Alice Leal

Is the Glass Half Empty or Half Full? Reflections on Translation Theory and Practice in Brazil

Transkulturalität – Translation – Transfer, Band 12
Herausgegeben von
Dörte Andres / Martina Behr / Larisa Schippel / Cornelia Zwischenberger

Alice Leal

Is the Glass Half Empty or Half Full? Reflections on Translation Theory and Practice in Brazil

Frank & Timme
Verlag für wissenschaftliche Literatur

Umschlagabbildung: © Angélica Leal

ISBN 978-3-7329-0068-8
ISSN 2196-2405

Herstellung durch das atelier eilenberger, Taucha bei Leipzig.
Printed in Germany.
Gedruckt auf säurefreiem, alterungsbeständigem Papier.

www.frank-timme.de

É sempre bom lembrar
Que um copo vazio
Está cheio de ar

Gilberto Gil

To my parents, Maria and Rosalvo.

Contents

Preface and Acknowledgements

I met Mary Snell-Hornby in Brazil at the Universidade Federal de Santa Catarina when she was there as a guest professor in 2006. She encouraged me to finish my MA as quickly as possible to start a PhD at the Universität Wien under her supervision. So in October 2007 I moved to Vienna and started working on the thesis that is now being published as book. I wish to express my gratitude to Mary Snell-Hornby, without whose encouragement and support this work would not have been possible.

Back in Brazil, several colleagues at the Universidade Federal de Santa Catarina (UFSC) and the Universidade Federal do Paraná (UFPR) played a pivotal role in the reflections I then carried out for this work. I would like to thank especially my MA supervisor, Markus Weininger, together with Werner Heidermann, both from UFSC; my BA supervisor Luci Collin, as well as Maurício Cardozo from UFPR. His patient reading and careful feedback was of utmost importance in the realisation of this book.

In 2009 I was delighted to receive an EST (European Society for Translation Studies) grant to take part in the 2009 CETRA Summer School at the Katholieke Universiteit Leuven (K.U.Leuven). The support, feedback and encouragement kindly offered by the entire CETRA staff greatly contributed to my research. My warmest thanks to José Lambert (K.U.Leuven), Andrew Chesterman (University of Helsinki), Dirk Delabastita (FUNDP Namur), Lieven D'hulst (K.U.Leuven), Peter Flynn (Lessius, Antwerp), Yves Gambier (University of Turku), Daniel Gile (ESIT, Université Paris 3 Sorbonne Nouvelle), Reine Meylaerts (K.U.Leuven), Franz Pöchhacker (Universität Wien), Christina Schäffner (Aston University), Maria Tymoczko (University of Massachusetts) and Luc Van Doorslaer (Lessius, Antwerp). The late Martha Cheung, who was the chair professor of the summer school in 2009, also played a very important role in this learning process. Finally, I would like to express my gratitude to all fellow students who took part in the Summer School.

In summer 2010 I received financial support from the Universität Wien to carry out a part of my research in Brazil, at the Universidade Federal do Paraná. The Universität Wien awarded me the grant "kurzfristige wissenschaftliche Arbeiten im Ausland" for two months (July and August), which was of fundamental importance to my stay in Brazil and subsequently to this work. Many thanks to the Universität Wien for their support, and to the Universidade Federal do Paraná, particularly to Maurício Cardozo, for receiving me.

Finally, I would like to thank Daniel Stirrat not only for his careful reading and suggestions, but especially for his unconditional support; my parents, Rosalvo and Maria Leal, as well as my sisters, Aline and Angélica Leal, for being so reliable and encouraging; Cornelia Zwischenberger and Larisa Schippel, my last readers, for their trust, patience and support.

One last clarification before we move on to the Introduction below. When I came to Vienna my intention was to write about the interface between the German functional approach and contemporary thought. This initial idea was gradually substituted by a wish to write about translation studies in Brazil – since to my surprise everyone in Europe seemed so interested in Brazil. I then set out to write about contemporary translation studies in Brazil, but soon realised it was a far too vast subject for me to be able to cover in one thesis. The number of books, journals, and academic works and events on translation in Brazil is simply enormous. I would without doubt end up with a very partial overview of a few unfairly homogenised tendencies squeezed into convenient categories.

As I started my research, I began with the works quoted here and realised I had more than enough material for one thesis. Of course the choice of subject – theory versus practice – and the choice of perspective – poststructuralist – were by no means random. Rather, these are subjects that have marked my short career in translation studies the most.

Introduction and Objectives

A much discussed issue in the field of translation studies is the relationship or the conflict between translation theory and translation practice. Indeed, this conflict has a strong impact on the way one perceives translation studies in general, including its institutionalisation, the aim of translator/interpreter training, the purpose of translation theory, the relationship between original and translation, the notion of professionalism as far as translators and interpreters are concerned, the object of translation research, amongst various others.

One such conflict appears to stem from numerous different elements. In this book this conflict is understood as a symptom of at least two sets of factors. The first set of factors refers to one's standpoint as regards translation. The most typical standpoints in our area are those of the practitioner, the scholar and the student. In addition to standpoints, the question of whether one's affinities lie closer to a predominantly essentialist or a predominantly anti-essentialist way of thinking seems equally important, thus making up the second set of factors mentioned here.

The different possible combinations amongst these sets of factors give rise to several disparate views as for the questions raised in the first paragraph. While some, for example, will claim that the aim of translation theory should be to make translation practice more efficient, others will defend the awareness-raising nature of theoretical reflection. Likewise, while some will argue that the purpose of translator/interpreter training is to fulfil the needs of the market, others will say that higher education should have more to do with critical thinking than with acquiring technical competence. The discussions are indeed endless and manifold, and certainly go beyond the modest boundaries of translation studies to influence one's very notion of theory, research and professionalism.

In this work I set out to investigate the elements lying beneath these different views on translation theory and practice – together with the issues that arise from these views – trying to understand not only where they come from, but also their implications. I have limited the scope of my work to translation studies in Brazil, but am convinced that many of the issues relevant in Brazil today may also help to shed light on discussions taking place all over the world.

The first part of this book will be dedicated to the different notions of theory and practice (and their interaction) that appear to emerge from the two sets of factors mentioned previously. In addition to the questions of standpoints and (anti-)essentialist affinities, Part I will start exploring some of the questions raised in the first paragraph above, such as the purpose of higher education and the notion of research in translation.

Next, Parts II and III, dedicated to Brazilian scholar Rosemary Arrojo and to Brazilian poet, translator and translation professor Paulo Henriques Britto, respectively, will illustrate and further develop some of the issues raised in Part I. As Arrojo and Britto represent very different views as far as translation is concerned, the analysis of their standpoints and academic work will offer an overview particularly of the reception of poststructuralist thought in Brazil.

Finally, in Part IV I will come back to a number of questions raised in the previous parts, all of which surrounding the notions of theory and practice in translation studies together with the various issues that derive from these notions. But rather than provide pretentiously universal answers to the questions I raise, I hope to open a debate that is well aware of the ineluctable heterogeneity of translation studies.

All in all, my two main objectives are the following. Firstly to offer an overview of the works of two prominent translation studies scholars in Brazil. Their work should unveil not only some of the most discussed issues in Brazilian translation studies, but also the reception of foreign theories and tendencies in the country. Secondly to present a thorough discussion of the issues of translation theory and practice, as well as of the other questions that emerge from these issues. This analysis, though initially based mostly on works by Brazilian thinkers, should reveal itself to be fruitful for translation studies in general.

Because the present book draws inspiration from poststructuralist thought, my intention is by no means to follow a strictly "scientific" methodology, overlooking heterogeneity, flattening differences and proposing, in the end, absolute and universally applicable answers to the questions I raise. Instead, I understand that controversy and fuzzy taxonomy are inevitable. But even more importantly than that, I understand that the questions raised in this book only make sense when embedded in a context and viewed through a particular perspective. In this sense, this book embraces heterogeneity and rules out the will for consensus and unanimity.

Part I: A Theoretical Practice and a Practical Practice

In Brazil 60-80 per cent of new titles consist of translations.

UNESCO 2008

In 2007 the second edition of the book *Conversas com Tradutores*[1] (BENEDETTI & SOBRAL 2003/2007)[2], or Interviews with Translators, was published in Brazil. According to the blurb, the objective of the book is to provide an overview of translation in Brazil from the translators' points of view. The organisers of the book came up with nine rather general questions about the translation market in Brazil, the basic competences translators must master, the influence of globalisation on translation, the concept of error in translation, the future perspectives in the field of translation, the role of translation criticism and translators' pay, amongst other issues. These nine questions were then sent to nineteen "dos nomes mais importantes da tradução em nosso país, na atualidade", or nineteen of the most important names in the field of translation in Brazil today[3] (BENEDETTI & SOBRAL 2003/2007: 31), who then wrote relatively short answers (the book has a total of 214 pages). Leaving aside the fact that the questions are not only overly general, but also identical for all nineteen translators – who, in turn, work in completely different areas – the Introduction (by Francis H. Aubert, professor of translation studies at the Universidade de São Paulo), Preface (by Ivone C. Benedetti, professional translator) and some of the interviews themselves seem to point towards the apparent "divórcio", or divorce of translation theory and translation practice in Brazil, which is perceived by most involved as problematic.

1 Italics will be used to indicate book titles and to convey emphasis in English. In indented quotations emphasis will be marked in boldface.

2 Because I will quote various excerpts from different interviews, both here and in I, 3, I will simply refer to the book as a whole, and not to the individual interviews. The only exceptions will be Aubert's Introduction and Benedetti's Preface. Similarly, in the References, the book will feature as a whole and the individual interviews will not be mentioned, except for the Introduction and Preface.

3 Unless otherwise indicated, all translations were done by me.

AUBERT (2003/2007: 10) claims that this divorce is due to the fact that

> *(...) para um bom número de tradutores, a boa teoria é aquela que proporciona soluções diretas e límpidas para o cotidiano do fazer tradutório. O anseio, ainda que inconfesso, é pelo modelo do dicionário, do manual de estilo (...).*
>
> [(...) for a large number of translators, a good theory should provide clear and direct solutions for their everyday translation tasks. Although secretly, what they long for is a dictionary-like model, a stylistic manual (...).]

Benedetti seems to be one of these translators as she asserts that "para as próprias pessoas envolvidas com tradução é muito clara a divisão entre teoria e prática", or for the people involved with translation, there is a very clear division between theory and practice (BENEDETTI 2003/2007: 25). Furthermore, she maintains that

> *(...) seria muito salutar uma interação produtiva entre teoria e prática (...). Os tradutores práticos, ou praticantes, raramente lêem teoria. (...) a teorização sem consideração da prática corre o risco de pecar pela generalização apressada, pelas soluções idealísticas (...)* (BENEDETTI 2003/2007: 25-26).
>
> [(...) a productive interaction between theory and practice would be extremely salutary. (...) Translators or practitioners rarely read theory. (...) theorisation that does not take practice into account risks being overly generalised, filled with idealistic solutions (...).]

One of the interviewees, Paulo Henriques Britto, professor of translation studies at the Pontifícia Universidade Católica do Rio de Janeiro, translation theorist, translator and poet (see Part III), also seems to corroborate this idea: "no campo da teoria de tradução atual, há alguns posicionamentos teóricos que são francamente irreais, e que não parecem resistir à prova dos nove da aplicação à prática", or in the field of translation theory today, there are a few theoretical standpoints that are simply unreal, standpoints which are not applicable to translation practice once you have gone through them with a fine-tooth comb (BENEDETTI & SOBRAL 2003/2007: 92-93).

While some emphasise that translation theory should be practice-oriented, thus implying that a marriage of theory and practice would be ideal, others stress the awareness-raising character of theoretical reflection quite independently from practical aims. Francis Aubert, for instance, is an example of the latter:

É da teoria, ou da teorização, que derivam as práticas conscientes, lúcidas, capazes, a qualquer tempo, de se justificarem, de se defenderem, de se imporem (...). Da teorização nasce a conscientização (awareness). É a partir da teorização que se faz uma prática verdadeiramente profissional (...) (AUBERT 2003/2007: 14-15).

[Conscientious, lucid practices derive from theory and from theorisation; practices that can always be justified, defended, imposed (...). The act of theorising gives rise to awareness (...). And it is through awareness that one can then proceed to a truly professional practice (...).]

Similarly, Heloísa Gonçalvez Barbosa, professor of translation studies at the Universidade Federal do Rio de Janeiro, asserts that

Embora muitos tradutores profissionais tenham alguma desconfiança da teoria, é ela quem me dá segurança (...). A teoria é importante na formação do tradutor, porque lhe confere um poder de ***reflexão*** *sobre sua vida profissional. Dá-lhe mais segurança nas tomadas de decisão e nos posicionamentos profissionais que toma. Ao mesmo tempo, a teoria ajuda o tradutor a encontrar seu lugar no mundo, na história* (BENEDETTI & SOBRAL 2003/2007: 59 – my emphasis).

[Theory is what gives me confidence (...), although many professional translators find it questionable. Theory is extremely important in translator training because it grants translators the power to *reflect* upon their work. It provides them with confidence in decision-making and helps them to find a professional stand. In addition to that, theory helps translators to find their place in the world, in history.]

In this sense, it seems clear that what Benedetti, Britto, Aubert and Barbosa mean by translation theory (and practice) is quite different. At first sight,

Benedetti and Britto seem to be speaking of translation theory as a set of precepts that should, ideally, govern over practice, systematising it and making it easier; a notion of theory, therefore, that is strictly practice-oriented and whose success or failure derives from its usefulness as far as translation practice is concerned, from the interaction between translation theory and translating. Indeed, in Aubert's opinion, this appears to be the idea and expectation "a large number of translators" (AUBERT 2003/2007: 10) have of translation theory. In this case, one could speak of an ideal "marriage" of theory and practice – that ends up in "divorce" when the interaction is not as successful as expected. Aubert and Barbosa, on the other hand, seem to speak here of translation theory as a source of awareness that should ideally lead to more "lucid, conscientious practices", to "reflect[ion]" (AUBERT 2003/2007: 14-15; BENEDETTI & SOBRAL 2003/2007: 59). In this light, the objective of translation theory, rather than to govern over practice, would be to give rise to more aware translators. As our discussions advance, it is essential to keep this difference in mind as regards the notions of translation theory.

Ruth Bohunovsky investigates, in a 2001 paper[4] (BOHUNOVSKY 2001), the relationship between contemporary translation theories and contemporary translation practices in Brazil, and her findings are rather curious. On the one hand we have translation theorists – most of whom scholars, professors at universities – that keep up with the latest tendencies in the field, bringing contributions academically recognised the world over. On the other hand we have a number of translators who have nothing to do with academia and who claim that their translations are faithful and neutral, deprived of any ideology or even interpretation – which in the eyes of certain members of academia, particularly through a more postmodern perspective, is not only regrettable, but also impossible.

Though significantly less comprehensive, Bohunovsky's research is somewhat similar to the one recently carried out by David Katan (2009). In his paper "Translation Theory and Professional Practice: A Global Survey of the Great Divide"[5] he presents the results of a survey carried out with 1000 practicing translators and interpreters from 25 different countries and with various educational backgrounds – though most went through some kind of interpret-

4 "A (Im)possibilidade da 'Invisibilidade' do Tradutor e da sua 'Fidelidade': Por um Diálogo entre a Teoria e a Prática de Tradução" – The (Im)possibility of Translators' "Invisibility" and their "Faithfulness": Towards a Dialogue between Translation Theory and Practice.

5 However large Katan's sample size may be, his methods have been deemed questionable, casting doubt on the results obtained (cf. ZWISCHENBERGER 2013: 115).

er/translator training. His findings are indeed very similar to Bohunovsky's: (i) most participants claimed that their loyalty lies with source texts – rather than with the reader or listener, the commissioner or client, or the translators/interpreters themselves (KATAN 2009: 138-139); (ii) most agree that "ideally a translator/interpreter should be invisible" (KATAN 2009: 140-141); (iii) most claimed that translation theory is only a "useful" part of translator/interpreter training – as opposed to an "essential" or "important" part – placing it at number 8 in importance amongst 12 elements – behind "practice", "strategies", "electronic tools", "subject specific knowledge", "contrastive grammar/linguistics", "the translator's / interpreter's profession", and finally "intercultural theory and practice", in this order (KATAN 2009: 142-147).

Therefore, one can say that Katan's findings, together with Bohunovsky's, appear to reinforce the aforementioned "divorce", or divide of translation theory and practice. And here one could look back at Aubert's first indented quotation above and generalise that the subjects of these surveys are practitioners and, therefore, their overall expectation as far as translation theory is concerned tends to be, indeed, the marriage of theory and practice – i.e. the interaction that makes practice easier, more automatic and systematised. But for some reason, as hinted at by Aubert (cf. AUBERT 2003/2007: 10) and clearly stated by Britto (cf. BENEDETTI & SOBRAL 2003/2007: 92-93), this marriage is not always prosperous, thus leading to a divorce.

Still trying to shed light on this "divorce", the 1998 book *Tradução: Teoria e Prática* [Translation: Theory and Practice] has a promising title. The professor of English at the Universidade de São Paulo, John Milton, presents us with a broad overview of the field of translation studies, particularly of literary translation. As regards the possible contributions of his book, he asserts in the introduction that it examines the area [of literary translation] through a new perspective, contrasting its recent developments with traditional approaches – "ele examin[a] a área de uma nova perspectiva, contrastando desenvolvimentos recentes com abordagens tradicionais" (MILTON 1998: 10). Indeed, chapter by chapter the book dissects the opinions of the most renowned literary translators and translation scholars from the times of Cicero and Jerome until the late 20th century.

The last chapter of Milton's book is dedicated to literary translation theory in Brazil. However, Milton warns us in his introduction that besides the Concretistas (i.e. mainly Augusto and Haroldo de Campos), the chapter "chama a atenção para outros estudos aleatórios, sem encontrar, no entanto, alguma outra escola de tradução com linhas definidas no Brasil", or mentions other random studies, but nevertheless finds no other well-defined translation school in Brazil (MILTON 1998: 9). The chapter is divided into four sections, as follows: section one is about the Campos brothers; section two is about José Paulo Paes, distinguished Brazilian poet, literary translator and translation theorist; section three is dedicated to "Outros grupos de tradutores brasileiros", or other Brazilian translation groups (MILTON 1998: 214), including writers from two different literary movements (Brazilian Modernism and the 1945 Generation); and, finally, section four examines "Outros trabalhos sobre a tradução literária no Brasil", or other works about literary translation in Brazil (MILTON 1998: 217).

In this last section, Milton insists that it is impossible to distinguish other literary translation schools in Brazil – "não é possível distinguir nenhuma outra escola de tradução literária" (MILTON 1998: 217). He then divides the recent material on literary translation published in Brazil ("material recente sobre a tradução literária publicado no Brasil" – MILTON 1998: 203) into four different types. The first type is what Milton calls "conselhos para o futuro tradutor", or advising future translators (MILTON 1998: 217-218), whereby authors caution future translators against translation traps and false cognate words. The books written by Paulo Rónai[6] in the mid-1970s (*Guia Prático da Tradução Francesa* [Translating into French: A Practical Guide] and *Escola de Tradutores*[7] [Translation School]) are instances of this first type. The second type comprises translation memoirs, such as *A Tradução Vivida* [Experiencing Translation] (1981), again by Paulo Rónai. The third type is what Milton calls comparative translation and mainly consists of papers, published in several media, comparing a number of different translations of the same work. According to Milton, these works tend to be extremely prescriptive, such as Walter C. Costa's article on the different translations of Emily Dickinson's poems ("Emily Dickinson Brasileira" or Brazilian Emily Dickinson).

Finally, the fourth type of recent publication in the field of translation studies includes books, papers, theses and dissertations on translation theory. And here once again John Milton reminds his readers of the "quadro bastante con-

6 Paulo Rónai is perceived as the first author to have written about translation studies in Brazil (*Escola de Tradutores* [Translation School], 1952). Born in Hungary in 1907, he first went to Brazil in 1940 to escape the Second World War. There he worked as a French and Latin teacher and published a number of translations and books on the art and craft of the translator. In 1974 he founded the ABRATES, Associação Brasileira de Tradutores or Association of Brazilian Translators.

7 See footnote 83 on page 153.

fuso", or quite confusing outlook (MILTON 1998: 218) of translation theory in Brazil. He argues that most works on translation in Brazil took no interest in abstract ideas, but rather in translation practice – "a maioria dos trabalhos sobre a tradução no Brasil não se interessou por idéias abstratas, mas pela tradução prática" (MILTON 1998: 219). Moreover, he asserts that very little has been written in Brazil on literary translation from a historic or descriptive perspective – "muito pouco foi escrito em termos de uma abordagem histórica ou descritiva da tradução literária no Brasil" (MILTON 1998: 222). He adds that "falta uma história da tradução literária no Brasil", or we need a history of literary translation in Brazil (MILTON 1998: 226).

In fact, lately Milton has become increasingly more associated with a line of research that could be called historiographical-descriptive, mostly aiming at describing translation schools and tendencies across times – his 1998 book (mentioned earlier) being a good example of it, as well as his 2002 book, *O Clube do Livro e a Tradução* (MILTON 2002). It is nonetheless interesting to remark that Milton appears to be seeking translation theory mostly within translation schools, thus denoting a very specific notion of translation theory, namely one that is strictly linked with translation models, and/or large and institutionalised translation movements – hence his remarks on the "confusing outlook", lack of "literary translation schools" or lack of interest in "abstract ideas" in Brazil (see above). In this sense, like Bohunovsky's and Katan's works, John Milton's work seems to strengthen the divorce of translation theory and practice in Brazilian translation studies – at least as far as this model-like notion of translation theory is concerned. And once again one could speak of "a divorce" because beforehand "a marriage" was presupposed.

Perhaps for this reason Milton decided to only briefly mention Rosemary Arrojo, as well as the reception of deconstruction in Brazil, which by 1998 had already established itself quite firmly. Though rather abundant by 1998,[8] these works labelled "poststructuralist" or "deconstructionist"[9] never constituted a

8 It is important to point out, however, that this 1998 book was actually published for the first time in 1993 under the title *O Poder da Tradução* [The Power of Translation], and the 1998 edition did not suffer substantial changes. This may explain the only brief references to the reception of deconstruction in Brazil, which in the early 1990s was certainly a far cry from what it became in the late 1990s (see Part II).

9 We can understand "deconstructionist" here as a poststructuralist tendency, as it became known particularly in the United States but also in Brazil. Nevertheless, Derrida seems to dislike this label, favouring instead an idea of antistructuralism, whereby both structuralism and a movement against it (hence "anti") are practised simultaneously (cf. Derrida's "Letter to a Japanese Friend" – DERRIDA 1985/1988 – translated by David Wood and Andrew Benjamin). In the present book, references to deconstruction and poststructuralism can be understood as explained by Christo-

hegemonic whole in Brazil, never leading to the establishment of so-called translation schools and models. In any case, it is clear that the scope of Milton's work was not translation studies in Brazil specifically. I will come back to the question of this apparent lack of works dedicated to translation theory in Brazil at the end of Part I.

Maybe what triggers the divorce (and the marriage, for that matter) of theory and practice in translation studies is the fact that what is meant by the words "theory" and "practice" varies greatly from theorist to theorist, professor to professor, translator to translator, interpreter to interpreter – as already briefly discussed. In this part of the present book, I will look into how different translation scholars and translators/interpreters perceive the concepts of theory and practice and their (lack of) interaction. The main objective of this first part is to carve the contours of the different notions of theory and practice to which translation theorists and practitioners appear to refer. This will show that they usually speak from utterly different standpoints and hence have completely disparate expectations as far as the interaction theory-practice is concerned. These different standpoints and perspectives are intimately related to the analogy used in the title, namely "is the glass half empty or half full?", and will be crucial for the following chapters, when the very definition of these standpoints will help us to understand dissimilar points of view on translation.

By the end of Part I, an overview of the different concepts of theory and practice predominant in translation studies (mostly in Brazil, but not exclusively), as well as of the symptoms and motivations behind these different ideas, will have been drawn. This overview will be both in terms of a more *essentialist* and a more *anti-essentialist* way of thinking, and in terms of different *standpoints*. And here it is important to warn the reader that I have no taxonomic intentions as such, and that these factors proposed here – i.e. standpoints and (anti-)essentialism – should by no means imply watertight categories. In fact, the conflict between translation theory and translation practice involves numerous factors, depending on the point of view one takes, and I have simply chosen to concentrate on and flesh out these two issues – for reasons that will become clear later in Part I. Moreover, these factors not only have rather blurry boundaries, but they often intersect each other, giving rise

pher Norris: "Deconstruction is avowedly 'post-structuralist' in its refusal to accept the idea of structure as in any sense given or objectively 'there' in a text. Above all, it questions the assumption (…) that structures of meaning correspond to some deep-laid mental 'set' or pattern of response which determines the limits of intelligibility" (NORRIS 1982/2002: 3). For more on this, see the Interchapter.

to various possible combinations. In summary, the present book is marked by a postmodern way of thinking – postmodern in the sense that it has no pretensions to repress contradiction or to erase or flatten difference and heterogeneity in favour of unanimity and homogeneity. It is as John McGowan puts it (McGowan 1991: 19-20):

Since reason's divisive strategies are taken as its means towards achieving domination, postmodernism attacks any number of traditional differentiations, including those between literary and other types of discourse, between high and low, between artistic and critic, and between signifier and signified. In each case the goal is to unsettle a privilege that accrues to one side of their pair and that can be maintained only by a logic of separation. (…) Against the modernist obsession with purity, now interpreted as part and parcel of the fundamental flaw of Western reason, we find postmodernism's celebration of heterogeneity.

Therefore, it is in this sense described by McGowan that I ask the reader to take this book as a work with a postmodern orientation.

INTERCHAPTER: The "Posts" and "Antis" of this Book

Operationally, essentialism is
the failure to allow for variation.

Stephan Fuchs

As addressed in footnote 9 on pages 21-22, terms such as "poststructuralism" and "deconstruction" will be extremely important in this book, so this brief interchapter shall clarify the terminology henceforth used. I understand that any attempt to group and categorise may run into generalisations and belittlement. Nevertheless, it is important to at least try and draw the general contours of the terminology used here, even though they are by no means absolute or inflexible, and exceptions and intersections do apply.

In very general lines, and largely based on the works by Norris (1982/2002), Best and Kellner (1991), and Sarup (1988/1993), deconstruction is understood here as a poststructuralist tendency. According to these authors, generally speaking one can understand poststructuralism as a movement that is not only similar, but that in many cases overlaps postmodernism,[10] which in turn is slightly more comprehensive than poststructuralism.

As Sarup explains (SARUP 1988/1993: 143-144), after World War II a new kind of society began to emerge – a society labelled as "post-industrial", "consumer society", "society of the spectacle", etc. For this society, Marxist theory was "outmoded", and the main question at its heart was whether or not the projects and aspirations of the Enlightenment – epitomised in the wish for "objective science, universal morality and law and autonomous art" – had failed, and whether or not we should stick to them. For Sarup, both postmodernists and poststructuralists would claim that the projects of the Enlightenment *have* indeed failed, and should therefore be fiercely criticised. In his view, in comparison with postmodernism, poststructuralism has attacked the Enlightenment project "less stridently, but with more intellectual sharpness".

10 And here we could go further and make a distinction between postmodernism and postmodernity, whereby the former refers to the "cultural sphere, especially literature, philosophy, and the various arts, including architecture", whereas the latter comprises "the geopolitical scheme" or "world process" (HASSAN 2000). In this light, postmodernity would be even more inclusive than postmodernism, embracing "postmodernism in the arts, poststructuralism in philosophy, feminism in social discourse, postcolonial and cultural studies in academia, but also multi-national capitalism, cybertechnologies, international terrorism, assorted separatist, ethnic, nationalist, and religious movements (…)" (HASSAN 2000).

Furthermore, he maintains that "post-structuralists like Foucault, Derrida and Lyotard are postmodernists" because "there are so many similarities between post-structuralist theories and postmodern practices that it is difficult to make a clear distinction between them".

Likewise, Best and Kellner assert that "while the term postmodern was occasionally used in the 1940s and 1950s to describe new forms of architecture or poetry, it was not widely used in the field of cultural theory to describe artefacts that opposed and/or came after modernism until the 1960s and 1970s" (BEST & KELLNER 1991: 9-10). For them, by the 1970s postmodern discourses had been disseminated the world over; nevertheless, it was not until then that the "most significant developments of postmodern theory" (BEST & KELLNER 1991: 16) took place. They refer here mainly to a so-called poststructuralist movement that erupted mostly in postwar France. In their view, the rapid changes that happened in social and economic spheres in the 1950s and 1960s led to dramatic changes in the "world of theory" (BEST & KELLNER 1991: 18), whereby the domination exerted by Marxism, existentialism and phenomenology was gradually replaced by the "linguistically-oriented discourses of structuralism" (BEST & KELLNER 1991: 18). These structuralist discourses aimed at "objectivity, coherence, rigour and truth" (BEST & KELLNER 1991: 19), and therefore claimed that their work was strictly scientific, free from subjectivity and bias.

Poststructuralism emerged, then, as an attack to these "scientific pretensions" which not only "attempted to create a scientific basis for the study of culture", but also "strove for the standard modern goals of foundation, truth, objectivity, certainty and system" (BEST & KELLNER 1991: 20). According to Best and Kellner, in its first moment, the poststructuralist critique was articulated in numerous texts chiefly by Jacques Derrida, Michel Foucault, Julia Kristeva, Roland Barthes and Jean-François Lyotard, which in turn "produced an atmosphere of intense theoretical upheaval that helped to form postmodern theory" (BEST & KELLNER 1991: 20). For them, these critiques pervaded literary, philosophical, sociological and political spheres firstly in France in the late 1960s and 1970s, and then in other countries as well, having a "decisive impact on postmodern theory" (BEST & KELLNER 1991: 20).

Indeed, as far as the relationship between poststructuralism and postmodernism is concerned, Best and Kellner believe that (BEST & KELLNER 1991: 25)

Poststructuralism forms part of the matrix of postmodern theory, and while the theoretical breaks described as postmodern are directly related to poststructuralist critiques, we shall interpret Poststructuralism as a subset of a broader range of theoretical, cultural and social tendencies which constitute postmodern discourses. (...) postmodern theory is a more inclusive phenomenon than Poststructuralism which we interpret as a critique of modern theory and a production of new models of thought, writing, and subjectivity, some of which are later taken up by postmodern theory.

Moreover, the authors quite rightly stress that the prefix "post" conveys an ambiguity that bears explaining. On the one hand, "post" implies a negation, a rupture which, in this sense, is similar to the prefix "anti". "Postmodern" and "poststructuralist" would hence mean "antimodern" and "antistructuralist", both in the sense of liberation from "old constraints and oppressive conditions" (BEST & KELLNER 1991: 29), and as an assertion of new ideas, new developments, new discourses. On the other hand, however, the prefix "post" may convey an idea of dependence, continuity and complementariness that by no means implies full rupture and negation. In this light, postmodernism and poststructuralism would be similar to "hypermodernism" or "hyperstructuralism", as in modernism and structuralism taken to their extremes. The challenge lies precisely in taking the prefix "post" in both the meanings described above, as opposed to favouring one over the other. As will be discussed at length in the chapters that follow, these "posts" can be understood simultaneously as negation and complementariness, constituting a paradox of great importance. In fact, the same paradox can be read in the term "deconstruction", whereby the ideas of "undoing" and "decomposing" exist side by side with the ideas of "reconstructing" and "redoing", and whereby sheer "destruction" or "demolition" are out of the question (cf. "Letter to a Japanese Friend" – DERRIDA 1985/1988 – translated by David Wood and Andrew Benjamin).

In this book, the terms "postmodern" and "poststructuralist" shall thus be used in the sense described here. Even though in many cases they could be used interchangeably, "postmodern" should be understood as more comprehensive and general, whereas "poststructuralist" as more specific and intellectually or scholarly-minded. The intention is not to propose a rigid, watertight classification to deal with this issue – firstly because one such classification would defeat the very objective of postmodernism, and secondly because grouping a large number of theorists and ideas under rigid epithets always risks erasing their differences and specificities (as already mentioned). In this sense, these terms should be taken more as approximations used for the sake of the argumentation than as watertight, absolute categories.

It is also crucial to take into consideration the fact that these postmodern and poststructuralist tendencies are by no means homogeneous and entail in no way some kind of consensus or unanimity. Quite the contrary: there are probably just as many quarrels *within* these so-called poststructuralist circles as there are quarrels between these thinkers and those outside their circles. It is beyond the scope of this work to explore the heterogeneity inherent to postmodern thought – though a few examples will repeatedly point to this lack of consensus. In any case, I will return to the question of heterogeneity and consensus in IV, 4. For now I would simply like to caution the reader against this initial impression that poststructuralist thought constitutes a harmonic whole.

In addition to "deconstruction", "poststructuralism" and "postmodernism", another term will be repeated often, namely "essentialism". Essentialism (as well as "anti-essentialism" or "nonessentialism") is understood here as an even broader term than postmodernism, since there are various lines of thought that oppose essentialism without being postmodern or poststructuralist as such – as is the case of hermeneutics and new criticism, for instance, or even Darwinism. In other words, while the poststructuralist perspective is in its multiplicity largely marked by a movement of anti-essentialism, one cannot say that only poststructuralist perspectives oppose essentialism. In this sense, poststructuralism draws a lot of inspiration from anti-essentialist ideas, whereas structuralism can be said to be marked by a predominantly essentiallist view – especially when one compares it to poststructuralism.

What I mean by essentialism in the present context can be synthesised through Brian Ellis' words (ELLIS 2001: 178):

If you are a scientific essentialist, than you must believe that the laws of nature are grounded in the properties and structures of things. They are intrinsic to things in the world, and not imposed on them by God or anything else. You will also believe that things belonging to natural kinds must behave as they do because this is how they are essentially.

And though Ellis concedes that "human laws, institutions, social structures, cultures, political organizations, and so on are *not* members of natural kinds"

(ELLIS 2001: 178 – my emphasis), he insists that there can be laws – through an essentialist perspective – governing over both human beings and human undertakings.

Another typical trait of an essentialist way of thinking is dichotomic classifications, polar oppositions, either-or distinctions, rather than "variable distinctions in degree" or "empirical continua", as Stephan Fuchs explains (FUCHS 2001: 13). Accordingly, Fuchs asserts that the "preferred mode of operation [of essentialism] is static typologies and rigid classifications, whose grids separate things that are everywhere, under all circumstances (…)" (FUCHS 2001: 15). If we look back at the epigraph above, we will remember that "[e]ssentialism is the failure to allow for variation" (FUCHS 2001: 15); in other words, from an essentialist perspective, both social and natural phenomena can be fit into watertight, universal, unchanging categories.

Therefore, while references in this book to "essentialism" (together with "anti-essentialism" or "nonessentialism") and "modernism" (together with "postmodernism") should be taken as more general and inclusive, as traits common to countless lines of thought, mentions of "structuralism" (and "poststructuralism") ought to be understood as more specific, related chiefly to language and discourse. From this perspective, "deconstruction" is even more specific because it can be taken as one of the many tendencies within poststructuralist thought.

1. Who's Afraid of Theoretical Reflection?

Grau, teurer Freund, ist alle Theorie
Und grün des Lebens goldner Baum.

Johann Wolfgang von Goethe

Although this first and compact chapter may strike one as less "scholarly" and more "impressionistic", it offers an efficient preamble to this book. Moreover, it will unveil some of the motivations behind it, shedding light on the bias underneath the surface of the text. Most of the content presented here stems from my personal experience as a BA student at the Universidade Federal do Paraná, then as an MA student at the Universidade Federal de Santa Catarina (both in Brazil) and, finally, as a PhD student and as a lecturer at the Universität Wien, in Austria.

Ever since I started studying translation, back in 2001, I have been through the following experience countless times. A certain professor is giving a lecture on theoretical (sometimes even philosophical) aspects of translating, when suddenly s/he is interrupted by an impatient student with a specific, practical translation question – such as "what should I do when I come across a typically Brazilian/Austrian word in the text I am translating into English?" Professors have responded in various ways over the years, but one aspect remains unaltered: there have always been a number of students who simply do not see the point in abstraction, in theorising. Most would probably rather be given some sort of formula or model which could solve their pressing problem immediately, even if this formula were so limited that they might only be able to apply it to this one case. Indeed, this attitude, on the part of the students, seems coherent with the profile expected of students nowadays, restlessly flipping pages of virtual books and hastily shifting from one website to another.[11] Regardless of whether the reader thinks this is positive or negative, let us keep

11 Isaiah Berlin's 1953 paper "The Hedgehog and the Fox" has become increasingly popular in recent times to describe the profile of students, particularly after the internet. In a nutshell, the idea is that people tend to perceive knowledge and approach information in either of the following ways. Either one is like a hedgehog and pursues a single idea thoroughly, or one is like a fox and pursues numerous ideas at once, without spending much time on each one. Applied to the contemporary context of education, this idea describes two predominant student profiles, with that of the fox prevailing by far over that of the hedgehog. Some claim that this has to do with the internet and its nonlinear structure, so young people today tend to swiftly skip from one piece of information to another in a nonlinear fashion (cf. Rasmussen & Ludvigsen 2009).

in mind this initial theory-practice conflict I have experienced throughout my academic life while we read the following paragraphs.

As will be shown particularly in I, 2, one of the key questions surrounding Brazilian universities – and certainly not only *Brazilian* universities – is indeed whether greater emphasis should be placed on *practice*, i.e. on exercising *technical competences* that will be needed later in the marketplace, or on *theory*, i.e. on *reflecting* and exercising *critical thinking* within a given field, mostly with a view to producing researchers and educators. Of course these two movements need not exclude each other; nevertheless, there does appear to be a tendency (be it within an entire university, a specific department, a major or even a lecture) towards one or the other, which more often than not is a source of conflict.

Take the Universidade Federal do Paraná, in Southern Brazil, for example. The Department of Languages is divided into two subdepartments: one of Linguistics, Classical Languages and Portuguese (henceforth Department of Portuguese), and the other one of Modern Foreign Languages (henceforth Department of Foreign Languages). The Department of Portuguese offers the so-called theoretical courses – mostly on Literary Theory and Linguistic Theory, in addition to various different courses depending on the major one takes. The Department of Foreign Languages can then take this theoretical foundation for granted and concentrate on teaching language and literature – most students do start learning foreign languages there from scratch. Recently, however, a conflict broke out and threatened to do away with this structure. Concerned about the ever growing need for teachers of Japanese and Polish in the state schools of Paraná, the Department of Foreign Languages proposed a plan to offer these languages within a major taught independently from the Department of Portuguese, i.e. without (at least officially) including much theoretical reflection. Their claim was that this kind of exclusively theoretical activity was superfluous to a certain extent, particularly considering that they wanted to train teachers for primary and secondary education only. In fact their plan was to offer all foreign language majors in this regime in the future – whereby any need for theoretical discussions would be fulfilled within their own chiefly practice-oriented classes.

By the same token, the BA in translation studies offered by the same university has been the target of similar criticism. In Brazil, the study of languages, linguistics and literature (and more often than not translation, too) takes place within one course called Letras, or letters. The major I took, for example, comprised nine semesters of Portuguese and English (linguistics and literature) with an emphasis on translation studies – which in turn is a joint effort of the Department of Portuguese and the Department of Foreign Languages, since teachers from both departments teach in it. The available BA emphases are translation studies, literary studies and linguistic studies. As it is a BA, its main aim is to train researchers and scholars, i.e. professionals that will most likely work at university. Nevertheless, many of these BA graduates end up pursuing various different careers that involve work with texts – proofreaders, editors, literary critics, writers, journalists, translators, interpreters, amongst many others. As for those who wish to work in primary and secondary education, instead of taking a BA they take a "Licenciatura", i.e. a teaching degree, a major that will grant them a licence to teach in primary and secondary schools. The principal differences between this course and the BA are the fact that it (the teaching degree) offers no emphases and that it focuses on classroom-oriented pedagogy and psychology rather than on research in linguistics, literature and translation studies.

With the ever-growing need for translators and interpreters, particularly in Curitiba (where the Universidade Federal do Paraná is located), a city home to various multinationals such as Renault and Bosch, the BA emphasis on translation studies has been attracting increasingly more attention. As I started explaining in the previous paragraph, for obvious reasons this emphasis has not escaped the question of whether or not it should be more market-oriented, more practice-oriented, including less *reflecting* and more *doing*. As it currently stands, the emphasis covers seven different courses or lectures, most of which are designed to stimulate reflection on the issues surrounding translation *studies*, with translator *training* playing a secondary role.

The merits of this curriculum aside, around Curitiba and in fact almost everywhere in Brazil there are no alternatives as far as translator and interpreter training are concerned. This is the major argument of those who defend the creation of a new translation curriculum for the Universidade Federal do Paraná with a view to training translators and interpreters. Those who oppose this and defend the current curriculum contend that a nine semester BA cannot provide fully professionalising training, particularly considering the fact that most students begin learning a foreign language there. Furthermore, the question of whether the university should be the institution to fulfil market demands remains a crucial issue at the heart of this debate. These discussions (and severe conflicts) remain a sensitive issue, and indeed they illustrate our discussion about the role of "theory" in higher education. Some members of the faculty seem to believe that theoretical reflection is vital, especially considering that it is, after all, a *BA* in translation *studies*. Others, in contrast, seem to perceive this theoretical tendency as a waste of time precisely because of its lack of practical application, hence confirming an image of the university as a professionalising institution.

Similarly, at the Centre for Translation Studies at the Universität Wien, the curricula of both the BA (Transcultural Communication) and the MA (Translation or Interpreting) have recently undergone changes that once again illustrate our debate. Because of their alleged lack of theoretical content, the former curricula were transformed so as to accommodate more translation and interpreting theory – instead of translating and interpreting. As one of the main mentors behind this curricular change, Mary Snell-Hornby explains that the intention is to provide a more solid theoretical foundation (as a source of reflection and critical thinking) so that they can build their practical competence upon it afterwards. Nonetheless, some claim that the courses have become overly theoretical, not preparing them well enough to face the marketplace.[12] As students enter the course with good knowledge (level B2 or C1 of the Common European Framework) of at least two foreign languages, many believe that the focus at BA level should already lie on *transcultural communication*, and not so much on theorising. As the example of the Universität Wien illustrates yet again, the question of whether the role of the university should be to fulfil market needs remains fundamental. What seems even more crucial at the moment, however, is how and to what extent this conflict theory-practice is established at university, as well as how it fits the perceived role of university.

But let us focus here on translation studies in Brazil. From a formal, academic point of view, this theory-practice conflict seems to be firmly established in translation courses, and so does a view of translation research as strictly applied, empirical or practice-oriented. Take the Universidade Federal de Santa Catarina,[13] for instance, the first federal university to ever have a specific programme for translation studies *strictu sensu*, first at MA level in

12 I say this based on a recent survey (2013) carried out by the student council with the current MA students, which revealed that most participants wish their curriculum were more practice and market-oriented and included fewer so-called theoretical courses (the results will be published on http://www.stv-translation.at/ in 2014).

13 All the information about the Universidade Federal de Santa Catarina was taken from my own experience as their MA student in 2006-2007, as well as from their official website – www.pget.ufsc.br (including the theses).

2003, and then at PhD level in 2009. The programme is divided into two lines of research, namely "Lexicography, translation and language teaching" and "Translation theory, criticism and history". In the second semester of 2008, the following courses were offered: Translation Theories, Translation Criticism, Corpora and Translation, Translation Practice, Literary Translation, Translation and Culture, and finally Translation and Rupture (originally in Portuguese). Once the students have successfully concluded a number of credits they must write their dissertations. Students are not allowed to write an entirely "practical" work, nor can they write an entirely "theoretical" dissertation; ideally they should always combine both, aiming to show their interaction – we could speak again of an ideal marriage of theory and practice.

Particularly in the second line of research, "Translation theory, criticism and history", supervisors tend to encourage students to balance theory and practice, analysing a certain theoretical approach and applying it to a certain practice. Quite often, however, it seemed to me that some of the results were not as satisfactory as they could have been precisely because most students did not find (or struggled to find) any avowed link between a particular theory and the practice they wished to carry out. This practice consisted mostly of translation criticism or comparative translations, as well as of annotated translations done by the students themselves. As a result, students have used Catford's theory applied to literary translation, or Venuti's ideas of foreignising translation applied to the translation of canonical literature in English into Brazilian Portuguese, just to mention two remarkable examples.

I vividly remember that as I was writing my project, I wanted to write a fully theoretical dissertation about (roughly summarised) the relationship between the German functional approach and other translation theories or reflections (initially I had Nida's, Berman's and Venuti's). In other words, my object of study was theoretical discourse itself, as in the present book. Most of my peers, in contrast, seemed to have been motivated to take an MA in translation studies by more "practical" issues. They felt that the Portuguese translation of a particular work was not good enough and therefore wanted to redo and/or criticise it. Others wanted to write the very first Portuguese translation of a given work. Narceli Piuco, for example, both criticised an existing translation and proposed a new translation of Madame de Stäel's *Corinne ou l'Italie* (cf. PIUCO 2008); Théo Moosburger, Nana Coutinho, Silvânia Carvalho and Fabrício Coelho all presented annotated translations of the following works: verses 1 to 609 of the epic *Vasileios Digenis Akritis*, Bernard Shaw's *Widowers'*

Houses, Abel Steven's *Madame de Stäel: A Study of her Life and Times*, and Alexander von Humboldt's "Einleitende Betrachtungen über die Verschiedenartigkeit des Naturgenusses und eine Wissenschaftliche Ergründung der Weltgesetze", respectively (see footnote 13 on page 32 – cf. MOOSBURGER 2008; COUTINHO 2008; CARVALHO 2008; COELHO 2008).

As we advanced our studies, however, it became clear that both they and I would have to change our projects so as to accommodate theory and practice simultaneously, and "promote" their marriage, so to speak. In other words, in my case a fully theoretical dissertation might be deemed purposeless without a practical application; as for my peers, entirely practical dissertations would be incongruent with the research programme (indeed translation *studies*), which in turn called for theoretical reflections. In the end we all adapted our projects and the annotated translations and translation criticism mentioned earlier all included some sort of theoretical discussion – mostly so as to define and justify criticism and translation strategies. As for me, I changed my initial project completely, but kept the German functional approach as my main theoretical framework (cf. LEAL 2007a).

At that time, I remember debating with peers and professors about the intended aim of the MA. As the majority of my colleagues had more practical motivations in mind, different professors in various situations warned them that the MA was *not* professionalising and that only a translation or only translation criticism could not make up an MA dissertation. As we had a study group together, I followed my peers' struggles from up close; indeed, one of our objectives was to help one another to find a theoretical framework in which to embed the practice that they wanted to carry out.

As for me, the nature of my struggles was quite different since the practice I wished to carry out was *theoretical*. I never quite understood the near obsession with the Popperian model of falsifiability and testability as the pillar of the scientific method – which is by no means exclusive to the Translation Studies Programme at the Universidade Federal de Santa Catarina, but rather a tendency that seems to have swept across the entire Western world.[14]. To me, the idea of testing hypotheses on something as abstract and subjective as literary translation seemed beyond my capacities as a student. As I came to realise later, mostly through the works of those involved in the Science Wars of 1990s

14 Indeed, according to Niranjana, "a rapid survey of translation studies would show that for the past few centuries thinking on translation has remained within an empiricist-idealist conceptual framework that, structured by what Gramsci would call 'common sense', in turn uphold the premises of humanism" (NIRANJANA 1992: 50).

(see I, 2), my reservations might not have been so unreasonable after all. In Aronowitz's opinion, for example, "the attempt to model social sciences on the methods of the old natural sciences is entirely misplaced" (ARONOWITZ 1996: 186). Further, he holds that *method* is the *dogma* underlying the success of *science* ever since the 17th century. One of the main elements of this dogma of method can be explained as follows: "the way to legitimate and reliable knowledge is through the experimental method, the basis of which is our ability to make both observational and falsifiable statements" (ARONOWITZ 1996: 179 – quoting POPPER 1959/2002).

In addition to this question of method, another issue that puzzled me during my MA was why the vast majority of my peers appeared to confirm that initial impression I described at the beginning of this chapter, namely that students seem indifferent – not to say completely apathetic – to theorising, as most of their interests and motivations to take an MA were largely practice-oriented. As I have argued elsewhere (cf. LEAL 2010b), there does appear to be a strong tendency for students in translation studies to derive research topics from a given practice or practical issue, and then (sometimes unwillingly) search for a theoretical framework in which to embed this practice – as already mentioned. A similar trend seems verifiable at the Universität Wien, amongst Snell-Hornby's MA and PhD students, for instance. The vast majority are either working on literary translation criticism – for reasons and interests similar to those of my MA peers in Brazil – or on media translation – mostly the dubbing of popular American TV-series in Austria.[15]

In any case, I shall come back to the issue of translation research and the Science Wars (see penultimate paragraph above) in I, 2. Before we move on to it, though, let us quickly summarise the points (mainly three) raised in this chapter. These three issues have strongly marked my path in translation studies and exerted an enormous influence in the composition of the present book. Firstly, as already mentioned in the introduction to Part I, there seems to be a lack of disposition on the part of some involved with translating and interpreting to discuss more abstract, theoretical issues – issues that do not directly refer to practical matters. In fact, for these people theory only seems interesting when its interaction with practice is successful. This attitude had come up before, as we analysed Bohunovsky's paper and Katan's survey with practition-

15 The following database offers an overview of all BA, MA and PhD theses presented at the Centre for Translation Studies since 1998: transvienna.univie.ac.at/forschung/datenbank-wiss-arbeiten/ (last accessed in August 2013).

ers, but also as a few translation scholars and translators (mostly Benedetti and Britto) expressed their opinion about translation theory. In the present chapter, this attitude appeared mainly through the unwillingness or lack of interest of many of my colleagues to engage in theoretical debates, mostly because their interests lie with practice. Secondly, numerous academic conflicts – be it at faculty level or amongst students – appear to stem from the question of whether the university should fulfil market needs and therefore tailor its courses to make them more practice-oriented. Thirdly and finally, research in translation studies in Brazil (but certainly not only there) seems to be predominantly applied and empirical, with Karl Popper's model of the scientific method at its forefront. I will come back to these questions repeatedly in what follows.

So far these personal impressions and comments on my short experience as a translation student and teacher have hardly contributed to understanding the nature of the questions raised here. In fact, I believe that they have somewhat emphasised this confusing outlook without much clarification. The next chapter, dedicated to the role of higher education, will throw some light on these issues.

2. The Inevitable End of Higher Education

Beware of what opens the university
to the outside and the bottomless,
but also of what, closing it in on itself,
would create only an illusion of closure,
would make the university available
to any sort of interest, or else
render it perfectly useless. Beware of ends;
but what would a university be without ends?

Jacques Derrida
(translated by Catherine Porter
and Edward P. Morris)

Derrida's words invite us to reflect upon the role of higher education in terms of "ends" – as in "goals", "applications", "orientation". One should on the one hand beware of "what opens the university to the outside" and, on the other hand, prevent it from "clos[ing] in on itself"; one should at once beware of "ends" and embrace "ends" as the way to the future.

As we have briefly seen in I, 1, while some will claim that the purpose of higher education is to teach professional competences to be exercised mostly outside the university, thus fulfilling the needs of the market, others will feel that the role of the university is first and foremost to stimulate reflection and critical thinking within a given field – particularly as far as the Humanities are concerned. Caught in this dilemma, the university does not seem to know whether it should do away with or embrace "ends", to use Derrida's words.

Transformations and debates like the ones happening at the Universidade Federal do Paraná or the Universität Wien (see I, 1) are probably happening at many different institutions for similar reasons. And the role that theory plays within these debates remains manifold. When theory is perceived as a source of reflection without direct practical applications, some will deem it essential, crucial to the establishment of critical, well-informed students and professionals; others, in contrast, will consider it superfluous and vague mainly because, once in the market, these students should primarily be functional and efficient. Likewise, when theory is understood as practice-oriented, a means to govern over practice, a set of rules and precepts to be followed, some will find it indispensable to the teaching of a profession, while others will claim that it is overly minimising and simplistic, leading to the production of uncritical students with merely mechanical, technical competences.

Jacques Derrida explores these ideas in his 1983 essay "The Principle of Reason: The University in the Eyes of its Pupils" (DERRIDA 1983 – translated by Catherine Porter and Edward P. Morris[16]). He begins by going back to Aristotle and analysing how he appeared to understand the notions of knowledge and theory, concluding that, for the Greek philosopher, the human drive for knowledge and theoretical thinking is mostly purposeless: "the desire to know for the sake of knowing, the desire for knowledge with no practical purpose" (DERRIDA 1983: 4). Derrida identifies more or less the same viewpoint as far as Kant, Nietzsche and Heidegger are concerned: "In modern times, Kant, Nietzsche, Heidegger and numerous others have all said as much, quite unmistakably: the essential feature of academic responsibility must not be professional education" (DERRIDA 1983: 18). Yet modern times, Derrida maintains, have witnessed the emergence and establishment of oriented, applied research. In other words,

> *research that is programmed, focused, organized in an authoritarian fashion in view of its utilization (in view of "ta khreia", Aristotle would say), whether we are talking about technology, economy, medicine, psychosociology, or military power – and in fact we are talking about all of these at once* (DERRIDA 1983: 11).

This kind of research would be overshadowing – in terms of both obtaining funds and getting attention – "disinterested research with aims that would not be pledged in advance to some utilitarian purpose" (DERRIDA 1983: 11).

Derrida then takes a more political path, going over Kant's attempt to define the limits between "the essential and noble ends of reason that give rise to a fundamental science versus the incidental and empirical ends which can be systematized only in terms of technical schemas and necessities" (DERRIDA 1983: 11). He arrives at the conclusion that nowadays, as far as research and theorising are concerned, we have a different, far more omnipresent kind of censorship than ever before:

16 The epigraph that opens the present chapter was taken from this essay.

Today, in the Western democracies, that form of [traditional] censorship has almost entirely disappeared. The prohibiting limitations function through multiple channels that are decentralized, difficult to bring together into a system. The unacceptability of a discourse, the noncertification of a research project, the illegitimacy of a course offering are declared by evaluative actions: studying such evaluations is, it seems to me, one of the tasks most indispensable to the exercise of academic responsibility, most urgent for the maintenance of its dignity (DERRIDA 1983: 13).

The so-called Science Wars and Culture Wars are a corollary of Derrida's remarks. Stanley Aronowitz explains, for example, that although the awards granted to research projects by the US-American Department of Defence did not have "utility" as their main criterion, the vast majority of "funding proposals justified their requests on practical grounds" (ARONOWITZ 1996: 189). Further, he asserts that "the choice of investigative objects and their promised results is ineluctably designed to persuade the funder that the payoff is worth the money" (ARONOWITZ 1996: 189). Dorothy Nelkin understands this phenomenon in terms of a "social contract" or a "marriage" between science (or scientific research) and the state that has been in effect since World War II:

This contract included a set of both tacit and open agreements about the autonomy of science. The government would provide research support, relatively unfettered by requirements for accountability, if scientists would work in the interest of public progress and conscientiously administer and regulate themselves. (...) This relationship was often described as a marriage, implying shared assumptions and mutual trust. The unusual degree of autonomy granted to science reflected its apolitical image, the reputation of scientists as unbiased and "disinterested" and therefore reliable as a source of truth (NELKIN 1996: 95).

In this light, the Science Wars appear to have been triggered by a strong desire to demystify science and scientific discourse, showing it is subject to numerous influences – such as ideology, culture, context and bias – just as any other type of discourse (cf. ARONOWITZ 1996: 178; FRANKLIN 1996: 141). This process of deconstruction of science led to the questioning of everything surrounding it: its methods, its validity, its institutions, its motivations, its neutrality. Although the issue of the Science Wars is a side issue here, it shows how far the conflicts underlying the purpose of research and higher education have gone. "Application" is key: application as far as theory is concerned, the application of science and reason, the application of university education. Be it in terms of producing professionals, producing theories or producing technologies, whether or not to directly and openly respond to the needs of the market is probably the crucial question surrounding the university at the moment – and in fact whether not responding is even a possibility (see epigraph above). And as Derrida very aptly explains, it is a mistake to think that this kind of ethical responsibility is relevant only in Natural Sciences (DERRIDA 1983: 13):

From now on, so long as it has the means, a military budget can invest in anything at all, in view of deferred profits: "basic" scientific theory, the humanities, literary theory and philosophy. The compartment of philosophy which covered all this, and which Kant thought ought to be kept unavailable to any utilitarian purpose and to the orders of any power whatsoever in its search for truth, can no longer lay claim to such autonomy. What is produced in this field can always be used.

Yet he concedes that "desiring to remove the university from 'useful' programs and from professional ends, one may always, willingly or not, find oneself serving unrecognized ends, reconstituting powers of caste, class, or corporation" (DERRIDA 1983: 18 – see epigraph above). For him, a possible way out of the dilemma lies in what he calls "thought" – a concept that has very much to do with the idea of theorising as a source of awareness, reflection and critical thought. Having been asked by the French government to help to set up a College of Philosophy in the 1980s (today the Collège International de Philosophie), Derrida insisted upon this dimension of "thought" – something "not reducible to technique, nor to science, nor to philosophy" (DERRIDA 1983: 16). By stimulating and promoting thought, this institution would be able to "provid[e] a place to work on the value and meaning of the basic, the fundamental, on its opposition to goal-orientation, on the ruses of orientation in all its domains" (DERRIDA 1983: 18). In addition to give rise to criticism of philosophy itself, this dimension of thought should also

be accompanied at least by a movement of suspicion, even of rejection with respect to the professionalization of the university in these two senses, and especially in the first, which regulates university life according to the supply and demand of the marketplace and according to a purely technical ideal of competence (DERRIDA 1983: 17).

The "two senses" to which he refers here are the following: (i) "does the university have as its essential mission that of producing professional competencies, which may sometimes be external to the university?" and (ii) "is the task of the university to ensure within itself – and under what conditions – the reproduction of professional competence by preparing professors for pedagogy and for research who have respect for a certain code?" (DERRIDA 1983: 17).

As far as reason is concerned – and here reason should be understood as the principle lying beneath science – this dimension of thought should lead to the "infinite task" of unmasking "all the ruses of end-orienting reason, the paths by which apparently disinterested research can find itself indirectly reappropriated, reinvested by programs of all sorts" (DERRIDA 1983: 16). And here Derrida stresses that "ends", "orientation", "application" should not necessarily be perceived as the enemy, but rather that a need has arisen "for a new way of educating students that will prepare them to undertake new analyses in order to evaluate these ends and to choose, when possible, among them all" (DERRIDA 1983: 16). In other words – and let us look back at the epigraph above – the idea would not be to do away with "ends" and "external investment", but rather to be more critical or careful about the consequences of these ends and investments.[17]

Notwithstanding this emphasis on "thought" and criticism, the College of Philosophy should be a "normal" institution. "Normal" in the sense that it cannot fiercely reject and discard everything it criticises. As he writes his essay, for example, Derrida is prepared to submit himself to the principle of reason for the sake of communication, which does not mean that he is not prepared to try and transform communication, language, academic relationships and approaches to pedagogy (DERRIDA 1983: 16). Accordingly, he demands this double gesture of the university as well, i.e. "to ensure professional competence and the most serious tradition of the university even while going as far as pos-

17 And here, if we embrace this dimension of "thought", we might already have a possible answer to the third question asked at the end of I, 1, namely about the predominance of the methods of applied research in translation studies. If the university is to be perceived as a place of "thought", an institution that fosters the deconstruction of its own premises, then applied research should no longer be a must in any area. In any case, we will come back to this question towards the end of this book, in IV, 3.

sible, theoretically and practically, in the most directly underground thinking about the abyss beneath the university" (DERRIDA 1983: 16).

Indeed, in face of the university "crisis" (in a way for many of the reasons that culminated in the Science Wars of the 1990s), Derrida clings to this double gesture (DERRIDA 1983: 20):

In a period of "crisis", as we say, a period of decadence and renewal, when the institution is "on the blink", provocation to think brings together in the same instant the desire for memory and exposure to the future, the fidelity of a guardian faithful enough to want to keep even the chance of a future, in other words the singular responsibility of what he does not have and of what is not yet.

This double gesture is very much in line with his idea of "double bind", as a sort of tension that stems from resistance and gives rise "neither to an analysis nor to a synthesis, neither to an analytic nor to a dialectic. It provokes both the analytic and the dialectic to infinity, but in order to resist them *absolutely*" (DERRIDA 1996/1998: 26 – translated by Kamuf, Brault and Naas – his emphasis).

In two later works dedicated to the future of the university, Derrida develops this idea further and places great emphasis on the need for a "university without condition", in other words a university that is "capable of deconstruction, beginning with the deconstruction of their own history and their own axioms" (DERRIDA 2001a: 26). This unconditional university would be a place where everyone had "the principal right to say everything, whether it be under the heading of fiction and the experimentation of knowledge, and the right to say it publically, to publish it" (DERRIDA 2001a: 26); it would be the "ultimate place" of "critical resistance" to "all the powers of dogmatic and unjust appropriation" (DERRIDA 2001a: 25-26). One such "more than critical" university would welcome all sorts of critical questioning: of "the history of the concept of man", "the history even of the notion of critique", "the form and the authority of the question" and "the interrogative form of thought" (DERRIDA 2001a: 25-26).

More specifically as far as the Humanities are concerned, Derrida believes in that double gesture again: preserving and changing at once; rethinking the concepts underlying the Humanities but, at the same time, protecting their canons at all costs (DERRIDA 2001b: 22). For him, the concept of interdisciplinarity plays a pivotal role, as the Humanities must, on the one hand, go beyond the borders that delimit their departments and, on the other hand, still keep their specificity (DERRIDA 2001b: 65). Dilek Dizdar understands Derrida's advice, as well as the role of "thought" in the Humanities as follows:

(…) Derrida [verteidigt die] Notwendigkeit einer Widerstandskraft mit Widerstandsprinzip. Für diese Widerstandskraft und die Unbedingtheitsprinzip eignen sich die Wissenschaften vom Menschen, die ***Humanities****, ganz besonders. Denn sie (…) öffnen den Raum für eine Diskussion und Neubestimmung des Prinzips. Diese Bewegung stützt sich auf die Möglichkeit der Kritik, des grundlegenden Fragens und der Dekonstruktion; denn in den* ***Humanities*** *geht es um den Menschen (…). Die Universität und darin besonders die* ***Humanities*** *sind also der Ort, der ein (selbst) kritisches Denken, den zivilen Ungehorsam pflegen und entwickeln sollte* (DIZDAR 2006: 364-365 – her emphasis).

Derrida's double gesture does not, therefore, imply substantial structural changes, but rather a change of attitude on the part of those involved with the university. As far as teachers are concerned, Rosemary Arrojo puts this change of attitude as follows (ARROJO 1996b: 102):

Armed with such an awareness and with their compromise to unmask the power relations that determine the meanings and the hierarchies that constitute the "truths" that shape their classroom, community and, ultimately, also their own history, postmodern teachers will start their courses by the very deconstruction of the practices and contents to be taught, as well as of the "official" goals to be met. (…) They will not teach their students to follow "universal", "objective" translation rules. Instead, they will show them how "rules" are always local and unstable (…). In other words, postmodern teachers will not give up their "textual authority", but will make an effort to make it as transparent as possible, showing their students that this kind of authority, like any other, both inside and outside the school, is a form of power that can be overcome.

The debate about the future and the purpose of the university is, undoubtedly, a multifaceted, complex one. The suggestions made by Derrida (and Dizdar, and Arrojo) are certainly feasible, and are probably being carried out by teachers with postmodern affinities regardless of the disposition of their colleagues and institutions. And in accordance with Derrida's double gesture or double bind one could still respond to the demands of the marketplace (to go back to the question raised previously about the role of higher education) and be critical of this response, reject it, strive to change it. Somehow, however, this does not appear to be the gesture behind measures such as the ones taken by the Department of Modern Foreign Languages at the Universidade Federal do Paraná, for example (see I, 1), or the motivations behind POSI, "Praxisorientierte Studieninhalte für die Ausbildung von Übersetzern und Dolmetschern" (see next paragraph). Instead, the need to fulfil market needs seems to set itself against the questioning, the "more than critical" attitude, the deconstruction of these very demands and their implications.

Indeed, as we will see in the next few chapters, many of those involved with translation (be they practitioners, scholars or students) seem to perceive the move towards university professionalisation as a sign of progress, evolution, a step closer to true science. Let us take POSI (see previous paragraph) as an example. According to Anderman and Rogers, the establishment of POSI had a lot to do with "the problem created by the discrepancy that appeared to exist between the priorities of university training programmes and the rapidly changing demands placed on translators in the market place" (ANDERMAN & ROGERS 2000: 64). However, as discussions between academics and practitioners began, it became clear that "the perspective of the employers/professionals still cannot be easily reconciled with that of the trainers", even though both sides have "better practice-oriented training for translators and interpreters as a shared goal" (ANDERMAN & ROGERS 2000: 64). According to the authors, universities have to overcome two different "constraints": the heterogeneous nature of the student body in terms of their professional future and "broader considerations of *their perceived role in society as educators in the humanistic tradition*" (ANDERMAN & ROGERS 2000: 64 – my emphasis).

The question of whether education should be professionalising, thus meeting the demands of the market, has been relevant ever since the onset of the idea of "education". According to Frijhoff, "Aristotle [asked] what exactly was the purpose of education of his age: to produce learned men, to educate in virtue, or to satisfy the needs of society. Learning, virtue, utility (…)" (FRIJHOFF 1996/2003: 43). In Brazil, this debate became particularly heated in the 1970s, mostly due to the gradual softening of the military regime. Alfredo

Bosi, one of Brazil's most important literary critics and historians, would write in 1979:

O ***mundo do receituário*** *é a forma formada da cultura dominante e vigora em todas as carreiras a que a Universidade dá acesso. É particularmente deprimente quando se pensa na passagem, em geral entrópica, da cultura universitária para o meio secundário. O que se transmite aos alunos do ginásio (...), o que se estratifica em termos de instrução fundamental, é, quase sempre, a fórmula final, reduzida, retificada, da antepenúltima tendência da cultura superior* (BOSI 1992/2006: 317 – his emphasis).

[The *world of formulae* is the way of the dominant culture, manifesting itself in every single university major. This is particularly depressing when one thinks about the usually entropic passage from university culture to the secondary sector. What is transmitted to secondary school pupils (...), what becomes fixed in their minds in terms of basic instruction is almost always the final, reduced, rectified formula of an already out-dated tendency dictated by the dominant culture.]

His bitter remarks are a response to two main changes that had been taking place more or less at that time: firstly the fact that philosophy had been abruptly cut out of the curricula of secondary school (BOSI 1992/2006: 314), and secondly the fact that Latin and Greek were becoming ever so unpopular at university level – unlike the previous decades that had seen their golden years. Later in this same essay Bosi asserts that university is the place where culture becomes prematurely formalising and professionalising; this culture hastily develops formulae and nurtures itself on them until new formulae prevail and replace them – "A Universidade é o lugar em que a cultura se formaliza e se profissionaliza precocemente. (...) essa cultura chega logo à cunhagem de fórmulas e se nutre dessas fórmulas até que sobrevenham outras que a substituam" (BOSI 1992/2006: 320).

Going back to the question about whether education should be practice-oriented, market-oriented, Bosi replies that education should be (BOSI 1992/2006: 340).

(...) uma introdução larga ao conhecimento do Homem e da Sociedade, uma ocasião constante de desenvolvimento da própria linguagem, como

expressão subjetiva e comunicação intersubjetiva (...). Este ideal, que forma o ser consciente das conquistas do gênero humano, não pode ser barateado nem trocado por esquemas inertes ou migalhas de uma informação científica ou histórica.

[(...) a large introduction to the knowledge of Man and Society, a constant opportunity to develop one's own language both as subjective expression and intersubjective communication. (...) This ideal results in the awareness of human achievements and hence cannot be cheapened or exchanged for inert schemas or scraps of scientific or historical knowledge.]

Indeed, as already argued, if we accept that theory is something generally produced in academic settings, for an academic public ("academic" here as opposed to "for laymen", including students as well), the role of the university becomes crucial in this debate about the role and purpose of theory. In Brazil, the concept of university is closely related to the concept of research (MENDONÇA 2000: 148):

(...) a lei 5.540 afirmava explicitamente constituir-se a universidade na forma ideal de organização do ensino superior, na sua tríplice função de ensino, pesquisa e extensão, enfatizando-se a indissolubilidade entre essas funções, particularmente entre ensino e pesquisa, sendo esta última o próprio distintivo da universidade.

[(...) law number 5.540 clearly stated that the university was the ideal structure in which to organise higher education, and that it should fulfil three main functions: teaching, research and extracurricular activities. These functions were indissoluble – particularly teaching and research, the latter being the distinctive feature of university.]

Although this law was revoked by a new one in 1996 (law 9.394), the role of research remained the same.[18]

Between the early 1990s, when Bosi published his essay, and today, Brazil has seen on the one hand a return to philosophy, Latin and Greek, but on the other hand a significant change as far as the "indissolubility" (see last indented

18 For more on this, see http://www.planalto.gov.br/ccivil_03/Leis/L9394.htm#art92 (last accessed in September 2010).

quotation) of research and university is concerned, mainly as a result of the establishment of numerous private universities across the country. Most of these new universities tend to employ underqualified teachers (at least in comparison to public universities), who in turn have different jobs at the same time and thus cannot dedicate their time to research.

And speaking of research, needless to say that the political and ethical struggles behind both the Science Wars and Derrida's concerns about the future of university (expressed particularly in DERRIDA 1983, 13-15) have also become a reality in Brazil. This is how Leyla Perrone-Moisés roughly summarises the current situation of the Brazilian university:

> *A situação atual é a seguinte: a universidade está exposta, à venda, ou, para usar de um eufemismo, sponsorisée, com objetivos comerciais e industriais. As Humanidades são reféns dos departamentos de ciências puras e aplicadas, que concentram os capitais rentáveis vindos de fora da universidade* (PERRONE-MOISÉS 2006: 28).
>
> [The current situation is the following: the university is exposed, for sale or, to use a euphemism, sponsorisée for commercial and industrial purposes. The Humanities are hostage to the departments of natural and applied sciences, which in turn receive all the investment from outside the university.]

In this sense, a lot of the research being done at the moment is funded through private investments and is directly and openly oriented to serving "utilitarian purposes" (cf. DERRIDA 1983: 13). Furthermore, conflicts in the Humanities – as the one at the Universidade Federal do Paraná – have become increasingly recurrent, whereby the move towards full professionalisation appears to represent a goal to be reached (see references to POSI above).

All in all, the picture drawn in the present chapter is marked by the idea of difference. Difference in terms of the way the concepts of theory and practice are understood, differences as far as the purpose of higher education is concerned, differences regarding the wherefores of research. In fact, this entire part and even this entire book is permeated by the notion of difference, and I certainly have no intention of erasing or reconciling these differences. More important than reconciliation – because it is not possible, nor is it desirable – would be to understand these differences, and this is how I would like to end the present chapter.

Derrida's suggestion of a dimension of "thought", he himself remarks, "cannot be produced outside of certain historical, techno-economic, politico-institutional and linguistic conditions" (DERRIDA 1983: 17), which might lead to the reappropriation of his idea by forces with which he does not necessarily agree – a risk he, as a "faithful guardian", is nevertheless willing to take as it is the risk of the future itself (DERRIDA 1983: 17). Indeed, all these differences mentioned in the previous paragraph are products of "historical, techno-economic, politico-institutional and linguistic conditions", and they all constantly risk being – and are – reappropriated. Understanding the disparate standpoints behind these differences as well as these movements of reappropriation is understanding the conflicts that derive from them – the famous dichotomies such as theory versus practice, oriented or applied versus basic research, as well as university for "virtue" versus university for "utility" (cf. FRIJHOFF 1996/2003: 43).

As already argued in the introduction to Part I, and as the example of POSI illustrates very well, the different outlooks one may have on the dichotomies that end the previous paragraph are closely related to one's standpoint – amongst other factors, of course. As Anderman and Rogers (2000) explain, academics are "still" reluctant to accept fully professionalising translation and interpreting courses because of "their perceived role in society as educators in the humanistic tradition" (ANDERMAN & ROGERS 2000: 65), though *some* members of academia are certainly not reluctant at all, and they will have their own motivations to do so. By the same token, employers and professionals of translation and interpreting are far more concerned about practice, hence their wish to make translation training fully professionalising and market-oriented. In fact, it makes sense that businesses in general should prefer their future employees to receive professionalising training at university, so that they do not have to invest in it afterwards – and this concern is as legitimate as the concern of the faithful guardian stood at the university gates, to use Derrida's allegory.

In any case, the question of standpoints will be raised again in I, 4. Let us now advance to Chapter 3, in which we will mostly go back to the first question raised in Chapter 1 about the apparent lack of disposition, particularly on the part of practitioners and students, to deal with "theory" as a source of reflection with no "utilitarian purpose" in mind, or in other words their keenness on a notion of translation theory that fulfils all their practical needs as practicing translators and interpreters.

3. Translation Theory and Practice: Married, Divorced or Nonrelated?

By all means marry; if you get a good wife,
you'll be happy; if you get a bad one,
you'll become a philosopher.

Socrates

As already mentioned in the introduction to Part I, the book *Conversas com Tradutores* (BENEDETTI & SOBRAL 2003/2007), or Interviews with Translators, reveals that the relationship the different interviewees have with the idea of translation theory is manifold. In fact, one of the nine interview questions goes straight to the heart of the matter. It reads as follows: "Em sua opinião, que tipo de tradutor (formação, dons pessoais, etc.) exige a tradução em geral e a tradução em sua área? Qual a função da teoria da tradução na formação do tradutor?", or in your opinion, what kind of translator (in terms of qualifications, personal qualities, etc.) is most fit for translating in general and for translating in your area? What is the role of translation theory in translator training? (BENEDETTI & SOBRAL 2003/2007). Out of the nineteen interviewees (most of whom are experienced translators), eight are directly involved with university, mostly as language or translation professors. The other eleven are practitioners, some of whom with specific translation training (a few are translation masters and doctors), and others self-taught translators.

Most of the translators who also work at university seem to have a positive outlook on the role of theory in translator training, with the words "reflection" and "awareness" echoing throughout the book. João Azenha Jr., for example, claims that "ela nos ensina a ordenar a experiência, a recuperar informações valiosas, a criar argumentos (...) para justificar as escolhas que são, impreterivelmente, diferentes de tradutor para tradutor", or it [translation theory] teaches us to organise our own experience, to recall valuable information, to come up with arguments (...) to justify the choices that will inevitably vary from translator to translator (BENEDETTI & SOBRAL 2003/2007: 48). Similarly, as already quoted on page 17, Heloísa Gonçalvez Barbosa maintains that

Embora muitos tradutores profissionais tenham alguma desconfiança da teoria, é ela quem me dá segurança (...). A teoria é importante na

formação do tradutor, porque lhe confere um poder de ***reflexão*** *sobre sua vida profissional. Dá-lhe mais segurança nas tomadas de decisão e nos posicionamentos profissionais que toma. Ao mesmo tempo, a teoria ajuda o tradutor a encontrar seu lugar no mundo, na história* (BENEDETTI & SOBRAL 2003/2007: 59 – my emphasis).

[Theory is what gives me confidence (...), although many professional translators find it questionable. Theory is extremely important in translator training because it grants translators the power to *reflect* upon their work. It provides them with confidence in decision-making and helps them to find a professional stand. In addition to that, theory helps translators to find their place in the world, in history.]

Quite in the same way, Lúcia Helena França places emphasis on the awareness-raising nature of theoretical reflection:

[A teoria] É extremamente importante, pois ***conscientiza*** *o aluno-tradutor sobre os processos envolvidos no ato tradutório, dando-lhe uma maior* ***consciência*** *da sua atuação e da importância de aprimorar seu desempenho, levando-o a refletir sobre o seu trabalho* (BENEDETTI & SOBRAL 2003/2007: 105 – my emphasis).

[It [theory] is extremely important because it raises the students-translators' *awareness* of the processes involved in translation, providing them with greater *awareness* both of their work and of the importance of improving their performance, hence causing them to reflect upon their work.]

Lia Wyler,[19] on the other hand, argues that studying theory can only be fruitful when two principles are followed. First of all, translation teachers must take into account "as expectativas do mercado editorial e do leitor brasileiro", or the expectations both of the publishing market and of the Brazilian reader (BENEDETTI & SOBRAL 2003/2007: 196). Secondly, she adds that ideally university students and teachers should evaluate together how the translation theories

19 Although Wyler is not an academic as such, but has become increasingly more involved in academic activities – such as publishing papers on translation and teaching translation courses at university. In Brazil she is best-known for her translations of *Harry Potter*. This division proposed here between "academics" and "nonacademics" is rather flexible as the intention is simply to point to those interviewees who have *some involvement* with academic activities.

they studied can be applied to the Brazilian culture and language (BENEDETTI & SOBRAL 2003/2007: 196). She concludes her answer to this question by saying that

A desatenção a esses pressupostos não tem produzido bons frutos: o tradutor profissional não tem se interessado muito por teorias, e o graduado em tradução não tem encontrado no mercado muita receptividade para o que estudou (BENEDETTI & SOBRAL 2003/2007: 196).

[Not following these principles has led to two negative consequences: the professional translator has not had much interest in theories and the translation major has had great difficulty finding receptivity in the market for what s/he has learnt.]

Except for one translator-professor, all the other eight "academic" interviewees (see footnote 19 on page 50) do not mention any particular theories, nor do they announce affiliations to a certain theoretical standpoint. The exception is Paulo Henriques Britto, whose work as a translation theorist and literary translator will be more closely analysed at a later point (see Part III). Even though he believes that theory plays an important role in translator training, he has one major restriction. He explains that "a reflexão teórica é uma coisa boa para o tradutor literário", or theoretical reflection is a good thing for the literary translator (BENEDETTI & SOBRAL 2003/2007: 92), but concedes that the relationships between theory and practice within the field of literary translation are not exactly ideal – "as relações entre teoria e prática no campo da tradução literária não são as melhores possíveis" (BENEDETTI & SOBRAL 2003/2007: 92). In Britto's point of view,

(...) é preciso reconhecer que a antipatia que muitos tradutores literários nutrem pela teoria não é completamente sem razão de ser. De fato, no campo da teoria de tradução atual, há alguns posicionamentos teóricos que são francamente irreais, e que não parecem resistir à prova dos nove da aplicação à prática. Refiro-me em particular a algumas teorias radicais que surgem no campo da desconstrução e do feminismo. Por exemplo, alguns teóricos sustentam que a noção de original em tradução é um simples mito, e que o tradutor literário deveria ter o mesmo status que o autor do suposto "original" (palavra que eles sempre colocam entre aspas); a posição secundária da tradução em relação ao original seria

uma decorrência do capitalismo, do machismo, do logocentrismo, da opressão das culturas européias sobre as periféricas, etc. Os tradutores contra-argumentam, com toda razão, que os mesmos teóricos que pontificam sobre a impossibilidade de se distinguir entre original e tradução, sobre a impossibilidade de se ter acesso às intenções do autor, sobre a relatividade radical de todo julgamento de valor referente às traduções, quando se defrontam com a tradução de um texto por eles escrito, se comportam tal como os autores logocêntricos, eurocêntricos e falocêntricos: querem que o tradutor reproduza em suas traduções exatamente o que eles queriam dizer, exigindo a mesma noção de "fidelidade" ao "original" que eles tanto criticam em seus pronunciamentos teóricos. Não admira que, diante de tais pronunciamentos, alguns tradutores literários apressadamente concluam que a teoria é uma perda de tempo (...). É importante que os tradutores literários percam seu preconceito contra a teoria e acompanhem com interesse um campo que é um dos mais produtivos e interessantes na área das ciências humanas na atualidade; mas também acho importante que alguns teóricos mais radicais façam uma autocrítica e reconheçam que há, realmente, uma boa dose de irrealismo em algumas propostas que têm circulado nos meios acadêmicos em décadas recentes (BENEDETTI & SOBRAL 2003/2007: 92-93).

[(...) one must acknowledge the fact that the antipathy felt by literary translators towards translation theory is not entirely unjustifiable. Indeed, in the field of translation theory today, there are a few theoretical standpoints that are simply unreal, standpoints which are not applicable to translation practice once you have gone through them with a fine-tooth comb. And here I mean particularly those theories derived from deconstruction and feminism. Some theorists maintain, for example, that the notion of original in translation is a myth, that literary translators should enjoy the same status as the author of the allegedly "original" work (and in their texts the word "original" always comes between inverted commas); the inferior role of a translation when compared to the original text would be a product of capitalism, chauvinism, logocentrism, of the oppression imposed by European cultures on peripheral cultures, etc. As a response, translators quite rightly claim that these very same theorists who defend the impossibility of distinguishing between translation and original text, the impossibility of accessing the author's intentions, the radical relativity of all and every value judgement of a translation, these theorists, when faced with a translation of their own texts, behave in the exact same way as the logocentric, Eurocentric, phallocentric authors they criticise: they expect their translators to reproduce in their translations exactly what they meant to say, demanding the very same notion of "fidelity" to the "original" of which they so fiercely disapprove. Unsurprisingly, faced with this, some literary translators hastily arrive at the conclusion that theory is a waste of time (...). It is important that literary translators reconsider their prejudice against translation theory and follow this that is nowadays one of the most interesting and productive fields in the Humanities. But I also think it is important for certain radical theorists to be more self-critical and recognise that in the past few decades there have been a great number of unrealistic ideas in the theories discussed in academic circles.]

Most of Britto's questioning will be addressed at length in Parts III and IV. What seems curious, however, is the way he and Wyler, though in completely different fashions, find the relationship theory-practice so problematic. For Wyler, the role of the translation teacher is to evaluate to what extent foreign translation theories may be "applicable" to the Brazilian context, publishing market, culture and language. Leaving aside the fact that she does not account for Brazilian theories, she concludes her reflection by reinforcing the divorce of theory and practice mentioned here repeatedly. She seems to believe that for this reason, "professional translators" have little to do with theory, whereas those who studied translation theory have difficulty "applying" it to the translation market. Here two questions seem relevant, namely to what extent can someone be a professional (in whatever area) when s/he has no specific education in this area? And should translation students be able to directly apply the theories they have learnt to the translation market of their country? In any case, let us keep in mind that these two interviewees, Britto and Wyler, are well-known translators in Brazil – Britto has translated more than 100 volumes of prose and poetry, whereas Wyler translated, amongst other works, all the *Harry Potter* books (I will come back to their standpoints in the next chapter).

Before we come back to these two questions and to Britto's long quotation, let us look at what the other eleven interviewees, i.e. the ones who have no involvement with academia, have to say about the "role of translation theory in translator training" (BENEDETTI & SOBRAL 2003/2007). Erik Borten and Claudia Berliner, for example, admit that studying theory may be fruitful, but personal qualities and specific knowledge of the area in which one works are far more important (BENEDETTI & SOBRAL 2003/2007: 74-75; 82-83). Heloísa Martins Costa seems to have a similar opinion: "Na minha área, que é técnica, é necessário gostar de pesquisar e saber usar as ferramentas que a internet oferece", or in my area, which is a technical one, one must like to research and know how to use the tools the internet offers (BENEDETTI & SOBRAL 2003/2007: 100).

Other interviewees emphasise the requirements that translation theories have to fulfil in order to be valid. Maria Stela Gonçalves asserts that a translated text must be

> *uma produção oriunda da* ***língua de chegada*** *e não (...) algo cristalizado, que parece dizer "Sou uma tradução desta ou daquela língua". Só acredito no que se designa como teoria da tradução se esta puder dar conta dessas condições* (BENEDETTI & SOBRAL 2003/2007: 114 – her emphasis).
>
> [a production typical of the *target language*, rather than (...) something crystallised, something that appears to say "I am a translation from this or that language". I only believe in what is called translation theory if it fulfils these requirements.]

Renato Rosenberg defends that "toda teoria fica ótima quando é respaldada pela prática", or every theory can be great when supported by practice (BENEDETTI & SOBRAL 2003/2007: 165). Nilson Louzada corroborates this idea by saying that good knowledge of translation theory helps one to discuss with and persuade editors about translation choices, but regrets the fact that there are not many Brazilian productions in this area (BENEDETTI & SOBRAL 2003/2007: 131).

Haroldo Netto connects theory and practice by stating that "se perguntarem a um tradutor (...) em plena execução do seu trabalho, o que está aplicando ali da teoria de tradução aprendida, muito provavelmente ele não saberá responder de pronto", or if you ask a translator which translation theories he is applying to a certain text he is translating at the moment, he would certainly not be able to reply right away (BENEDETTI & SOBRAL 2003/2007: 137). He quotes a colleague, Robert Bonono (BENEDETTI & SOBRAL 2003/2007: 138), to explain his relationship with translation theory:

Acho que, em termos gerais, os "tradutores praticantes" não percebem a relevância da teoria para o seu trabalho. Suas preocupações primordiais têm a ver com os detalhes dos textos (terminologia, significados obscuros) e questões gerais de comércio, como a obtenção de trabalho.

[I think that, in general, "translation practitioners" do not see the relevance of theory to their work. Their main concerns relate to text details (such as terminology, obscure meanings) and general business issues, like finding work.]

There are those who concede that although they were not specifically trained as translators, this theoretical knowledge is (or would be) very welcome. Alfredo de Lemos, for instance, claims that he wishes he had a better "base teórica", or theoretical foundation, because theory is what helps us to organise and systematise our knowledge – "organizar e sistematizar nosso conhecimento" (BENEDETTI & SOBRAL 2003/2007: 126). Similarly, Mauro Sobhie perceives translation theory as a way to avoid "longas horas de sofrimento em dilemas existenciais", or hours agonising over existential dilemmas (BENEDETTI & SOBRAL 2003/2007: 175). Based on his own experience, he feels that reading translation theory motivates us to reflect upon our role in translation as well as in society ("nosso papel na tradução e dentro da sociedade" – BENEDETTI & SOBRAL 2003/2007: 175).

Finally, Vera Pereira admits that she finds it a difficult question seeing as she was not trained as a translator at all, but rather learnt by practising, "errando muito e acertando às vezes", or often making mistakes and occasionnally getting things right (BENEDETTI & SOBRAL 2003/2007: 149). She confesses that, when faced with two different translations of the same text, they never seem to be the same and that she is unable to tell which one is *the* right one. Pereira suggests it might be up to the author to decide but then confesses, quite puzzled, that she herself has already found translation solutions far better than the ones suggested by the author of a certain text. She concludes her reflection by saying that "como autodidata em tradução, sinto falta de conhecimentos teóricos (...) e tenho dificuldades que suponho serem menores para quem passou anos (...) estudando para o exercício dessa função", or as a self-taught translator, I wish I had more theoretical knowledge, and I probably have more difficulties than those who spent years studying to become a translator (BENEDETTI & SOBRAL 2003/2007: 149-150).

Indeed, Pereira's example of the two different translations of the same text and the unproven authority of the author illustrates the comments of the other interviewees who defend the awareness-raising character of translation theory. Had Vera Pereira studied translation theory, perhaps she would not have to go through "hours agonising over existential dilemmas", and she would certainly have better insights as for her "role in translation as well as in society" (see the previous paragraphs). However, this does not necessarily mean that she is a worse translator because of this lack of theoretical knowledge. I will come back to this question (of whether "truly" professional translators must have some sort of education in the area) at the end of the next chapter.[20]

Another work that offers an interesting dialogue between "practicing translators" and "academia" is Andrew Chesterman's and Emma Wagner's book *Can Theory Help Translators?* (CHESTERMAN & WAGNER 2002), raising similar questions to the ones discussed in *Conversas com Tradutores*. At the very beginning of the book, Wagner, representing practicing translators as an EU translation manager, states that

Most of us had a brief brush with theory in our student days, when we absorbed whatever was needed to get us through our exams... and then proceeded to forget it, as we got to grips with the realities of learning how to do the job. There can be few professions with such a yawning gap between theory and practice (CHESTERMAN & WAGNER 2002: 1).

She goes on to ask "Can it [translation theory] help us to become better translators and give us a feeling of self-esteem?" (CHESTERMAN & WAGNER 2002: 1), to which Chesterman replies that "translation theory SHOULD have this aim" (CHESTERMAN & WAGNER 2002: 1 – his emphasis). Wagner's urge is clearly similar to the wishes of the translators just mentioned above, particularly the ones who do not belong to academia – but not solely. It is hard to tell whether she and her colleagues really have "forgotten" what they learnt about translation theory, or whether they have simply assimilated it and stopped consciously thinking about it – based on the book, I would say the latter.

In any case, Chesterman's response also seems to leave a few questions unanswered – "does he believe in practice-oriented theory only?" is probably the most important one. Judging by his apparent despise of postmodern thought, I

20 Pereira's example will come up again IV, 1, where I will close this discussion about what makes a translator professional.

believe so: "I will bypass here the deconstructionist argument that there is no objective (…) in the first place" (CHESTERMAN & WAGNER 2002: 9-10),

> *[s]ome postmodern/deconstructivist thinkers (…) stress that (…) the author is dead (…). Outside texts: nothing. Further: meanings are not fixed but endlessly shifting and deferred, all is indeterminate, everyone interprets a text in their own way (…)* (CHESTERMAN & WAGNER 2002: 24).[21]

All in all, in Wagner and Chesterman's debate it seems quite clear that Wagner does not really see a point in translation theory as "it is", and wishes it would at last directly and objectively solve her everyday hitches as an EU translator. One could speak of Fish's theory hope here: "the hope that our claims to knowledge can be justified on the basis of some objective method of assessing such claims rather than on the basis of the individual beliefs that have been derived from the accidents of education and experience" (FISH 1989/1999: 322). In other words, the hope that a theory will one day emerge and establish "a range of standard solutions" to all sorts of translation problems everywhere in the world (CHESTERMAN & WAGNER 2002: 6) – a view very much in line with the one that a few scholars and several practitioners interviewed in *Conversas com Tradutores* showed. As for Chesterman, though he does not appear to suffer from "theory hope", he makes his reservations clear as far as postmodern or openly prescriptive approaches are concerned. His notion of translation theory seems to be a "perspective from which to contemplate something, so as to understand" (CHESTERMAN & WAGNER 2002: 2), but at the same time something that can be directly applied to translation practice (CHESTERMAN & WAGNER 2002: especially 131-136).

By looking at all these opinions on translation theory and practice by people with completely different backgrounds and occupations, it seems once again evident that what they call "theory" varies greatly from person to person, and so does their notion of "practice". For those who stress its awareness-raising nature, theory seems to be a practice justifiable in itself. It is not supposed to provide correct answers, applicable to all contexts, all cultures and all publishing markets; it is not supposed to directly help one to persuade editors or mathematically determine which translation of a particular text is *the* correct one.

21 I will come back to Chesterman's criticism to postmodern tendencies in IV, 4.

Nevertheless, for those who stress the immediate bonds between theory and practice, theory is supposed to fulfil all the demands made by practice, regardless of how manifold or specific this practice may be. Theories must consist of manuals, prescriptive or descriptive models applicable to any situation anywhere, as products of large and institutionalised translation schools. Perhaps this is due to the fact that a number of so-called translation theories in fact attempt to be translation manuals or models – a simple Google search shows over 35.000 results for the expression "translation model". Indeed, I can still recall my enthusiasm when I first started reading Christiane Nord's model of text analysis in translation (NORD 1988/2005). In the Introduction she explains that

> *Translation-oriented text analysis should not only ensure* ***full comprehension and correct interpretation*** *of the text or explain its linguistic and textual structures and their relationship with the system and norms of the source language (SL). It should also provide a reliable foundation for* ***each and every decision*** *which the translator has to make in a particular translation process* (NORD 1988/2005: 1 – my emphasis).

Words such as "full comprehension", "correct interpretation", "reliable foundation for each and every decision", and all this for every possible text type, culture and translation context (NORD 1988/2005: 2-3), certainly sound promising. In addition to that, they seem to justify why translators expect so much of translation theory, as well as why they can only understand the relevance of a certain theory when it is directly applicable to translation practice. For them, there should always be a happy marriage between theory and practice. Perhaps it is also because of words like Nord's that some of the translators mentioned above feel frustrated and suspicious of translation theories, or claim that there have hardly been any Brazilian contributions to the field of translation theory – precisely because there are indeed hardly any Brazilian translation models in these terms. While it is true that it is not the scope of the present book to analyse to what extent Nord's model fulfils the requirements she sets (cf. LEAL 2006, 2007a), it is also true that even the most essentialist reader will find the expressions "full comprehension" and "correct interpretation" extremely ambitious, to say the least.

These different perceptions of theory and practice will be clarified in the following chapters, firstly by looking, in Chapter 4, at the question of *standpoints* and how they appear to influence one's outlook as regards translation theory and translation practice. So far, according to the perspective proposed in this book, there seems to be a predominance of two notions of translation theory, mentioned various times, and whose boundaries often overlap. The first notion presupposes the marriage of theory and practice, whereby the former attempts to describe, regulate and govern over the latter. Should these attempts fail, then the divorce of theory and practice ensues. This first notion seems to be more closely related (but not exclusively) to an idea of translation models, as well as of large and institutionalised translation schools. Furthermore, this first outlook on translation theory seems to yield a notion of practice that, in turn, is stable and homogeneous enough to be fully controlled by theory. Another point of view that seems to be fairly in line with this first notion of translation theory is that of *professionalising* translator and interpreter training, within which this type of theory plays a crucial role – as it is through theory and its schemas and precepts that pupils will be properly trained to face the marketplace.

The second notion of theory that seems to be entertained by some of the voices quoted here is not practice-oriented, but rather awareness-raising. In fact, the idea of translating as a practice plays a smaller role in the big picture since the practice this second perspective appears to entail is, indeed, *theoretical*. For some of those who share this outlook, translating is too manifold and heterogeneous to be fully systematised by schemas and rules. For others, the focus should lie on critical thinking, on understanding one's role in history and society rather than on translating. More often than not, this notion of theory is associated with an idea of higher education that is not strictly professionalising, and that favours in its structures and departments first and foremost the establishment of critical pupils.

It is my own eyes as a translation researcher that reveal, or rather *build*, these two notions of translation theory and practice. Perhaps for a self-taught practitioner these two views proposed here are amalgamated into one, whereby some theories may strike him/her as more practice-oriented and hence potentially "useful", whereas others may strike him/her as more abstract and probably "less useful". If one claims that translation theory is pointless (as we have seen a few examples above), one may have the first or the second notion of theory in mind. In other words, one may have had contact with a model-like, outwardly practice-oriented theory that, in one's view, did not fulfil all the demands made by one's practice. Similarly, one may have felt interested in a theoretical reflection with no practical aims, but later felt disappointed that this reflection did not directly contribute to one's work.

All in all, these two notions discussed here are highly specific perspectives that could be implied by the words "theory" and "practice" every time they appear, and in this sense I warn the reader once again to be careful and take this constant ambiguity into account. In the next chapter this ambiguity will be explored as the symptoms of different standpoints, which in turn are marked not only by one's occupation, but also by one's culture.

4. To Each his Own Standpoint

Wenn ich mit meiner Relativitätstheorie recht behalte,
werden die Deutschen sagen, ich sei Deutscher,
und die Franzosen, ich sei Weltbürger.
Erweist sich meine Theorie als falsch,
werden die Franzosen sagen,
ich sei Deutscher und die Deutschen, ich sei Jude.

Albert Einstein

Let us begin the present chapter by looking at an allegory often used by Paulo Henriques Britto (see Part III) to defend a practice-oriented notion of translation theory. A few young people got lost in a forest. It is getting dark and they do not know how to get home, nor can they contact anyone to come and rescue them. They have no compass but know they must go towards the south. So one of them suggests climbing up a tree to see where the sun is setting. If one knows the sun sets in the west, one can easily find north and south. Suddenly, however, another member of the group calls the group's attention to the fact that north, west... those are all human inventions, illusions, a mental fabrication. At this point Britto usually asks something along the following lines: "But if one wants to get home, what does it matter if north and south are illusions? Which method is more efficient then, climbing up the tree or philosophising about cardinal points?"[22]

Indeed, Britto's question is very pertinent and goes straight to the heart of this chapter. If one must get home, reflecting on cardinal points is obviously not the best strategy – it is certainly *not* a strategy at all. Nevertheless, thinking that the cardinal points are absolute truths above and prior to mankind is not exactly ingenious either. From the point of view of getting home, acknowledging the fact that the cardinal points were invented by human beings and do not exist *as such* makes no difference at all. However, from the point of view of one's awareness of the world around one, as well as of all the implications of one's beliefs, this reflection is indispensable. In any case, asking someone who

22 Britto's allegory resembles Richard Dawkins' words: "Show me a relativist at 30,000 feet and I will show you a hypocrite" (apud FRANKLIN 1996: 143). In her paper, Franklin uses Dawkins' quotation to illustrate and criticise what she calls "the power of science" and of "scientific objectivity", and how it manifests itself through its claim to "instrumentalism, its practice, its efficacy" (FRANKLIN 1996: 143).

wants to get home to argue about this with someone who wants to heighten his/her awareness of the world is undoubtedly a bad idea, as these two people will speak from utterly different *standpoints*.

So far I have argued that the great "divorce" of translation theory and practice, mentioned time again by translation scholars, students and practitioners, may have some of its roots in the fact that what one usually means by theory and practice varies greatly. This variation may occur for a number of different reasons, depending on the person and on the situation. As far as translation studies is concerned, nevertheless, there is one issue that seems particularly crucial, and it can be summarised as follows: Where do all these different people speak from? What is their *standpoint* as far as their relationship with translation is concerned? Indeed, Britto's allegory provides a good illustration to this point.

If we look back at the previous chapters and remember discussions carried out in the books *Conversas com Tradutores* (BENEDETTI & SOBRAL 2003/2007) and *Can Theory Help Translators?* (CHESTERMAN & WAGNER 2002), mainly, we will remember that those voices emerge from utterly diverse places, hence circumscribing utterly diverse standpoints. Let us take Chesterman and Wagner, for instance. It is only natural that they should have different opinions as far as translation theory is concerned: Wagner is an EU translation manager and therefore is mostly concerned about nitty-gritty aspects of the profession, whereas Chesterman is a scholar, a university professor interested in translation as a discipline. However enlightening and thought-provoking their debate may be, one can hardly expect them to reconcile their differences – and there is no reason why one should.

A similar dynamics appears to take place in *Conversas com Tradutores*. No wonder most interviewees engaged in academia saw translation theory with better eyes, whereas those strictly involved with translation/interpreting sounded sceptical or indifferent. When one's bread and butter is to translate/interpret, one's concern will probably be practice-oriented. Nevertheless, when one's job is to reflect about translation/interpreting, to teach it, to establish it as a discipline and exercise it in academic settings, theory and theorising acquire a whole new dimension.

Moreover, all the "academic" interviewees who seemed to have a somewhat negative outlook on translation theory and/or imposed a number of obstacles to it (mainly Britto and Wyler) are also practitioners, both renowned translators (see I, 3). No wonder they emphasised the irrevocable bonds between translation theory and translation practice, and showed great concern for the direct applicability of translation theory to practice. Earlier we read that "the professional translator has not had much interest in theories and the translation major has had great difficulty finding receptivity in the market for what s/he has learnt", by Wyler (BENEDETTI & SOBRAL 2003/2007: 196); and "the relationships between theory and practice within the field of literary translation are not exactly ideal", by Britto (BENEDETTI & SOBRAL 2003/2007: 92).

Likewise, regarding the debate as to whether higher education should be professionalising or not (see I, 2), there seems to be a very general tendency for those who speak from inside the university to play the role of Derrida's "faithful guardian" (DERRIDA 1983: 17). In other words, however much one may be willing to open the university gates to external interests, one is certainly not willing to open them completely – particularly as far as the Humanities are concerned. On the other hand, those who speak from outside the university – in this case mainly businesses – would probably like to rely on the university to provide professionalising training to their future employees (as already argued), so as to exempt them from this extra "investment". Of course these two "categories" are far from antagonistic and self-excluding, and we know that numerous examples of intersections between the two *do* exist. Indeed, there are a number of departments within the Humanities that are now tailoring the curricula of their courses so as to better fulfil market demands – as we have seen through the examples of POSI and the Universidade Federal do Paraná. By the same token, surely there are businesses that are interested in taking on people with a broader humanistic education rather than with technical training. In any case, the general trend of each defending their own *standpoint* seems to prevail.

Most of the voices we listened to in Chapter 2 – including mine – are the voices of "faithful guardians", i.e. scholars, professors and thinkers dedicated to the Humanities. In this sense, it is not surprising that most of them will defend the autonomy of the university and will try to protect it against restricted and exclusive ends. Had we listened to different voices stemming from different standpoints (such as that of the businessman or the pupil,[23] we would most likely have had access to different points of view on the same issue. Taking POSI (see I, 2) again as an emblematic example of the dynamics between disparate standpoints, one can hardly expect academia and businesses to find *that*

23 In fact, the voices of a few pupils – mostly my BA and MA peers and my students at the Universität Wien –were heard in I, 1 and revealed an opposite point of view.

many common denominators precisely because their motivations, their interests and expectations – in other words, their *standpoints* – are utterly diverse, so any attempt to reconcile them completely can only be accompanied by a simultaneous movement of criticism of this attempt.

Another element that contributes to the diversity of the question of standpoints is the fact that they take different shapes in different cultures. All of us who have lived in a foreign country know that issues such as experience and education are valued completely differently depending on the culture. Following some of my friends' odyssey looking for jobs in Austria, for example, it is quite evident that professional experience is valued over education here, and this tendency certainly has a great impact on people's mindsets. Sworn translators, for instance, need not have any specific qualifications as the exam is open to anyone with at least five years' *experience* as a translator (cf. PYM et al. 2012: 28). Even at university academic qualifications are less relevant than, say, in Brazil. The teaching staff at the Centre for Translation Studies at the Universität Wien, for example, consists of 131 people, out of whom just over a third (45) have a PhD.[24] In addition to that, only a few have a fixed contract with the university, whereas the majority are "temporary lecturers". They work on a six-month-basis, teach a limited number of courses, get the same pay irrespective of their qualifications and do not usually do research.

Unlike in Austria, in Brazil people without a doctorate are only very rarely hired to teach at public universities. If we take the Programme for Translation Studies of the Universidade Federal de Santa Catarina, for example, all 40 teaching staff are doctors and enjoy a permanent contract with the university. They are expected to actively engage in research activities and do not usually work outside the university.[25] Qualifications are also very much valued outside university, with salaries tending to increase accordingly.

With these examples I do not mean to say that the situation in Brazil is better than it is in Austria, but rather to call the readers' attention to *difference*, to the fact that these questions are culture-specific. In other words, an average academic in Austria cannot be said to hold the same standpoint as an average academic in Brazil as far as, for example, the conflict theory-practice is con-

24 Cf. http://transvienna.univie.ac.at/lehre-lehrende/lehrende-alphabetisch/ (last accessed in August 2013). It must be added, nevertheless, that a few doctor titles mentioned in this page do not correspond to what is today considered a doctor title, which means that this figure (45) is, in reality, probably even lower.

25 Cf. http://www.pget.ufsc.br/curso/colegiado.php (last accessed in August 2013).

cerned. Indeed, the way each culture tends to perceive each of these notions is simply different.

The question of standpoints is influenced by yet another factor, a factor that has become increasingly common in Brazil, namely the intersection of occupations. Particularly in translation studies, the number of professionals engaged both in academia and translation is striking. The trend is particularly strong in literary translation and translation of nonfiction – nearly half of all translated volumes recently published by Companhia das Letras, one of the largest and most prestigious Brazilian publishers, was translated by members of academia.[26] Furthermore, in addition to being professors (and hence scholars and researchers) and translators, a few are also poets and novelists, Paulo Henriques Britto (see Part III) probably being one of the most renowned examples.

In this light, it is interesting to observe how the dynamics amongst these multiple standpoints work within the same person. As already mentioned above, amongst all "academic" interviewees from the book *Conversas com Tradutores* (BENEDETTI & SOBRAL 2003/2007), the only two to overtly stress the importance of the applicability of translation theory to translation practice were two well-established translators (Britto and Wyler – see I, 3). In this sense, the idea that the standpoint of the practitioner may occasionally exert a stronger influence on them than that of the academic is indeed clarifying.

Let us also consider the involvement of academics in administrative or political affairs, which gives rise to new sets of implications. Being engaged in political or institutional struggles – be they in terms of obtaining funds, establishing a new department, getting a new curriculum through, amongst various other possibilities – certainly sets new motivations and interests into motion. Witness, for example, the will to turn translation studies into an independent discipline, along with the will to grant the work of the interpreter and translator the status of a profession in the full sense of the term – both particularly strong in Europe during the 1980s, but still very much present. These movements towards "independence" and "professionalisation" seem to have brought about a notion of translator and interpreter training that is strictly applied, largely market-oriented. After all, if translation were to become a profession, it had to be taught and perceived as such, which in turn seems to have led to a process of professionalisation of university courses. The complete breakup

26 This piece of information is available on the publisher's website, under "Lançamentos" (new books) – http://www.companhiadasletras.com.br/. The remarks made here are based on the list published there in November 2010.

with the linguistics and literature departments of the university – which is the case not only of the Universität Wien, but also of several European universities that offer translation and interpreting courses – has also contributed to this tendency towards professionalisation. This is because the vast majority of the time and space that used to be dedicated to those disciplines within translation courses has been substituted by subjects deemed more "useful" for the pupils' future careers – such as the use of media, terminology, translation technology, project management, diversity management, amongst many others.[27]

In this sense, the process towards the acknowledgement and professionalisation of the work of the translator/interpreter (advertently or inadvertently) led to a change of focus as regards translator/interpreter training, making it more market-oriented, more practice-oriented – even after efforts such as the one recently made at the Universität Wien to make their curricula more theoretical, against the advice of POSI, for example (see I, 1 and 2).[28] Political agendas such as this one also have a strong impact on the dynamics of standpoints, as this example clearly illustrates. Surely many of the scholars behind these great European movements to grant independence and autonomy to translation studies were not primarily motivated by aspirations such as POSI's, for instance; on the contrary. However, the interests and motivations behind this political agenda – translated mostly through the wish to elevate the status both of the discipline and of the profession – brought about changes in translator/interpreter training as far as the tension theory-practice is concerned. New departments were founded and new curricula were passed so as to raise the status of the profession, and the question of "ends" (see I, 2) was somewhat overshadowed by this initial objective, remaining in the background. Different countries and different institutions dealt (and are currently dealing) with this question of theory-practice in translator/interpreter training in different man-

27 These subjects are all taught within the BA in Transcultural Communication of the Universität Wien – its full curriculum is available through the following link: http://transvienna.univie.ac.at/fileadmin/user_upload/fak_translationswissenschaft/ Studienprogrammleitung/Curr_2007_Aenderung_BA_Transkulturelle_Kommunikation.pdf (last accessed in December 2010). In 2011, a new curriculum was put in place (see http://transvienna.univie.ac.at/fileadmin/user_upload/fak_translationswissenschaft/Studieninformation/BA_Curriculum_2011.pdf - last accessed in February 2014).

28 For more on this, see II, 2.4. In Rosemary Arrojo's view (mostly in 1998a), this wish to establish an independent discipline for translation studies goes hand in hand with the belief in objective science and universally applicable theories, with science as a means to increase efficiency and with theory as a resource to systematise knowledge and make it more practice-oriented.

ners, and it is beyond the scope of this book to analyse these individual trends.[29]

In any case, the example of the will to elevate the status of translation both as a discipline and as a profession illustrates how political agendas exert a strong influence on one's standpoint, and in this case specifically, this influence changes the dynamics of the conflict theory-practice as well. I wonder what will happen in Brazil in the next few years, with increasingly more translation departments being planned to be founded quite independently from the linguistics and literature departments – though the vast majority of those involved in these changes were trained in literature and linguistics, as was once the case in Europe. Will the desire to establish an independent discipline and to grant professional status to the translator/interpreter gradually make translation courses more market-oriented? I will come back to this question in IV, 3.

All in all, the objective of this chapter was to call the readers' attention to the fact that one's standpoint exerts a great influence on how one perceives translation theory and practice. Additionally, I have argued that, within translation studies, these standpoints are marked by numerous factors, the most conspicuous of which being "occupation" – whether one is a scholar, a practitioner, a student, etc., considering that in some cases multiple simultaneous occupations apply. However, one can hardly say that those who share the same occupation will invariably share the same point of view as regards translation theory and translation practice – as we have seen through various examples in the previous chapters. For this reason, we have looked into two different factors that add more dimensions to the question of standpoints, namely *culture*, which is a highly specific and changeable factor to be taken into account, and *political or institutional interests and motivations*, which may lead to changes in standpoint. Surely they are additional elements that exert influence on one's standpoint; nonetheless, in light of the examples presented here, the elements explored in this chapter seem the most conspicuous.

Yet another factor seems to play a fundamental role in one's perception and expectations about translation theory and translation practice – a factor not accounted for in the present chapter. This factor refers to whether one's perspectives are marked by a predominantly essentialist way of thinking or a pre-

29 An important issue here that will be addressed in IV, 4 is that of a unified methodology for translation studies across the globe. Scholars such as José Lambert (cf. LAMBERT 2007) argue for the adoption of a universal translation methodology, regardless of the cultures and languages involved. His universalistic "model" has become increasingly influential in Brazil, particularly because of his recent work at the Universidade Federal de Santa Catarina as a guest professor.

dominantly anti-essentialist way of thinking. Indeed, in Chapter 5 we will look into Arrojo's argument of the essentialist versus the anti-essentialist, which will then be further developed in Chapter 6.

5. (Anti-)Essentialist Theory, (Anti-)Essentialist Practice?

Since language, which Saussure says is a classification, has not fallen from the sky, its differences have been produced, are produced effects, but they are effects which do not find their cause in a subject or a substance, in a thing in general, a being that is somewhere present thereby eluding the play of différance.

Jacques Derrida
(translated by Alan Bass)

In Rosemary Arrojo's view, what seems to be the distinctive feature of translation theories in general is whether they subscribe to a more essentialist or a more anti-essentialist line of thought, be it overtly or not. In her paper entitled "The Revision of the Traditional Gap between Theory and Practice and the Empowerment of Translation in Postmodern Times" (1998b), Arrojo offers great insight on this issue. For her, the ever growing visibility of translation studies in recent times has much to do with the "opening of new undergraduate and graduate programmes in universities worldwide" (ARROJO 1998b: 25), but it is also "closely related to the dissemination of a reflection generally labelled as postmodern, (...) anti-essentialist" (ARROJO 1998b: 25). Furthermore, she claims that this polarity between essentialist and anti-essentialist approaches has little to do with "the division between cultural studies and linguistic inspired theories of translation" (ARROJO 1998b: 26). I will develop these initial ideas in the following paragraphs.

What Arrojo very aptly describes in this paper is how "modern, essentialist approaches" (ARROJO 1998b: 27), such as Nida's, Catford's, Mounin's and Wilss', rely on what Stanley Fish calls "theory hope" (see I, 3), i.e. "'the promise that theory seems to offer' as a 'means of salvation' against relativism, chaos" (ARROJO 1998b: 29 – quoting Fish – cf. FISH 1989/1999: 342, 155). This "means of salvation" is strongly bound to a notion of translation practice as a "form of symmetrical, ideally neutral exchange or interaction between cultures" (ARROJO 1998b: 29), hence the firm belief of these so-called essentialist theorists in an *exemplary model or method that could show translators how to find adequate equivalents, (...) universally acceptable ethical terms (...), [a model or method that could] frame the relationships to be established between original and translation, no matter which languages, cultures and interests are involved (...). It is something a practitioner should ideally consult when he 'wishes to perform correctly'* (ARROJO 1998b: 28, 30 – quoting FISH 1989/1999).

Indeed, this perspective appears to be very much in line with one of the concepts of translation theory to which some of the voices quoted in this book seem to refer – particularly in I, 3. I have summarised this perspective of theory as a set of precepts that describes and governs over practice, making up a general and universally acceptable model. For this notion of theory, practice is stable, homogenous and hence fully susceptible to the theory that regulates it.

In her paper, Arrojo shows how the attempts in this direction made by theorists like Peter Newmark, Wolfram Wilss and Joseph Graham ended up being perceived as impossible by their mentors themselves. Newmark begins his 1981 *Approaches to Translation* by saying that his aim is to "determine appropriate translation methods for the widest possible range of texts and text-categories", as well as to provide a "framework of principles, restricted rules and hints for translating texts and criticizing translations" (NEWMARK 1981/1986: 19). Remarkably similarly, Wills, in his 1982 *The Science of Translation: Problems and Methods*, set out to write a "modern translation theory" whose main objective was to "determine the relationships between the individual constants of the translation process as explicitly as possible (...) and regardless of the languages involved" (apud ARROJO 1998b: 28-29).

Still at the beginning of the 1980s, Graham opened his article "Theory for Translation" by announcing the development of a "rigorous theory for translation" including "something like a practical evaluation procedure with criteria necessarily specific, though general nevertheless"; a theory "leaving no implication, no supposition, and thus demanding no interpretation at all"; a theory, in other words, "absolutely scientific" (apud ARROJO 1998b: 29-30). Therefore, his aims and expectations did not differ much from those of the authors mentioned in the previous paragraph. By the end of his attempt to write one such theory, however, he comes to the conclusion that the task is impossible because it is impossible to find "grammatical equivalents across languages". He then concludes the paper by asking himself "whether indeed translation really is a subject for theory after all" (apud ARROJO 1998b: 30). I will come back to this question towards the end of the present chapter.

In the aforementioned paper, Arrojo also examines Mounin's 1963 *Les Problèmes Théoriques de la Traduction*, and many of his ambitions and frustrations reveal themselves to be extraordinarily similar to his "essentialist" successors. Right at the beginning of his book, Mounin states that translating would be impossible if one took into account the latest developments in linguistics. He then concedes not only that translating *seems* possible, but also that translators and translations *do* exist and appear to work well, which leads him to the well-known conclusion that "l'existence de la traduction constitue le scandale de la linguistique contemporaine" (MOUNIN 1963/1998: 8). From his point of view, this "scandal" leaves one with two possibilities: either translation is theoretically condemned because of linguistics or linguistics is questioned in the name of translation:

> *(...) ou condamner la possibilité théorique de l'activité traduisante au nom de la linguistique (et rejeter ainsi l'activité traduisante dans la zone des opérations approximatives, non scientifiques, en fait de langage) ; ou mettre en cause la validité des théories linguistiques au nom de l'activité traduisante* (MOUNIN 1963/1998: 8-9).

Nevertheless, he chooses neither the former nor the latter and sets out to describe translation as a linguistic "problem". At the end of his work, though, he concludes that despite his efforts translation is a process that resists systematisation after all (MOUNIN 1963/1998: 271). Indeed, as Arrojo pointed out in the paper in question, already in the second chapter of the book, Mounin, commenting on Bloomfield's "radical" conclusion on this matter (MOUNIN 1963/1998: 27), had already provided great insight as to whether his task would be feasible or not. In Mounin's words,

> *(...) Bloomfield élimine, en premier lieu, tout recours aux mots* ***pensée, conscience, concept, image, impression, sentiment*** *comme autant de notions non encore vérifiées scientifiquement. Pour avoir le droit d'utiliser ces mots dans une sémantique scientifique (une science des significations) nous devions avoir une psychologie scientifique, c'est-à-dire, une explication totale des processus dans le cerveau du locuteur et le siège* (MOUNIN 1963/1998: 27 – his emphasis).

Yet, although he admits that in these terms translation is scientifically impossible, his "theory hope" (cf. FISH 1989/1999: 322)[30] persists as he believes that these theoretical problems will be resolved in the future, by fully scientific, universal theories. So does Bloomfield, as he regrets the fact that "nous en sommes *encore* très loin" (apud MOUNIN 1963/1998: 27 – my emphasis) from this "sémantique scientifique". This pointing to the future is, according to Arrojo, a typical trait of essentialist theorists who, even when faced with the impossibility of the tasks they assign themselves, still believe some day someone will be able to carry them out successfully. Indeed, scientism is an emblematic characteristic of essentialism – and vice-versa.

Arrojo then looks into more contemporary approaches which, from her point of view, are still "essentialist" and have the same "old dilemmas and expectations" (ARROJO 1998b: 34). She takes Mona Baker, for example, in her 1996 paper "Linguistics and Cultural Studies: Complementary or Competing Paradigms in Translation Studies?", and analyses how Baker announces her affiliation to an anti-essentialist perspective, but still derives her dilemmas and expectations from essentialism, very much in line with the authors quoted in the previous paragraphs. Although Baker admits that meaning is social and conventional, produced by history and ideology and therefore not intrinsic to texts, she very contradictorily criticises heterogeneity and multiplicity in translation studies. For Baker, this multiplicity and heterogeneity prevent the field of translation studies from establishing a truly scientific, united and universally accepted methodology (see footnote 29 on page 67). Baker is particularly concerned about cultural studies and the fact that it has "never had a distinct methodology of its own" since it tends to "reject absolute values", thus "encourag[ing] an amateurish, incoherent and less rigorous approach to its object of study" (apud ARROJO 1998: 35). Here again Fish's "theory hope" seems pertinent.

Arrojo, in contrast, relying on Derrida's ideas developed mainly in his essay "Différance", perceives this heterogeneity and multiplicity as only human:

If meaning and values are socially determined, and if different societies and cultures have different concerns, different groups within such societies and cultures will always be inclined towards heterogeneity rather than consensus. In such a strictly human (or Babelic) scenario, where "lan-

30 Fish's notion of "theory hope" permeates this entire book. I will continue mentioning it between quotation marks but will stop adding the references.

guage has not fallen from the sky" and, therefore, where there is no room for the absolute or the transcendental, "fragmentation" is not merely an unfortunate accident to be avoided at all costs (...), it is simply the very condition of its own possibility (ARROJO 1998: 35 – quoting Derrida 1972/1982).[31]

In other words, she perceives fragmentation as the result of our differences – in terms of values, cultures and societies. In this sense, "the absolute and the transcendental" would not only be unfeasible, but also undesirable. I will come back to the question of heterogeneity in translation studies later in this chapter, as well as in Part IV.

As for the role of theory in translator training, Baker asserts in her 1992 *In Other Words: A Coursebook on Translation* that "if translation is to become a profession in the full sense of the word, translators will need something other than the current mixture of intuition and practice to enable them to reflect on what they do and how they do it" (BAKER 1992: 4).[32] Indeed, this takes us back to the question asked in I, 3, i.e. to what extent can someone be a professional (in whatever area) when s/he has no specific training in this area? It seems that Baker's reply is reasonable enough, and she makes an interesting analogy between translation and medicine, saying that without theory "we will never be seen as anything but witch doctors and faith healers" (BAKER 1992: 4). When she expands on the possible scope of a translation theory, she quite suitably maintains that

Almost every aspect of life in general and of the interaction between speech communities in particular can be considered relevant to translation, a discipline which has to concern itself with how meaning is generated within and between various groups of people in various cultural settings (BAKER 1992: 4).

31 This is more or less the passage in Derrida's essay "Différance" to which Arrojo is referring (see the epigraph that opens this chapter): "In a language, in the *system* of language, there are only differences. (…) But (…) these differences *play*: in language, in speech too, and in the exchange between language and speech. On the other hand, these differences are themselves *effects*. They have not fallen from the sky fully formed, and are no more inscribed in a *topos noetos*, than are prescribed in the grey matter of the brain. (…) Since language, which Saussure says is a classification, has not fallen from the sky, its differences have been produced, are produced effects, but they are effects which do not find their cause in a subject or a substance, in a thing in general, a being that is somewhere present thereby eluding the play of *différance*" (DERRIDA 1972/1982: 11 – translated by Alan Bass – his emphasis).

32 I will comment on this allegory in IV, 1.

And here she asserts that linguistics is the discipline translators must learn so as to "acquire a sound knowledge of the raw material with which they work" (BAKER 1992: 4).[33] As Arrojo points out, this is where the problem lies. In Baker's context, theory and science are "of course only related to essentialist, logocentric, Eurocentric and Western conceptions of science which by no means take into account the desire, the history or the circumstances of the scientist" (ARROJO 1998b: 36). Furthermore, in Baker's reflection "theory is unquestionably only essentialist theory, just as translation is only that which such theory is equipped to establish or deal with" (ARROJO 1998b: 37). In this context, Arrojo quite rightly emphasises that if Baker's point is that translators must learn linguistics in order to be truly professional, then "we are not at all far from Mounin" (ARROJO 1998b: 37). Another interesting aspect of Baker's argumentation is the fact that she maintains that this "theoretical knowledge is itself of no value unless it is firmly grounded in practical experience" (BAKER 1992: 2). Therefore, her notions of translation theory and practice seem to be in line with those of her predecessors analysed previously.[34]

Taking an instance from cultural studies to show how the essentialist versus anti-essentialist dichotomy has nothing to do with the linguistics versus cultural studies dichotomy, Arrojo looks into Lefevere 1992 *Translating Literature: Practice and Theory in a Comparative Literature Context*. She comes to the conclusion that even though he proposed a revolutionary theoretical framework, advocating concepts like "rewriting", "manipulation" and "patronage" (which sound anti-essentialist), his concepts of meaning and text are not at all different from the ones used in essentialist, linguistic approaches. As an example, Rosemary Arrojo mentions Lefevere's "traditional separation between the semantic (locutionary) and the pragmatic (illocutionary) levels entertained by linguistics" (ARROJO 1998b: 27). I will come back to cultural studies and Lefevere's work in the following chapter. For now what is important is Arrojo's thesis that however revolutionary approaches as Lefevere's may appear to be, they still rely greatly on essentialist, linguistic principles. This confirms a notion of theory that is ultimately prescriptive, with universal aims, as well as a notion of practice that must be controlled by theory, that must be as rational and watertight as the linguistic categories that permeate it. Lefevere's apparent-

33 A second edition of Baker's *In Other Words: A Coursebook on Translation* came out in 2013. The excerpts quoted/glossed here remained fairly unaltered (see footnote 34 below).

34 It is beyond the scope of this book to analyse more recent works by Baker, for example, so as to verify whether Arrojo's argument of essentialist tendencies remains pertinent. In this sense, Arrojo's remarks (and mine) should be taken only as far as the works quoted here are concerned.

ly radical and innovative proposals, illustrated by words as the ones quoted by Arrojo, lose their strength when he bases his theories on the same traditional principles – without further questioning.

Another interesting approach with clearly anti-essentialist aspirations but still firmly influenced by an essentialist perspective is Peeter Torop's – former head of the Department of Semiotics at the University of Tartu, in Estonia, after Jurij Lotman. In 1995 Torop published his *Total'nyj perevod*, which was translated into Italian by Bruno Osimo as *Traduzione Totale* in 2000.[35] Very much in line with Toury's descriptive studies (whom he quotes repeatedly in his book), one of Torop's main objectives in his book is to "evitare qualsiasi approccio di carattere valutativo, attenendosi scrupolosamente ai principi dell'analisi descrittiva" as well as "porre al centro dell'analisi la descrizione del processo traduttivo", as Bruno Osimo comments in his Prefazione to the book (Osimo 2000). Particularly interesting is firstly his emphasis on the need for a thorough description of the translation process and, secondly, on the unquestionable importance of the translator's poetics. Because his aim is solely descriptive, Torop proposes a comprehensive model for translation criticism which, as explained by Osimo above, avoids value judgements and privileges descriptions of literary translation. Additionally, the translator plays a central role in his approach to translation description, as one of the main factors to be taken into account is indeed the translator – including his/her background, experience, style and poetics.

However, notwithstanding Torop's almost unprecedented and detailed acknowledgement (at least in so-called "translation models") of the translator's personal poetics as a key factor in literary translation, he contradictorily insists on the necessarily and exclusively scientific nature of translation criticism, thus excluding subjectivity and ideology. In Osimo's words, another main objective of Torop's work is to "accettare criteri di ricerca esclusivamente di tipo scientifico, e adottare un metalinguaggio comune, così che la traduttologia in futuro possa svilupparsi in modo produttivo" (Osimo 2000), thus pointing again to Fish's "theory hope". Furthermore, he makes use of traditional linguistic categories to refer to text and meaning, such as "livello sintattico", "livello semantico", "livello pragmatico", "piano del contenuto" and "piano dell'espressione" (Torop 1995/2000: 1.2, 2); yet, he maintains that "L'approccio linguistico (…)

35 This Italian translation is available online at www.logos.it (last accessed in May 2007). As the online page numbers are not the same as the pages in the actual book, when I quote from it I will mention the chapter and section only.

non può pretendere di avere un ruolo fondamentale nella traduttologia, poiché non abbraccia tutto il complesso dei problemi di traduzione". Another contradiction that reveals this clash between essentialist and anti-essentialist ideas in Torop's work is the fact that even though he acknowledges the importance of the literary translator's poetics, he believes in the possibility of a "pura ricodifica" as opposed to a "libera interpretazione" (TOROP 1995/2000: 2) – as if translating could be "pure recoding" of the "pure meanings" present in the text. In this sense, however relevant and innovative Torop's contributions to literary translation may be,[36] his approach is yet another example of a theory with anti-essentialist ambitious which, nevertheless, is still grounded in a predominantly essentialist basis.

Torop's or Lefevere's example (and Toury's too, as we will see in I, 6) illustrate Arrojo's thesis that although at first sight linguistic theories of translation appear to rely on an essentialist perspective, whereas cultural theories seem to announce their anti-essentialist affiliation, a closer look will reveal that the latter still relies on traditional linguistics, thus unveiling a largely essentialist outlook on language and translation. I will expand this argument in the next chapter.

In any case, this reliance on essentialist values can by no means invalidate these approaches, as already argued apropos of Torop's example. This is precisely what Arrojo claims in her 1998 paper before she turns to what she calls "Translation and Non-Essentialism" (ARROJO 1998b: 40). She maintains that her criticism of the approaches she calls "essentialist" should not "dismiss their overall validity" (ARROJO 1998b: 39). Her aim, she insists, is to

> *(...) point to the illusory character of their most ambitious pretensions and to the basically authoritarian, asymmetrical relationship they (wittingly or unwittingly) propose to establish between a supposedly comprehensive, rationalist theory and a practice mostly viewed as neutral and apolitical* (ARROJO 1998b: 39).

Moreover, she claims that these "scientifically-minded" theories, "in the name of rationality and of an allegedly universal ethics", simply repeat the "age-old precept according to which the translator's work should be invisible and ideally neutral" (ARROJO 1998b: 39). Equivalence also plays a pivotal role in these

36 For a thorough analysis of Torop's work, cf. LEAL (2007a).

theories marked by an essentialist perspective (as we will see in the following chapter), as Nida's allegory of the train very aptly illustrates. Certain "content" must be safely transported without any interference, change or loss, regardless of the background or ideology of the one who conducts this "transport", or the context in which it takes place. Nonetheless, this does not mean that theories such as these do not bring important and interesting contributions to the field of translation studies, as Arrojo points out quite rightly (cf. LEAL 2012b).

Finally, in the last section of her 1998 paper Arrojo moves on to anti-essentialist views on translation. For her, the main distinctive feature of the anti-essentialist perspective is that translation is no longer perceived as a "problem" or a "scandal" from which linguistics or translators must be saved. Similarly, nonessentialist thought does not rely on a "theory hope", an all-embracing theory that would finally establish universal truths about translation. In this sense, these perspectives she calls "nonessentialist" appear to be fairly in line with the second notion of translation theory and practice to which different voices quoted so far seem to refer (particularly in I, 3). I have summarised this notion of translation theory as not practice-oriented, but rather awareness-raising. Its aim is not to govern over practice – which is perceived as manifold and heterogeneous – but rather to stimulate reflection and critical thinking.

By taking the Saussurean postulates of the arbitrary and conventional sign to an extreme, these perspectives that Rosemary Arrojo calls "postmodern" question the "possibility of stable meanings that could be reproduced and recovered in their sameness" (Arrojo 1998b: 41), be it through translation or simply through reading and interpretation. In this new light, translating is perceived as "regulated transformation" ("transformation reglée" – DERRIDA 1972: 31) rather than "neutral transferral", which in turn raises the question of the translator's visibility not as a problem to be avoided, but rather as "the focal point of translation scholarship in postmodern times" (ARROJO 1998b: 42). In this sense, it would be impossible to ask *whether* translators "*should* ideally remain invisible" (KATAN 2009: 140-141), as visibility simply *is* a natural trait of transcultural communication.

Seeing translation in this new light takes us back to questions asked earlier, such as "should translation students be able to directly apply the theories they learnt to the translation market of their country?" (which I myself asked), or Graham's "[is] indeed translation really a subject for theory after all [?]"(apud ARROJO 1998b: 30). Here again what "theory" and what "practice" do we have in mind? Let us remember the double concept of theory and practice proposed in this book. If "theory" is prescriptive (or descriptive), model-like, with supposedly universally accepted, absolute truths; if "theory" relies on linguistic dichotomies such as form and content, signifier and signified[37] or meaning and form without questioning them, then "practice" is perceived as stable and homogeneous. In this context, practitioners expect theory to answer all their questions with absolute, indisputable, scientific answers, as we saw above. Their practice must ideally be controlled by this "all-embracing" theory – which does not usually happen. So the lack of success in the interaction between theory and practice may be due to the fact that, as already pointed out, theories marked by a predominantly essentialist perspective never appear to be able to fulfil all the tasks their authors assign themselves. This makes them hope for a better future (as Mounin does), or complain about the heterogeneity and multiplicity of translation studies (as Baker does), or about the lack of practical applicability of certain theoretical views (as Britto does).

Translation students hopelessly try to apply these essentialist approaches (and the anti-essentialist as well, for that matter) to the translation market of their countries (as I myself did several times), but often feel frustrated or, as Wyler puts it, they have "great difficulty finding receptivity in the market for what [they have] learnt" (BENEDETTI & SOBRAL 2003/2007: 196). Practitioners and self-taught translators feel these "theories" do not solve their problems, and therefore simply discard them for being far less important than talent and technical knowledge, as a few translators commented previously, or as Britto indicated. If translation theory, in Graham's words (apud ARROJO 1998b: 29-30), should "leav[e] no implication, no supposition, and thus deman[d] no interpretation at all", then perhaps it is not a subject for theorisation after all, since these demands seem impossible to be fulfilled, projecting its impossibility onto practice as well. In summary, from a more essentialist point of view, translation practice would always involve loss, relegating the translator to his position of traitor; theory, in turn, would never be able to comprise practice in its multiplicity "for the simple reason that, in order to be finally finished and absolutely accurate, it would also have to be infinite as it would have to reproduce everything that is, was or will ever be associated with translation, and, thus (...) would never be controllable or systematized" (ARROJO 1998b: 38).

37 For more on the nature of the criticism proposed here, see I, 6.

On the other hand, what Arrojo calls "anti-essentialist" perspectives – fairly akin to the second notion of translation theory discussed here – do not take meanings as stable and controllable, but rather as arbitrarily and conventionally produced within languages and cultures (or "interpretive communities", to borrow Stanley Fish's term – see FISH 1980). Therefore, from this point of view interpretation – both in general and in translation – become particularly crucial. In this light, the status of translation practice is elevated, as translators are perceived as autonomous professionals who *produce* rather than *recover* meanings. Theory is not designed to provide ubiquitous answers directly applicable to practice, nor is it supposed to govern over practice. Instead, theory provides insight, raises awareness, stimulates reflection on the role of translation in society and sheds light on questions about language and culture. It is as Gillian Lane-Mercier puts it (apud ARROJO 1998b: 43):

> *[a] translation theorist or practitioner cannot but position him or herself – aesthetically, politically, ideologically (...). Indeed, it is this very positioning, be it overt or covert, conscious or unconscious, avowed, unavowed or disavowed, that enables us to go beyond dualist conceptions of translation in order to bring to the fore the ethical stance which translation both entails and implies.*

In this sense, translation would be a very fruitful subject for theorisation; a subject which not only raises awareness about translation practice, but also sheds light on linguistic, cultural and social phenomena in general. Looking for applications to precise aspects of translation practice is no longer a relevant question, as the notion of "theory" loses its prescriptive nature. In other words, the act of theorising need not be done with a view to its "utilitarian purpose" (see I, 2), but rather constitutes a *practice* in itself, a *theoretical* practice.

Raising, once again, the question of what a professional translator is (and having Baker's medical analogy in mind), Baker's answer seems indeed quite pertinent: "if translation is to become a profession in the full sense of the word, translators will need *something* other than the current mixture of intuition and practice to enable them to reflect on what they do and how they do it" (BAKER 1992: 4 – my emphasis). Nevertheless, as Arrojo stresses in her 1998 paper, this "something" does not necessarily have to be linguistics since, as we have seen, there is far more to it than linguistics. Arrojo summarises this point as follows:

Since it emphasizes the need for translators to take full responsibility for their inevitably authorial intervention in the writing of the target text, this kind of anti-essentialist research on the ethical implications of the translator's visibility and on the notion of translation as "regulated transformation" has certainly begun to offer us a much needed instrument not only to raise awareness among practicing translators about the conflicting relationships they tend to establish both with "theory" or science and with their own work, but also to equip aspiring professionals with the critical background which will allow them to become fully responsible translators, well aware of their authorial voices. It is this awareness and the acceptance of its far-reaching consequences which will begin to allow translators to make the difficult transition from sensitive amateurs or talented craftsmen to self-conscious writers, who know about their fundamental role in the shaping of the cultural and social conditions of their work (ARROJO 1998b: 44).

From this point of view, translation theory is key to translator/interpreter training, but not necessarily as a ubiquitous, prescriptive or descriptive model that will ultimately (even if "inadvertently", in the case of more descriptive approaches) dictate the rules of practice, but rather as a potent stimulus for reflection. And whereas it is true that when it comes to professional translation this reflection is not indispensable, it is also true that it makes more aware, critical translators. This is what some practitioners very aptly pointed out in I, 3, or as Rosemary Arrojo summarises in this last quotation.[38]

One last question that seems pertinent – and to which I shall repeatedly return in this book – is whether these two different notions of translation theory and practice actually lead to different practices. Some translators will perceive their practice as inevitably doomed to failure, always involving loss and frustration, whereas others will see it as a natural process of transformation whereby change, bias and ideology are exercised. The question is whether these different perceptions of practice lead to remarkably, overtly different translation practices. In other words, do supposedly revolutionary or "radical" theories – as Britto puts it and as postmodern approaches are commonly la-

38 I will address this question of what makes a translator/interpreter professional at length in IV, 1 and 3.

belled (see I, 3) – entail revolutionary practices as well, whatever "radical"[39] means?

If one expects translation theory to describe a certain practice, it is quite obvious that one will also expect that *different* theories describe *different* practices. From this point of view, a more "conservative" translation theory would perceive translating as a more "conservative" practice. By the same token, a more "radical" theory would imply that the practice it describes is "radical", too. Nevertheless, if we discard the notion of applicability, very much in line with the contributions of poststructuralist thought, does this belief in the correspondence "radical" theory – "radical" practice, "conservative" theory – "conservative" practice make sense at all?

I have already hinted at this question earlier, and believe it is important to keep it in mind as our discussions advance. The answer to this question is also closely related to the analogy of the half empty or half full glass used in the title of the present book – but let us leave it aside for the time being, until the end of I, 6. In the next chapter, the different notions of theory and practice discussed here will be further clarified, accepting the hypothesis that essentialist beliefs and anti-essentialist beliefs lead to different notions of translation theory and translation practice, and discussing whether these differences circumscribe different paradigms in translation studies.

Before we move on to it, nonetheless, I would like to make a final remark concerning the association made in the present chapter between, on the one hand, the two notions of translation theory and translation practice mentioned repeatedly in this book and, on the other hand, Arrojo's "essentialist" and "anti-essentialist views". Similarly to the question of standpoints addressed in I, 4, these associations are both generalisations and approximations, which means that they in no way imply watertight, absolute categories. Even though the notion of theory as a set of precepts that rule over practice seems akin to a predominantly essentialist way of thinking, whereas the notion of theory as a source of reflection and awareness appears to be in line with a largely nonessentialist perspective, I do not mean to state that this association is clear-cut and outright. Nor do I intend to say, for example, that all those – quoted particularly in I, 3 – who perceive translation as a potent stimulus for reflection are strictly anti-essentialist thinkers. The intention is rather to show how one of the notions of translation theory and practice built from the voices of the

39 I will come back to this question of radicalism in Parts II and IV.

various practitioners, scholars and students quoted here entails hopes, motivations and interests similar to perspectives marked by an essentialist way of thinking. Likewise, the other notion of translation theory and practice proposed here involves hopes, motivations and interests fairly in line with those of so-called anti-essentialists views. Instead of some sort of affiliation to essentialism or anti-essentialism, this argument should be understood simply as symptoms of a more essentialist or more nonessentialist way of thinking.

6. The Question of Paradigms

There also seems to be a fourth type of progress.
I mean the one which looks like a zigzagging spiral,
advancing so to speak by leaps and bounds
but at the same time going round in a circle,
wasting a lot of breath and energy in fruitless repetitions,
but ultimately managing to come to a conclusion
some distance away from its starting point.[40]

Hans Vermeer

In this last chapter of Part I, I will develop Arrojo's argument of the essentialist and the anti-essentialist particularly in terms of their noncorrespondence to linguistic and cultural approaches to translation, respectively. By doing so, I intend to examine to what extent these approaches can be said to belong to the same paradigm in translation studies, as well as to what extent an anti-essentialist perspective of translation inaugurates a new paradigm in the discipline. As already argued a few times, a more essentialist perspective seems akin to a particular notion of translation theory and translation practice, whereas a perspective marked by anti-essentialism appears to entail a *different* notion of translation theory and translation practice. In this sense, the present chapter will help to close the discussion about the role that essentialist and anti-essentialist views play in translation studies, also helping to shed light on the diverse notions of translation theory and practice which, in turn, derive from these views.

In her *Tradução e Diferença* (1999), Cristina Rodrigues presents a thorough analysis of the impact of an essentialist way of thinking on different approaches to translation. In a nutshell, her work consists of an analysis of the relevance and the contours of the concept of equivalence in Catford's, Nida's, Lefevere's and Toury's translation theories. In spite of these authors' completely different departure points, objectives, contexts and backgrounds, Rodrigues shows how the notion of equivalence (and hence a tendency towards an essentialist view) permeates and, to some extent, works as the basis for their theories.

From a linguistic perspective (i.e. Catford and Nida), Rodrigues explains that "os vários conceitos de equivalência [são] a noção central para a definição

40 I first read this excerpt in Mary Snell-Hornby's *The Turns of Translation Studies* (2006: 2).

de tradução e para se determinar o que é uma boa tradução", or the various concepts of equivalence work as the central notion that defines translation and determines what a good translation is (RODRIGUES 1999: 163). As for the theorists who describe literary translation in their context of production (i.e. Lefevere and Toury), on the other hand, they question this notion of equivalence. However, they depart from the very same essentialist ideas of an original text in which all meanings have been intentionally placed by the author and whose effect can be measured and then transferred to another language (which, in other words, is a concept of equivalence, too). In this sense,

> *O paradigma em que os trabalhos se colocam é, portanto, semelhante, pois suas propostas decorrem do pressuposto de que há um sentido e uma mensagem presente nos textos que podem ser recuperados pelo tradutor ou pelo pesquisador e transmitidos por diferentes meios ou por diferentes línguas, sem que se afete sua integridade* (RODRIGUES 1999: 163).
>
> [The paradigm to which these works belong is similar, as their proposals derive from the presupposition that there is a meaning and message intrinsic to texts, which in turn can be recovered by translators and researchers and then transmitted in their entirety through different media or languages.]

Cristina Rodrigues asserts that Nida's and Catford's approaches are particularly similar in one respect, namely they both believe that source and target text have exactly the same meaning or, in other words, are equivalent. Therefore, the translation process would consist of a phase in which these fixed meanings are determined by the translator, and a second phase in which they are rendered, without substantial loss, into the target language (RODRIGUES 1999: 167). They perceive the source text as a whole containing all pertinent meanings and maintain that reading is a protective activity whose main objective is to recover these meanings correctly. Catford's essentialism is revealed through his belief that meanings exist before and outside languages, and that each language simply externalises a number of concepts which, in turn, are part of an "estoque universal", or "universal reservoir" (RODRIGUES 1999: 168) – hence the possibility of perfect equivalence. Nida, in contrast, emphasises the fact that meaning is placed in texts by their authors, and can then be recovered through a context and linguistic analysis of the text. For him, the key to a successful meaning recovery is to change the form so as to obtain the same effect as the original text (Rodrigues 1999: 168) or, in other words, so as to obtain an equivalent text.

Toury and Lefevere, on the other hand, reject this basic essentialist notion of meaning by not having the concept of equivalence as their departure point, as Rodrigues very aptly explains. Yet despite that, they both seem to believe in recoverable meanings placed in texts by their authors. Lefevere, in his aforementioned *Translating Literature: Practice and Theory in a Comparative Literature Context* (1992), entertains the same notion of protective reading since he repeatedly advises readers to seek the meaning intended by the author or to be careful not to misread the author's intentions (LEFEVERE 1992: 51, 63). Differently from Catford, Nida and Lefevere, Rodrigues points out, Toury is the only one to emphasise that source and target text are different texts, and that target texts work differently, in different contexts and different cultures compared to their source texts. Nevertheless, he places the source text in the very centre of his work as his *tertium comparationis,* i.e. an invariable element based on which translation norms can be fixed and target texts can be judged as acceptable or adequate. One such notion of source text can only imply the "correct" reading of a text, "correct" in the sense that it successfully recovers the meaning intended by the author. Therefore, Rodrigues asserts that

> *Assim como para Catford, Nida e Lefevere, também para Toury o "texto-fonte" acaba se colocando fora do jogo da linguagem, ou seja, o texto-fonte não estaria sujeito à interferência do leitor, do pesquisador, pois é considerado estável, portador de um conteúdo imanente* (RODRIGUES 1999: 169).
>
> [Just like for Catford, Nida and Lefevere, for Toury the "source text" is also located outside the dynamics of language; in other words, the source text is not subject to the reader's or researcher's interference as it is perceived as stable, as the receptacle of inherent content.]

Rodrigues' main criticism of the essentialist remainders of these four approaches stems from the fact that all four theorists entertain a notion of sign as the product of a stable relationship between signifier and signified, whereby form can be opposed to content (RODRIGUES 1999: 173). From this point of view, a sign would be the direct representation of something in its absence. Our thought would then be the untainted perception of the world, which in turn would mean that language represents things in their essence (RODRIGUES

1999: 173). This way, absolute equivalence between languages would be possible because each language would simply use different signifiers to refer to the exact same concept or thing. Taking into account some of Louis Kelly's considerations in his 1979 *The True Interpreter: A History of Translation Theory and Practice in the West*, Rodrigues argues that theories as Saussure's (1917) to a certain extent reinforced the abovementioned notion of sign and language, whereas theories as Quine's (1959) fiercely questioned it (RODRIGUES 1999: 174).

Largely based on linguistic experiments that focus on the question of reference, the American philosopher Williard Quine contends that there is not just one possible relationship between words and objects, but rather many, which leads Rodrigues to the conclusion that, when it comes to the objective comparison between languages,

> *Qualquer decisão que se toma sobre a referência se liga a aparatos conceituais arbitrários, pois o significado não pode ser uma entidade objetiva, na medida em que está vinculado ao comportamento dos usuários e à sociedade* (RODRIGUES 1999: 175).
>
> [Any decision that one may make about the question of reference is tied to arbitrary concepts because meaning is not an objective entity, but rather is always subject to the speaker's behaviour and to society.]

Indeed, this takes us back to Bloomfield's conclusion that we need scientific semantics before we can objectively compare languages (see I, 5), or even to Saussure's assigning the "problem" of reference to psychology (SAUSSURE 1917/1986: 15, 111). Despite (or because of) its potential impact on the notion of equivalence, or even on the possibility of a definite theory (shall we say science?) of language, Quine's ideas seem to have gone largely unnoticed – at least within translation studies – until the early 1990s, as Rodrigues points out (RODRIGUES 1999: 175).[41]

Concerning Saussure, Rodrigues maintains that he breaks away from the Aristotelian and Augustinian tradition of direct association between a word and the perception of reality in the world. As Susana Kampff Lages explains, "A concepção agostiniana partiria, pois, da idéia de que haveria dois planos da linguagem: o primeiro originário, dos nomes ou da pura nomeação e o segun-

41 Cf. STEINER (1975/1998, especially chapters 4 and 5).

do, do discurso, derivado do primeiro", or the Augustinian concept departed from the basic notion of two language levels: the first and original one was the plan of names and designations, whereas the second was the plan of speech, derivative of the first (LAGES 2002/2007: 211). Indeed, Saussure emphasises that languages are not lists of terms, and that there is no reality, no pre-established ideas before and outside language. In other words, there is no "essence" above words, governing them. From this point of view, each language would articulate its own meanings; hence it would be impossible to say that two words from two different languages refer to the same perception of the world, the same reality. In this sense, objective equivalence between languages would be impossible (RODRIGUES 1999: 186).

Another greatly significant element of Saussure's theory stressed by Rodrigues is the fact that signs cannot be taken individually, seeing as their very definition derives from their relationship to other signs and not from an image in the exterior world. This way, signs do not contain an essence that can define them, but rather can only be defined through their differences to other signs in the system (RODRIGUES 1999: 187). Once again, if there is no common essence, no bond between sign and referent, then one cannot attribute common values to words of different languages.

Yet somehow the very notion of equivalence in translation seems to have its roots in Saussurean structuralism. Rodrigues finds two possible reasons for this, the first one being the dichotomic distinction Saussure defends between form and substance. She uses John Lyons' words from his 1968 *Introduction to Theoretical Linguistics* to clarify it:

Just as the same lump of children's clay can be fashioned into objects of different shapes and sizes, so the ***substance*** *(or medium) within which distinctions and equivalences of meaning are drawn can be organized into a different form in different languages* (LYONS 1968/1995: 56 – his emphasis).

From this point of view, Rodrigues explains, there would be a "substance" of content before and outside languages, and each language would simply articulate it differently (RODRIGUES 1999: 187). For Saussure, this "substance" should be the object of study of psychology rather than of linguistics.

According to Cristina Rodrigues, the second reason why the claim for equivalence between languages is attributed to Saussure derives directly from this first reason, namely "a fala é a representação imediata do pensamento", or speech is the immediate representation of thought (RODRIGUES 1999: 188). This would allow for universalism, thus repressing the differences between languages in the name of the universality of reason. Indeed, this is what Derrida discusses in his *De la Grammatologie* (DERRIDA 1967/1997), a work of great importance for the argumentation developed by Rodrigues. For her (and Derrida), this universalism would guarantee the stability of meaning, allowing both for the mutual intelligibility between languages, and for perfect equivalence between languages, as most logocentric[42] theorists would claim.

As Derrida asserts, whereas on the one hand Saussure did break away from the then dominant notion of sign defined by Aristotle and Augustine, on the other hand he reinforced this same tradition by keeping watertight oppositions between signifier and signified, speech and writing. Furthermore, Saussure endorsed this age-old tradition by overvaluing speech and everything that binds a sign to its sound (cf. DERRIDA 1967/1997: 29-31). From Rodrigues' point of view, these are the reasons why one normally associates Saussure with the idea of a transcendental meaning, a "substance" that ensures the stability of meaning across languages (LYONS 1968/1995: 56; cf. RODRIGUES 1999: 187). By referring to Derrida, Rodrigues explains that the articulation between signifier and signified is what produces meaning, and that there is no such thing as a pure, material term opposed to a conceptual term, precisely because none of these terms can be perceptible or intelligible on their own (RODRIGUES 1999: 191).

Rodrigues concedes, however, that the deconstructionist criticism of Saussure is by no means radical, as the opposition between signifier and signified is necessary. In Derrida's words,

(...) la traduction pratique la différence entre signifié et signifiant. Mais, si cette différence n'est jamais pure, la traduction ne l'est pas davantage et, à la notion de traduction, il faudra substituer une notion de transformation: transformation réglée d'une langue par une autre, d'un texte par un autre. Nous n'aurons et n'avons en fait jamais eu affaire à quelque "transport" de signifiés purs que l'instrument — ou le "véhicule" — signi-

42 By "logocentric" I mean the derogatory term coined by Ludwig Klages in the 1920s and used mostly within deconstructionist thought (but also in critical theory) to refer to tendencies that centre all texts and discourses on the *logos*, as in the *word* or *reason*. Indeed, a corollary of logocentric thought is precisely the belief in a transparent and stable relationship between signifier and signified.

fiant laisserait vierge et inentamé, d'une langue à l'autre (...) (DERRIDA 1972: 31).

Therefore, what needs to be questioned is simply how "pure" and "stable" the relationship between signifier and signified is, and not the opposition itself. This is why structuralism and poststructuralism are simultaneously complementary and opposite, as already mentioned in the Interchapter. Arrojo puts it this way (ARROJO 1996a: 60):

(...) pode-se dizer que há, entre estruturalismo e pós-estruturalismo, uma relação paradoxal que se divide entre uma oposição radical e uma complementaridade. Na passagem (sem fronteiras claramente demarcadas) que marca a transição do estruturalismo para o pós-estruturalismo, há uma radicalização de insights e pressupostos que, de certa forma, permite pensar a reflexão pós-estruturalista como uma espécie de estruturalismo mais atento e menos iludido.

[(...) one can say that there is a paradoxical relationship between structuralism and poststructuralism, i.e. radical opposition and complementariness at the same time. In the transition (without clearly established boundaries) from structuralism to poststructuralism, a radicalisation of insights and presuppositions takes place, which in turn allows one to perceive poststructuralism as a more attentive and less naive kind of structuralism.]

By acknowledging that signifier and signified are different and that their relationship is by no means pure and stable, one can then perceive translation a sort of "regulated transformation" (cf. DERRIDA 1972: 31) whereby this difference, this unstable relationship between signifier and signified is practised.

In this sense (and to a certain extent in line with Rodrigues' arguments) these anti-essentialist perspectives *could* be said to inaugurate a new paradigm in translation studies; a paradigm that places linguistic and cultural approaches (to mention only the approaches discussed in the present chapter) side by side, as Arrojo argued in her 1998 paper examined in I, 5. But before we draw any conclusions, let us look at the issue of paradigms in our area. It is often said that translation studies went through a great paradigm shift in the 1980s, thanks particularly to Vermeer's Skopostheorie. Mary Snell-Hornby describes it as follows:

In his [Vermeer's] model, language is not an autonomous "system", but part of a culture. Hence the translator should not be only bilingual, but also bicultural. Similarly, the text is not a static and isolated linguistic fragment, but is dependent on its reception by the reader, and it invariably bears a relation to the extra-linguistic situation in which it is embedded, it is therefore "part of a world continuum". This approach relativises both text and translation: the one and only perfect translation does not exist, any translation is dependent on its skopos and its situation (...). With this approach a translation is seen in terms of how it serves its intended purpose, and the concept of translation, when set against the former criterion of SL [source-language] equivalence, is more differentiated and indeed closer to the realities of translation practice (SNELL-HORNBY 2006: 52-53).

Indeed, one could even claim that the turn in translation studies for which the German functional approach is (partly) responsible has something to do with the turn caused by anti-essentialist thought, as Vermeer himself admits in his Preface to Michaela Wolf's *Übersetzungswissenschaft in Brasilien: Beiträge zum Status von „Original" und Übersetzung*:

Seit dem Erscheinen der Oficina [de Tradução, Rosemary Arrojo] wurden weiteren Ansätze und Disziplinen, vor allem Dekonstruktivismus und die Psychoanalyse verstärkt eingezogen. Es war für mich eine Art Bestätigung meiner Gedanken, daß der Einbezug dieser Fächer meine brasilianische Kollegin [Arrojo] zu ähnlichen Folgerungen geführt hat, wie wir sie in Europa vertreten. Dabei gestehe ich gerne zu, daß Arrojos Ergebnisse die Schlüsse aus einem funktionalen Ansatz radikaler weiterführen, als dies bisher im alten Kontinent der Fall war (VERMEER 1997: 9).

Needless to say, Vermeer's work is largely marked by a hermeneutical perspective, which in turn is fiercely criticised by poststructuralist thought. Nevertheless, words such as "eine Art Bestätigung meiner Gedanken" or "Arrojos Ergebnisse [führen] die Schlüsse aus einem funktionalen Ansatz *radikaler* weiter" (my emphasis) seem to reveal a kinship between Arrojo's and Vermeer's works, however incipient this kinship may be.[43] Mary Snell-Hornby, for

43 For more on the possible intersections between Vermeer's ideas and deconstruction, cf. DIZDAR (2006).

example, goes as far as to say that Derrida's and Arrojo's ideas about translation as transformation and as an activity that involves meaning production rather than *re*production "correspon[d] to that of Vermeer, particularly with his image of 'dethroning' the source text, his rejection of the static concept of equivalence and his concept of the active role both of the translator and the target-text reader" (SNELL-HORNBY 2006: 61-62). In fact, she even speaks of a "mutual cooperation" (SNELL-HORNBY 2006: 62-63) amongst Vermeer's, Arrojo's and Holz-Mänttäri's works, with Wolf's book *Übersetzungswissenschaft in Brasilien* as an example of it.

Indeed, this movement towards more culture-oriented or *skopos*-oriented approaches can be perceived as an intermediate stage between linguistic approaches and anti-essentialist perspectives in translation, as Susana Kampff Lages summarises very well in her 2002 *Walter Benjamin: Tradução e Melancolia*:

> *(...) as reflexões funcionalistas (...) juntamente com os estudos desenvolvidos em Israel pela escola de Tel Aviv (...) aparecem igualmente como uma reação a concepções de tradução confinadas a limites estritamente lingüísticos, geralmente preocupadas com o conceito de fidelidade (...), voltadas sobretudo a uma adequação imediata ao texto de partida, que procura eliminar quaisquer marcas da intervenção do tradutor sobre o texto. Nesse sentido, elas ocupam um* ***espaço de transição entre a visão tradicional e as novas teorizações****, no sentido em que trazem para o primeiro plano (...) o contexto histórico-cultural da tradução. Elas realizam aquilo que Mary Snell-Hornby chama de "virada cultural" nos estudos sobre a tradução. Entretanto, essas correntes funcionalistas permanecem, por um lado, tributárias de uma visão segundo a qual seria possível compreender em sua totalidade as "intenções comunicativas" [Reiß e Nord] de determinado autor (...), [visando o] desenvolvimento de uma metodologia e de uma teoria que possibilite generalizações* (LAGES 2002/2007: 74-75 – my emphasis).
>
> [(...) the German functional approach, together with the studies developed in Tel Aviv, arises as a reaction to strictly linguistic translation approaches, which in turn are mainly concerned with fidelity and the immediate adequacy of target text when compared to source text, thus disregarding the translator's interference with the text. In this sense, they [functional approach, etc.] *represent the transition between the*

more traditional perspective [linguistic approaches] *and the newer theories* [anti-essentialist views] because they bring to the fore the historical-cultural context of the translation. They are responsible for what Mary Snell-Hornby calls the "cultural turn" in translation studies. Nevertheless, these functional theories remain, in a way, tributary to a notion according to which one can fully understand the "communicative intentions" [Reiß and Nord] of a given author, aiming at developing a methodology and a theory that allow for generalisation.]

Nevertheless, the question of whether the term "paradigm" is adequate to describe the contributions of anti-essentialist thought to translation studies remains unanswered. In *The Turns of Translation Studies: New Paradigms or Shifting Viewpoints?* (2006), Snell-Hornby defines "paradigm" as "groundbreaking contribution", whereas a "change in position on already established territory" would simply be a "shifting viewpoint" (SNELL-HORNBY 2006: ix). For her, the great paradigmatic changes in translation studies took place from the late 1970s to the mid-1980s, whereby the manipulation school and the German functional approach represented two different and more or less simultaneous paradigm shifts. On the one hand both break with linguistic dogmas and are target-oriented (as opposed to the then traditional source-oriented approach). On the other hand they differ in the sense that the manipulation school, unlike the German functional approach, claims to be strictly descriptive (SNELL-HORNBY 2006: 47-55).

In view of the issues addressed in I, 5 and in the present chapter, one could question (as Lages does – see indented quotation above) to what extent these approaches really broke with linguistic dogmas, as both the theories grouped under the manipulation school and under the German functional approach derive many of their methods and terminology from linguistics.[44] Similarly, the shift that these approaches provoked towards the target text – including the target culture and public – is questionable since, as Rodrigues points out (RODRIGUES 1999: 169), the source text remains to a large extent the *tertium comparationis* of these theories (see footnote 44 on this page). However, in view of the then dominant linguistic approaches to translation, of course the manipulation school and the German functional approach did represent a colossal

44 For time and space constraints I will not carry out a thorough analysis of the relationship between the German functional approach and so-called traditional linguistic dogmas. For more on this, cf. LEAL (2006, 2007a). I will take for granted here the conclusions drawn in these works.

change – no wonder Snell-Hornby and many others (SNELL-HORNBY 2006: 161-162; cf. LEAL 2012b) perceive these changes as paradigmatic in translation studies.

In this light, Lages' argument seems plausible enough – that between a predominantly essentialist perspective and a predominantly anti-essentialist perspective the so-called "cultural turn" of the 1980s represents an intermediate stage (LAGES 2002/2007: 74-75). So one could claim that these anti-essentialist contributions, embodied particularly by deconstruction, *do* make up a new paradigm in translation studies in the sense that their rejection of linguistic dogmas is taken to an extreme; i.e. it is carried forward in a more *radical* way (to recall Vermeer's words quoted above).

If we nonetheless consider deconstruction a poststructuralist movement, and if we bear in mind that the relationship between structuralism and poststructuralism is one of opposition and complementariness at the same time (see Arrojo's indented quotation above), the idea of a paradigm shift does *not* seem pertinent. It is as Norris puts it (NORRIS 1982/2002: 30):

> *It is not a question, [Derrida] repeats, of rejecting the entire Saussurean project or denying its historical significance. Rather it is a matter of driving that project to its ultimate conclusions and seeing where those conclusions work to challenge the project's conventional premises. (...) Saussure is thus not held up as one more exemplar of a blind and self-deceiving tradition. Derrida makes it clear that Structuralism, whatever its conceptual limits, was a necessary stage on the way to deconstruction.*

Given their close and rather unique relationship, the transition from structuralism to poststructuralism is hardly one of paradigm shift. Especially if one has in mind the notion of paradigm discussed in KUHN (1970), it is difficult to see how perspectives marked by poststructuralist thought could proceed to becoming hegemonic – firstly because this would require great willingness on the part of countless academics and scientific communities to fiercely criticise their own works, and secondly because hegemony would defeat the very objective of poststructuralist thought.[45]

All in all, poststructuralist thought cannot be said to inaugurate a new paradigm in translation studies because it remains rather close to structuralist

45 I will return to the issue of poststructuralist thought and hegemony in IV, 4.

views. Much as there appears to be a tendency for postmodern theories to be perceived as radical,[46] this is not at all the case since these theories do not propose new approaches, methodologies and *practices*. The contributions of poststructuralist thought are rather *awareness*-raising and *thought*-provoking, and hence do *not* extend to translation practice.

* * *

If we look back at the paradox described in the Interchapter, we will remember that the poststructuralist trend involves the negation of and the dependence on structuralism at the same time – or, as Derrida puts it as far as deconstruction is concerned, it practices structuralism and a movement against it simultaneously (see footnote 9 on pages 21-22). Likewise, one of the ideas underneath the term "deconstruction" appears to point to "undoing" and "redoing" at once. As we now approach the end of Part I, let us link this discussion of paradigms to the title of this book: "is the glass half empty of half full?" This paradoxical condition that seems crucial to poststructuralism and deconstruction is also fundamental to the title of this book, but perhaps before we try and understand the paradox, let us first look at how the present title *could* be understood through either-or oppositions.

In the aforementioned *Walter Benjamin: Tradução e Melancolia*, Susana Kampff Lages associates the symptoms of melancholia to the movements within translation studies. Based on Arrojo's ideas about the desperate search for meaning in translation, Lages asserts that what characterises "postmodern theories" is their acceptance of the fact that it is impossible to recover the allegedly "lost origin", Benjamin's Ursprache, transcendental meanings. "Modern theories", in contrast, still resent this loss and thus have not given up the search for the "lost origin" (LAGES 2002/2007: 169). In Lages' view, both (i.e. modernism and postmodernism) seem to fit the pathological description of melancholia, with modernism representing the state of severe depression, and postmodernism representing the manic, euphoric state (LAGES 2002/2007: 169).

What Lages calls "modern theories" – and Rodrigues and Arrojo call "essentialist theories"[47] – overtly or covertly rely on a notion of sign as a stable

46 See footnote 39 on page 81.

47 The point is not to state that all three thinkers refer to the exact same theories; my point is rather that Lages tends to group theories under the labels "modern" and "postmodern", whereas Arrojo and Rodrigues appear to prefer the labels "essentialist" and "nonessentialist" (see Interchapter).

and pure relationship between signifier and signified. Even when they admit that this stability or purity is not absolute – as would be the case of the German functional approach for instance, or culture-oriented approaches – they still somehow cling to the belief that the author's intention will provide the ultimate meaning, or perhaps the translation brief, or the text itself, its words will irrevocably announce what interpretation is more or less suitable. These approaches do not fully acknowledge translators and their autonomy, poetics and ideology. Projected onto practice, these essentialist views make translating border on impossibility as translators desperately seek these stable meanings, these textual meanings, these authorial intentions and yet do not seem to concretely find them, or are criticised for having found the *wrong* ones. In this vicious circle, theory promises to scientise and fully systematise practice, but still contains large amounts of subjectivity and leaves various questions unanswered. Practitioners have no or little interest in theory precisely because it does not seem to contribute to their practice; practice seems to insist on being frustrating, ambivalent, unscientific. Theory arises once again as the hope to put an end to subjectivity –let us not forget Fish's "theory hope" here.

In contrast, the approaches that Lages calls "postmodern" perceive ambivalence and subjectivity as natural and human. Translation studies will be subjective to the same extent that literature, linguistics, philosophy and history, for example, are subjective. In this light, the aim of theory is to stimulate reflection and raise awareness and not to prescribe formulae for success. Translators must no longer be semi-gods, possessing both encyclopaedic knowledge and the power to read people's minds and intentions; instead, translators are only human, subject to their background, context and ideology. Translation practice is an activity inevitably carried out within this framework (and not these "constraints") since no one can dispose of their ideology when reading or translating a text, for example.

In this light – and as already mentioned repeatedly – we could speak of theory and practice and their interaction in at least two different ways, which I will call "pessimistic" and "optimistic" for the sake of the analogy proposed in the title of the present work – and let us bear in mind that my intention here is to offer an *initial either-or interpretation* of the title. The "optimist" sees the glass half full and perceives translation theory as the (potential) answer to his/her questions. Theory is scientific, practice-oriented, objective and neutral – or if it is not, it will soon be (remember Fish's "theory hope"). From this point of view, practice is governed by this (seemingly) all-embracing theory, but precisely because theory is not as all-embracing as one would wish, practice seems frustrating, incomplete. Still, the optimist hopes that things will improve in the future, that science will advance and bring subjectivity to an end. And let us not forget the state of "depression" suggested by Lages, because this constant frustration and problematic interaction between a theory that does not fulfil its own objectives, and a practice that never meets the requirements imposed by theory, make the optimist hesitate. Some will claim theory is, for this reason, useless (when its marriage to practice is not prosperous); others will overcome depression and carry on hoping.

As for the "pessimist" (shall we say "realist"?), s/he has accepted that theory can only do so much for one, and that it cannot be neutral, objective and all-embracing. Science will never advance to substitute or erase subjectivity, and human undertakings will always be inevitably marked by bias. Practice is neither governed nor purely described by theory; instead, there is a constant movement of indirect and mutual influence between theory and practice. Rather than feeling that theory is missing something, or that practice is impossible and incomplete, this "pessimistic" point of view allows one to see possibility rather than impossibility in translation. Perhaps Lages' "manic" state may be associated precisely with this sense of possibility, of openness which, nevertheless, does not lead to "theory hope", but rather to the sense of resignation that human beings will never be able to overcome their human, and hence finite, subjective condition. Therefore, theoretical discourse has its roots in philosophical thinking and has no direct practical application, but may lead to a more aware, better-informed kind of practice.

In this sense, as far as a more essentialist and a more nonessentialist perspective of translation theory and practice are concerned, one could easily associate ideas marked by essentialism to optimism, to the glass half full. Accordingly, views influenced by nonessentialism could be perceived as pessimistic, the glass half empty. Indeed, associations between poststructuralist thought and nihilism and/or scepticism abound – cf., for example, GROSS & LEVITT (1994/1998: 71-106).[48] And these associations are more often than not combined with praising remarks of the essentialist precepts criticised by poststructuralist thought, as these precepts are useful, optimistic, cheerful and scientific (GROSS & LEVITT 1994/1998: especially 76 and 87-88) – hence "glass half full".

48 I will address the issue of poststructuralist thought and pessimism at length in IV, 2.

But let us bear in mind here that we are contemplating the *same* glass, and that the challenge – once again paradoxical – lies precisely in being able to perceive the glass as half full *and* half empty *at the same time*, as this is the challenge of postmodernism. If we agree that the contributions of poststructuralist thought *do not* inaugurate a new paradigm in translation studies because of its bonds with structuralism, then poststructuralist thinkers *do not* perceive the glass as constantly and solely half empty. In other words, as they practice structuralism (and the glass half full) and a movement against it (the glass half empty) *simultaneously*, they can but conceive both images *simultaneously*. If we take Lages' allegory of melancholia, we can understand the state of depression and the state of euphoria as aspects of the *same* illness, presenting similar symptoms.

This paradoxical duality advocated here may also be perceived in different ways. As someone whose affinities lie with poststructuralist thought, I would say that the numerous associations made between poststructuralist ideas and pessimism and/or nihilism are highly questionable. Particularly in translation studies, the constraints imposed by essentialist ideas strike me as extremely demotivating since according to them translators/interpreters are expected to be able to read people's minds and forget their own history and ideology in order to be successful – a success that, nonetheless, never seems attainable. In this light, the contributions made by a more poststructuralist perspective seem quite liberating, and would thus be more in line with an optimistic way of thinking. But I will come back to this criticism, especially to deconstruction, in IV, 2.

For now what is important is the conclusion that even though poststructuralist perspectives have brought countless contributions to translation studies, exerting influence on the notions of translation theory and translation practice, it did not inaugurate a new paradigm in the area because of its close relationship with structuralism. In this sense, I ask the reader to take the challenge of understanding the allegory in the title of this book through Derrida's double gesture or double bind (see I, 2), i.e. through a movement of embracing and rejection at the same time, through a perspective that allows one to see the glass *both* half empty *and* half full.

In Parts II and III we will look into the works of Rosemary Arrojo and Paulo Henriques Britto, respectively, as examples of the issues discussed here. Their work as translation scholars and, in the case of Britto, as a translator and poet, illustrates not only the question of different standpoints – addressed in I,

4 – but also the impact of a more essentialist, structuralist way of thinking as opposed to a more anti-essentialist, poststructuralist way of thinking – addressed in I, 5 and 6. Their debate reveals some of the most relevant questions at the moment in translation studies in Brazil (and certainly not only there). Also, it unveils very different concepts of translation theory and translation practice, thus leading to diverse notions of what the area itself, as well as education in the area, should be like. The intention in Parts II and III is hence to exemplify and flesh out some of the issues introduced in Part I through the works of Arrojo and Britto.

Part II: Rosemary Arrojo

Amável o senhor me ouviu, minha idéia confirmou:
que o Diabo não existe. Pois não?
O senhor é um homem soberano, circunspecto.
Amigos somos. Nonada. O diabo não há!
É o que eu digo, se for… Existe é homem humano. Travessia.

João Guimarães Rosa

Rosemary Arrojo is arguably the best-known Brazilian translation scholar – both in Brazil and abroad. Including innumerable papers and book chapters in Portuguese,[49] in addition to three books (one of which with five editions at the time of writing – see below), all of which on translation studies, Arrojo's work is unquestionably one of the most well-established in Brazilian translation studies. Not to mention her publications in English: dozens of papers and book chapters in the most varied translation journals and translation anthologies which, together with her participation in international conferences, have granted her the status of acclaimed translation scholar the world over. In fact, at this writing her works have already been translated several times into German and Spanish, but also into Catalan, Turkish, and Hungarian.

In this part we will firstly examine Arrojo's academic life, so as to try and sketch the contours of her standpoint in translation studies (Chapter 1). This initial reflection will help us to anticipate some of Arrojo's motivations and interests as far as translation is concerned, throwing light on her theoretical views, which in turn will be the object of Chapter 2. Given the vastness of her academic production, it would be impossible to address all her works here. Therefore, I have decided to divide Chapter 2 into four sections, so as to accommodate a selection of her publications. Sections 2.1 to 2.3 will be devoted to her three books, namely (1986/2002) *Oficina de Tradução: A Teoria na Prática* (Translation Workshop: Theory in Practice), (1992/2003) *O Signo Desconstruído: Implicações para a Tradução, Leitura e o Ensino* (The Decon-

49 By 2001, Arrojo had published about 80 papers – in specialised journals and proceedings of conferences – and book chapters – mostly in Brazilian and foreign anthologies on translation studies (see footnote 114 on pages 208-209).

structed Sign: Implications for Translation, Reading and Teaching), and (1993) *Tradução, Desconstrução, Psicanálise* (Translation, Deconstruction and Psychoanalysis). Finally, in Section 2.4 we will look into a selection of four of her more recent papers (2000a, 2000b, 2000 with Chesterman and 2004). The main objective of Chapter 2 will be to draft Arrojo's theoretical standpoint in translation studies – always in view of her standpoint as an academic.

As I analyse her works, I will constantly call the readers' attention to aspects that I find crucial to the debates I propose here. Several key questions will be raised in this Part II, many of which were already anticipated in Part I – such as the alleged radicalism of poststructuralist thought, the aim of translator and interpreter training, the wish to establish an independent discipline for translation studies, amongst various others. In order to keep the focus on Arrojo, I will address these questions at length in Part IV – always indicating here, in Part II, when I intend to do so.

One last clarification before we move on to Chapter 1: I have decided to present Arrojo before Britto (see Part III) for practical reasons. Seeing as Britto dedicates a part of his work to directly and openly criticising poststructuralist thought on translation – including declared critiques to Arrojo specifically – I thought this structure would facilitate the reading. This way, the reader will be well-acquainted with Arrojo's theoretical standpoint by the time we get to Britto's criticism.

1. Standpoint as an Academic

Antes do doutorado, fiz um mestrado em literatura na Inglaterra (Universidade de Essex), e o foco desse mestrado era precisamente a teoria e a prática da tradução literária. (...) Mas meu interesse primeiro sempre foi a literatura.

Rosemary Arrojo[50]

Rosemary Arrojo began her academic career in 1969 at the Universidade de São Paulo, where she obtained a double degree (a teaching degree and a BA) in English and Portuguese (Letras, or literature and linguistics) three years later. She then did an MA in theory and practice of literary translation under Arthur Terry at the University of Essex between 1975 and 1977. And here it is important to point out that Terry was arguably Britain's greatest Catalan literary studies specialist, having published numerous works, including literary translations (cf. HULME 2004). Indeed, his specialities are comparative literature and literary translation, hence the mutual interest between him and Arrojo. Her thesis was called "An Anthology of Brazilian Modernist Poetry in Translation" and, as far as I know, these translations have never been published, except for one long poem by Brazilian poet João Cabral de Melo Neto, "O Cão sem Plumas", which came out under the title "The Dog without Plumes" in *Mundus Artium: A Journal of International Literature and the Arts*, vol. XVI, in 1983.

Between 1982 and 1983 Arrojo obtained two further Master's degrees from Johns Hopkins University, one in Humanities and one in Spanish. At that time she was also doing her PhD in Humanities under Eduardo González at the same university, for which she received a Fulbright Scholarship. She concluded her doctorate in 1984 with the thesis "Jorge Luis Borges' Labyrinths and João Guimarães Rosa's Sertão: Images of Reality as Text". When she started her PhD, González had only recently been admitted to Johns Hopkins University (1981) as a professor of Latin American literature and cinema. Having published five books mostly on Hispanic America, he is currently the head of the

50 This excerpt is Arrojo's reply to an interview question about her interest in translation theory (see footnote 53 on page105). It translates as follows: Before the doctorate, I did an MA in literature in England (University of Essex) dedicated precisely to the theory and practice of literary translation. (...) But my primary interest has always been literature.

Spanish and Latin American Subdivision of the Department of Romance and German Languages and Literatures.

But let us concentrate for a moment on Arrojo's PhD thesis. The title and key words of this work reveal subjects with which she would work for decades to come: Jorge Luis Borges (see II, 2.2 and 2.4), João Guimarães Rosa (see II, 2.3) and deconstruction.[51] Even though this text is not available and has not been published, one can speculate about the motivations behind it. Having worked with two specialists in Hispanic Literature, Terry and González (see above), Arrojo may have been introduced to Borges' works by one of them. As she had a degree in Brazilian Literature, Rosa was certainly a part of her reading repertoire by then, and deconstruction must have come into play as the possible interface between the two works. Let us not forget that by the early 1980s the impact of poststructuralist tendencies in the Humanities was far from negligible in the USA, a far cry from the situation in Brazil. Indeed, nearly a decade later Arrojo would write in the preface of one of her books (ARROJO 1993: 11):

> *No início dos anos 80, nos Estados Unidos, quando procurava um tema e um álibi aceitáveis para uma dissertação de doutorado, o reencontro com Pierre Menard, sua curiosa biblioteca e sua óbvia modernidade, à luz e sob a sedução das estratégias desconstrutivistas de Derrida, foi uma revelação que não cessa de ser renovada. Além de uma tese, o Pierre Menard de Borges me propiciou ainda o ingresso "oficial" na reflexão sobre tradução praticada no País.*
>
> [As I was looking for an acceptable subject and alibi for a PhD thesis in the early 80s in the USA, the reencounter with Pierre Menard, his curious library and his obvious Modernism – both from the point of view and under the seduction of Derrida's deconstructionist strategies – was an endlessly renewing revelation. In addition to a thesis, Pierre Menard provided me with the "official" ingress into the reflections on translation taking place in Brazil at the time.]

And here one could speculate further and say that two of her papers published between 1984 and 1986 in specialised journals in Brazil derived from her PhD

51 These key words were taken from her CV and include "labyrinth" in addition to "deconstruction", "Jorge Luis Borges" and "João Guimarães Rosa" (see footnote 114 on pages 208-209).

thesis. The papers were named "'Pierre Menard, Autor Del Quijote': Esboço de Uma Poética da Tradução via Borges" ("Pierre Menard, Autor del Quijote": A Sketch of Translation Poetics via Borges) and "O Labirinto Inescapável da Linguagem: Matriz-temática da Obra de J. L. Borges" (The Inescapable Labyrinth of Language: The Leitmotif of Borges' Work). Together with two other papers published in the same period – "A Tradução como Reescritura: O Texto/Palimpsesto e um Novo Conceito de Fidelidade" (Translation as Rewriting: The Text/Palimpsest and a New Concept of Faithfulness) and "Um Áporo e suas Aporias: Reflexões sobre um Poema de Carlos Drummond de Andrade" (An *Áporo* and its Aporias: Reflections on a Poem by Drummond) – these papers (or at least versions of these papers) would later make up a part of the book *Oficina de Tradução* (see II, 2.2), first published in 1986, a book in which Arrojo lays out the foundation for her work on deconstruction and translation.

In 1984 Arrojo began working as a professor and researcher. She had already worked as a teaching assistant at Johns Hopkins between 1983 and 1984, as a PhD student. There she taught Portuguese and Spanish at undergraduate level. Even before then, between 1977 and 1979 – i.e. before her PhD and straight after her MA in Literary Translation – she had worked as a teaching assistant at the Pontifícia Universidade Católica de São Paulo (PUC/SP), where she taught the following undergraduate courses: Contrastive Analysis, English Literature, North-American Literature, Translation Practice, English Conversation, and Intensive Reading. In 1984 she was hired once again by the same university (PUC/SP) as an assistant professor to teach more or less the same courses, together with a course named Translation Theory – a job she abandoned four years later. Nevertheless, in 1984 she had also started working as a professor at the Universidade Estadual de Campinas (UNICAMP), where she remained for about 20 years. Her main research area was translation theory and teaching, and she taught the following courses at undergraduate level: English Language, Literary Translation Workshop, Translation Practice, Translation Theory, English Literature, and North-American Literature. From 1984 she was engaged in graduate programmes as well, being in charge of the courses Introduction to Translation Studies, Translation Theories, Translation Topics, Advanced Seminars on Translation, Pragmatics, Topics in A-language, Reading in A-language, Grammatical Theories and Pedagogical Grammars, Introduction to the Theoretical Problems of Translation, and Topics in Discourse Analysis.

Between 1984, when she started working as a full-time professor in translation studies, and the early 2000s, she published just over 60 papers and book chapters, as well as two more books, *O Signo Desconstruído* (1992) and *Tradução, Desconstrução e Psicanálise* (1993) (in addition to the abovementioned *Oficina de Tradução*, in 1986). Moreover, another nearly 20 papers were published in the proceedings of academic events on translation, mostly in Brazil but also abroad. All her works are directly dedicated to the issue of translation, the vast majority of which through an openly poststructuralist view. Only four of her over 30 papers published in specialised journals in Brazil in this period include a specific literary corpus that works as the basis of her argumentation. These literary texts include the novel *O Grande Sertão: Veredas* (by Rosa), the poem "Áporo" (by Drummond), the short-story "Pierre Menard, Autor del Quijote" (by Borges), and Ítalo Calvino's novel *Se una Notte D'Inverno un Viaggiatore*. All the other 30 papers or so are dedicated to "more theoretical" questions surrounding translation, such as the issue of literariness, the question of faithfulness, translation ethics, translator training but mainly contemporary thought on translation (poststructuralism, psychoanalysis, deconstruction, feminism and nonessentialism). In fact, even in those four papers in which a specific literary text seems to be in the spotlight, the discussion is always oriented towards a more theoretical issue. In other words, the focus of Arrojo's work does not usually seem to lie on translating – as it will become clear in the next chapter – but rather on theoretical issues and discourses about translation. I will come back to this question later in this chapter.

In this period between the mid-1980s and the early 2000s, she also supervised four PhD theses on translation, and all of them approached translation from a poststructuralist perspective.[52] As a matter of fact, her four PhD students went on to become professors at Brazilian universities, three of whom work specifically on translation studies through a poststructuralist perspective – Cristina Carneiro Rodrigues (see I, 6), Maria Paula Frota (see IV, 1) and Silene Moreno. In the same period she supervised nine MA theses, all of which were devoted specifically to translation through a poststructuralist perspective. Amongst these dissertations, four were predominantly theoretical, dealing

52 Most of the information presented here was taken from her national CV (see footnote 114 on pages 208-209), which was last updated in 2001, when Arrojo was still working in Brazil. However, a 2009 version of her CV, which she kindly sent me upon request, mentions eleven MA and seven PhD theses. As this newer version of her CV unfortunately does not include detailed information on these works, I have not taken them into account in the analysis here.

with the following issues: Antoine Berman's theory of translation criticism, the dichotomy theory versus practice in translator training, the impossibility of machine translations, and the influence of linguistics in the teaching of reading. The other five theses, in contrast, incorporated specific corpora, in this case literary texts: *Hamlet*, *A Brief History of Time*, *The Catcher in the Rye*, *Macbeth* and *On the Road*. Amongst her MA students, the majority went on to do a PhD and to work as scholars at different Brazilian universities, some of whom specifically in translation studies, others in cognate areas. A few, too, dedicate their work to the interface between translation studies and poststructuralist thought.

At the time of writing, Arrojo is the director of the Translation Programme at Binghamton University, State University of New York, where she has been working as a comparative literature professor since 2003. In fact, this programme is the only one dedicated specifically to translation in the USA, and Arrojo was taken on precisely to help to set it up.[53] Arrojo has recently shifted the focus of her research slightly to representations of translation in fiction. Although her works still explore the interface between translation and contemporary thought (mostly deconstruction, but also psychoanalysis), they do not strike one as militant poststructuralist works any longer – I will come back to this question in what follows and in II, 2.4. Different works from the late 1990s and early 2000s already signal this slight change of interest, such as her 1999 paper "Interpretation and Possessive Love: Hélène Cixous, Clarice Lispector and the Ambivalence of Fidelity", included in the anthology *Post-Colonial Translation: Theory and Practice* (edited by Bassnett and Trivedi), and her 2002 paper "Writing, Interpreting and the Power Struggle for the Control of Meaning: Scenes from Kafka, Borges and Kosztolányi", published in Tymoczko and Gentzler's *Translation and Power*.

Unfortunately there is not much information available on her work at Binghamton University, so I cannot depict this last decade with as much detail

53 In an interview given to Susana Kampff Lages for the *Jornal da UNICAMP*, Arrojo comments that "creio que o fato de ter auxiliado a implementar um programa de pós-graduação relacionado à tradução no IEL contribuiu para que a Binghamton University me contratasse para implementar o primeiro PhD em estudos da tradução dos EUA" – I believe that the fact that I helped to set up a postgraduation programme related to translation at the Institute of Language Studies [at UNICAMP] motivated Binghamton University to hire me to set up the first PhD Programme in Translation Studies in the USA. Unfortunately the year of the interview is not displayed in the newspaper, and neither is the year in which that particular issue of the newspaper came out (cf. http://www.unicamp.br/unicamp/unicamp_hoje/jornalPDF/237pag0607.pdf - last accessed in August 2013).

as I did the previous two.[54] According to the webpage of the university, Arrojo has been in charge of various translation courses, seminars and workshops, amongst which are Borges and Calvino on Translation and Reading; Introduction to Theories of Interpretation; Methods and Masterpieces in Comparative Literature; Translation Theory; Introduction to Translation Studies; Postcolonial Translation Studies; Translation Workshop (literary and non-literary); Translation and Fiction; Representations of Translation in Latin American Fiction; The Translators as Objects of Study and the Interests of the Discipline; Translation and Power; Teaching Deconstructively; Politics of Translation; The Translator's Role in Contemporary Approaches to Translator Training; Translation and Ideology; and Translation and Globalisation – some at undergraduate and others at graduate level.

Her recent publications include the following papers and book chapters, all published in English: "Philosophy and Translation" (*Handbook of Translation Studies*, John Benjamins, 2010); "Translation and Impropriety: A Reading of Claude Bleton's *Les Nègres du Traducteur*" (*TIS: Translation and Interpreting Studies*, vol. 1, number 2, Fall 2006); "Translation, Transference, and the Attraction to Otherness: Borges, Menard, Whitman" (*Diacritics: A Review of Contemporary Criticism*, fall-winter 2004, vol. 34, published in 2006 – see II, 2.4); "Tradition and the Resistance to Translation" (*Kultur, Interpretation, Translation*, Peter Lang, 2006); "The Ethics of Translation in Contemporary Approaches to Translators Training" (*Training for the New Millennium: Pedagogies for Translation and Interpreting*, John Benjamins, 2005); "The Gendering of Translation in Fiction: Translators, Authors, and Women/Texts in Scliar and Calvino" (*Gender, Sex and Translation*, St. Jerome, 2005); "Lessons Learned from Babel" (*Target* vol. 14, number 1, 2002) – in addition to various others, as well as a few book reviews.

During this last decade Arrojo has worked as a guest professor and keynote speaker at numerous universities in Europe and the Americas, in countries like Spain, Germany, Austria, Turkey, Denmark, the Czech Republic, the United States, Canada, Brazil and Argentina. She has also been a member of the advisory and/or editorial board of several translation studies journals and anthologies, including the Benjamins Translation Library, *Translation and Interpreting*

54 The information presented here was taken from her personal CV (see penultimate footnote above) and mainly from the following websites: http://www2.binghamton.edu/comparative-literature/index.html and http://pic.binghamton.edu/index.htm. Unfortunately I do not know exactly how many publications and how many academic works she has supervised in this last decade, but I do know she is very much active in the USA.

Studies (journal by John Benjamins), *TranScribe* (journal on translation teaching from the University of Nevada), *TTR* (*Translation, Terminology, Writing* – journal of the Canadian Association for Translation Studies), *Transfer* (*Revista Electrónica sobre Traducción e Interculturalidad* – Universitat de Barcelona), in addition to the Brazilian journals *D.E.L.T.A.*, *Tradução e Comunicação*, *Tradterm*, *Cadernos de Tradução*, *CROP*, *Trabalhos em Lingüística Aplicada*, and *Alfa*. In addition to that, between 2003 and 2006 she was a member of the Executive Committee of the MLA's (Modern Language Association) Discussion Group on Translation.

As a comparative literature professor at Binghamton University, she has directed three PhD theses since 2003 – other theses are currently underway. All three works have recently been published under the titles *Gregory Rabassa's Latin American Literature: A Translator's Visible Legacy* (Maria Constanza Guzman); *Delectable Bodies and Their Clothes: Plato, Nietzsche, and the Translation of Latin America* (Ben Van Wyke); and *Blackness, Translation and the (In)Visible: Harryette Mullen's Poetry in Brazilian Portuguese* (Lauro Amorim). All three PhD students are currently engaged at universities in Canada, the USA and Brazil, respectively, where they work with translation and different interfaces with contemporary thought.

Rosemary Arrojo is certainly one of the best-known, most-read Brazilian translation scholars. Although most translation scholars in Brazil are probably not too keen on poststructuralist thought, her works remain canonical, making up the syllabi of perhaps all translation courses in the country. And here a number of different factors must have contributed to the elevated status she enjoys today. Arrojo started writing about translation in the mid-1980s, when the area was far from established in academia – one cannot say it is well-established today, with the first translation courses emerging only very recently at federal universities, so never mind over 25 years ago. As a matter of fact, the whole time that Arrojo worked as a translation professor, she was engaged in the Programme in Applied Linguistics of UNICAMP which, by the way, is probably one of the most reputable universities in Brazil. So Arrojo is arguably one of the pioneers of translation studies in Brazil, possessing one of the most complete, mature oeuvres of the area.

Furthermore, her international acknowledgement decisively contributed to her recognition in Brazil. There certainly are not many Brazilian scholars whose work not only is known abroad, but has also been translated into several languages (cf. Leal 2010b). The fact that she went abroad for her MA and

PhD was also a key factor for her career and subsequent reputation in Brazil. Surely some of Arrojo's references whilst she was abroad first entered Brazil through her work, awakening the interest of various local scholars. If we take Jacques Derrida, for instance, by the mid-1980s only a couple of his works had been translated into Portuguese – the vast majority of the translations of Derrida's early works available today stem from the 1990s. In view of this fact, it seems pertinent to assume that numerous translation scholars may have first read about Derrida, for example, in Arrojo's works.[55]

On account of the brief overview drawn in the present chapter, Rosemary Arrojo strikes one as a typical scholar. She obtained her PhD early in her career and established, to a certain extent, the main subjects on which she would work for decades to come. This focus, not only on translation studies but also on its interface with poststructuralist thought, has been clear from the beginning, underlying her choice of universities for her postgraduate education, as well as the courses she teaches both at undergraduate and graduate levels, and the academic works she supervises. Her publications were and remain a corollary of her academic interests and motivations which, by the way, are far from mainstream in translation studies. As already hinted at above, Arrojo faced, in the course of her career, far more criticism and disagreement than assentation, which in turn may justify a certain radicalism of which her works are constantly accused (see Part III).

On the one hand, the fact that she worked in the Programme in Applied Linguistics of the Institute of Language Studies at the prestigious UNICAMP may have been a source of difficulty and conflict for her – hence her relentless outcry for the independence of translation studies particularly from linguistics. On the other hand, UNICAMP is renowned for the reception of deconstruction in Brazil, which in turn is a great source of inspiration in Arrojo's work.[56] Her colleague then, the late Paulo Ottoni, was arguably one of the most representative Derrida specialists in Brazil, with nearly 10 books published on deconstruction, along with dozens of papers and book chapters. Furthermore,

55 I do not mean to belabour this issue much further as it exceeds the scope of this book. I believe Silviano Santiago might have been the first whose works can be said to represent the reception of deconstruction in Brazil, back in the 1970s. In this sense, Arrojo was certainly not the first, but *one* of the first – and her work introduced not only Derrida to numerous Brazilian readers, but also other key thinkers of poststructuralist thought.

56 In the same interview quoted in footnote 53 on page 105, Arrojo explains that she had the opportunity and privilege to set up an area of concentration within the Postgraduation Programme in Applied Linguistics dedicated to translation. In this sense, though "applied linguistics" may initially sound like a constraint, Arrojo did have a "safe niche" for translation inside it.

from the early 1990s to mid-2000s Ottoni coordinated seven different research groups on the interface between translation and deconstruction, thus establishing a solid place in Brazil for the reception of deconstruction.[57] Moreover, the Institute of Language Studies at UNICAMP is renowned for its competent translators and interesting translation projects, particularly in the area of classical languages and literatures. A few examples include Odorico Mendes, Trajano Vieira and Joaquim Brasil Fontes. In this sense, the role that UNICAMP played in Arrojo's career may be said to be twofold; perhaps their openness to deconstruction and translation was a reason to stay in 1988, and the affiliation to linguistics was a reason to go in 2003. And let us not forget that she is now engaged in the Comparative Literature Department of Binghamton University. But this is of course pure speculation.

So as far as her academic standpoint is concerned, a few elements appear to be particularly noteworthy. As already mentioned above, Arrojo seems to have determined the areas on which she would concentrate – poststructuralist thought and literary translation – at a very early stage in her career. These two issues are unquestionably very close to her heart, as she has dedicated over 30 years of her life to them – if we take her MA in literary translation as a starting point. Moreover, she appears to have devoted everything to an academic career – and not *any* academic career, but one in translation, in literary translation.[58] She left the country in the late 1970s to do her MA, and then again in the mid-1980s to obtain two more Master's degrees, in addition to a PhD. At that time, leaving Brazil for postgraduate education was by no means common, and probably indicated a wish to pursue an academic career. Her subsequent work confirms this trend – Arrojo has certainly dedicated her life to academia. From the beginning, she put a lot of effort and time into publishing papers, book chapters and books, as well as into taking part in academic events.[59] Her work with MA and PhD students also unveils her special dedication to academia, as many of them went on to pursue an academic career in the exact same area as

57 Having said that, it is important to emphasise that Arrojo was the one to inaugurate this place for the reception of deconstruction at UNICAMP, with Ottoni joining her only years later – he obtained his PhD from the same university in 1990. My point here is by no means to imply that they worked closely with each other or even that their reception of deconstruction went in the same direction. I rather want to contextualise UNICAMP and its renowned relationship with deconstruction and poststructuralist thought as a whole.

58 Though "literary translation" is probably too narrow a category, as Arrojo has been working on various areas surrounding translation and language as a whole.

59 Unlike some of her peers, as we will see in III, 2. At that time, striving for an active academic life was not nearly as common as it is today.

well – I believe not by mere chance. Though she has worked as a freelance translator (starting in 1977), this does not seem to have taken her fancy. In fact, as far as I know she has never translated and published a book, for example. She does translate excerpts from texts as examples in her own books and papers, and she translates her own papers into English and Portuguese, but she never seems to have invested in a career as a literary translator as such. In view of her vast academic production, it is quite clear that she preferred the academic path over the career as a translator.

In this light, it seems quite clear that the main focus of Arrojo's work is by no means practice-oriented, and hence has far more to do with that notion of translation theory as a source of awareness mentioned repeatedly in Part I. Her theoretical affinities, together with her career path, appear to reveal first and foremost an interest to *reflect* about translation. And since this interest of hers is far from mainstream in translation studies, a certain militant attitude on her part is to be expected – an attitude that, as a matter of fact, is also very much in line with poststructuralist thought as a whole. And let us not rule out the possibility that, precisely because of these long decades of theoretical struggle with her peers, it is only natural that the focus of her work should now have taken a slight turn towards representations of translation in fiction. Indeed, within this new research object she can still work on the interface between translation and contemporary thought, though not in such a militant fashion anymore, since she can take much of her previous work for granted. And, by the way, her PhD thesis, defended over 25 years ago, already pointed in this direction (of representations of translation in fiction). But perhaps back then she felt a greater need to introduce and draw the contours of contemporary thought – which remain obscure to many in translation studies today, so never mind back in the 1980s. And although this argument is rather speculative, I believe it is important to contextualise and try to understand this shift in her research. So let us now move on to Chapter 2 and analyse to what extent her academic standpoint, briefly sketched in this chapter, confirms our expectations as for her books and papers.[60]

60 I will come back to the question of her allegedly "militant" or "radical attitude" throughout Part II and in IV, 2 as well. As for this shift in her research, I will address it again in II, 2.4.

2. Works

Ihre [von brasilianischen Wissenschaftlerinnen und Wissenschaftlern] Beiträge sollten auf verschiedene Fragen eine Antwort versuchen: Was zeichnet die Ansätze in der brasilianischen Übersetzungswissenschaft aus, wie sie sich im Kontext des Poststrukturalismus im vergangenen Jahrzehnt entwickelt hat? Welche Schulen, welche Literatur, welche philosophischen Zugänge werden dabei rezipiert? Wie wird das „Andere" im postkolonialen Text repräsentiert? Und wie spiegeln sich diese theoretischen Überlegungen in der Praxis des Übersetzens wider?

Michaela Wolf[61]

In light of Arrojo's standpoint as a translation scholar, briefly sketched in II, 1, let us now examine some of her most important works, including her three books (Sections 2.1 to 2.3) and a selection of four of her more recent papers (Section 2.4). Although technically *Oficina de Tradução* appeared before *O Signo Desconstruído* – in 1986 and 1992, respectively – I have decided to analyse her 1992 book first because it is more exemplary of Arrojo's work and, therefore, will help to lay the foundation for her theoretical views. Except for that, her other works will be presented chronologically, as follows: Section 2.1 *O Signo Desconstruído* (1992/2003), 2.2 *Oficina de Tradução* (1986/2002), 2.3 *Tradução, Desconstrução e Psicanálise* (1993), and 2.4 Selected Papers (2000a, 2000b, 2000 with Chesterman and 2004).

As for the translations of Arrojo's works, I will point out whenever one of the papers that make up the following sections has been published in different languages as well. In fact, Arrojo's three books consist largely of collections of papers, the majority of which had been published in Brazilian specialised journals before the publication of the books – or at least very similar versions of the papers. Some of these papers were later translated into various languages and published mostly as book chapters in translation anthologies.

61 This excerpt was taken from Michaela Wolf's Vorbemerkungen in her *Übersetzungswissenschaft in Brasilien* (Wolf 1997: 7). Though these words do not refer exclusively to Arrojo, they refer to her as well. I believe that in the present chapter these questions suggested by Wolf will also be of utmost importance.

The main objective of the present chapter is to evaluate to what extent the expectations drawn in the previous chapter – as far as Arrojo's standpoint as an academic is concerned – are met by her works on translation studies. Before we move on to Section 2.1, let us not forget that one of her papers (1998b) has already been thoroughly analysed in I, 5, which means that the reader is probably already aware of some of her theoretical affinities.

2.1 O Signo Desconstruído

Apesar de dividido entre o senso moral imposto pela sociedade e a força do inconsciente, o homem ocidental, forjado no culto ao racionalismo, ilude-se com a sua suposta autonomia "consciente" – que não passa de uma instância derivada de processos inconscientes – e crê poder separar-se do "real", ou seja, crê poder olhar o real e o outro com olhos neutros; crê, em suma, poder "descobrir" "verdades" que não sejam construídas por ele mesmo.

Rosemary Arrojo[62]

O Signo Desconstruído (The Deconstructed Sign) consists of a collection of seventeen short papers, most of which were written by Arrojo herself, on deconstruction, "a palavra-chave em torno da qual se reúnem os artigos desta coletânea", or the key-word around which these seventeen texts are organised (ARROJO 1992/2003: 9).[63] According to the author, the deconstructionist project devised by Derrida is devoted to the nearly obsessive unmasking of the moments of aporia, blind spots and subliminal contradictions underlying all and every dichotomy or hierarchy based on which we elaborate our sciences, theories and world-views – "desmascaramento quase obsessivo dos momentos de aporia, dos pontos cegos e das contradições subliminares que se instalam

62 This excerpt is quoted and translated later in this section (see pages 120-121).

63 Most of the papers analysed in this section were written by Arrojo alone, with the exception of four, which were written together with her UNICAMP colleague Kanavillil Rajagopalan. Therefore, his name will obviously be mentioned every time I refer to one of these four papers, but in order to keep the references short, I will mention Arrojo's name only, as I will quote the book as a whole rather than all ten papers individually. Rajagopalan's individual work will not be analysed here, but his works with Rosemary Arrojo will make up an important part of the present section.

nas bases de qualquer dicotomia ou hierarquia a partir das quais elaboramos nossas ciências, nossas teorias e nossas visões de mundo" (ARROJO 1992/2003: 9). Amongst these dichotomies and hierarchies, she highlights two which the precision of Derrida's deconstructing scalpel has affected ("que a precisão do bisturi desconstrutor de Derrida tem atingido" – ARROJO 1992/2003: 10), namely subject and object, and signifier and signified. The main aim of the papers is thus to examine the impact of the deconstruction of these dichotomies on translation, reading and teaching.

In the present section, I will examine all the papers written by Arrojo herself, as well as those written in cooperation with Kanavillil Rajagopalan (see footnote 63 on page 112) – 10 papers in all. The analysis of these essays will help to unveil the foundation of Arrojo's work as a translation scholar, which will then be developed in Sections 2.2 and 2.3.

In the paper entitled "O Ensino da Leitura e a Escamoteação da Ideologia" (Reading in School and the Theft of Ideology), Rosemary Arrojo, together with her colleague from the Universidade Estadual de Campinas, Kanavillil Rajagopalan (see footnote 63 on page 112), analyses the impact of the deconstruction of the dichotomies subject-object and signifier-signified on reading. Curiously, they make use of another dichotomy to get to the heart of the matter, namely poststructuralism and what Derrida calls logocentrism (see footnote 42 on page 88). By opposing these two theoretical points of view, the authors assert that when faced with a reading task, readers have two options. Their first option is to believe that the meaning of the text, or the meanings of the words in the text are "independentes do sujeito, da história e das circunstâncias de leitura" or independent of reader, history and reading circumstances (ARROJO 1992/2003: 88). The second option, in contrast, is to believe that meaning is inevitably imposed, attributed to objects rather than discovered in them – "significados [são] inevitavelmente atribuídos, impostos aos objetos e não descobertos neles" (ARROJO 1992/2003: 88).

The first option refers to a point of view largely derived from Saussurean ideas, which

> *pressupõe (...) a possibilidade da neutralidade e da ausência de perspectiva como marca da relação leitor texto; pressupõe, em outras palavras, a possibilidade de que um leitor, no exato momento da leitura, pudesse se esquecer de tudo que o constitui como sujeito: seu inconsciente, sua história, sua cultura, sua ideologia* (ARROJO 1992/2003: 88).

[presupposes the possibility of both neutrality and absence of perspective as marks of the relationship reader-text; in other words, it presupposes that readers could, throughout the course of the reading process, forget everything that constitutes them as subjects, i.e. their unconscious, their history, their culture and their ideology.]

The opposite perspective (i.e. the second option presented above), largely taken from Derridean ideas, maintains that

(...) o significado não se esconde nem se embute no texto à espera que um leitor o decifre, compreenda ou resgate; o significado é produzido pelo leitor a partir de suas circunstâncias e das convenções que organizam e delimitam suas instituições, inclusive a linguagem (ARROJO 1992/2003: 88).

[(...) meaning is not hidden or inherent to texts, waiting for readers to decipher, understand or recover it; meaning is produced by readers according to the circumstances and the conventions that shape their institutions, including language.]

From the latter point of view, a particular interpretation cannot be deemed fully correct or incorrect, but simply acceptable or unacceptable according to a given situation or perspective.

Based on this Derridean view, the authors argue that a new teaching method must be adopted in schools, whereby teachers become aware both of their own ideology – which will inevitably shape their choices, interpretations and judgements – and their important role as meaning-shapers in the classroom. This last issue is particularly important since "no âmbito de sala de aula, cabe ao professor determinar quais significados e qual perspectiva serão aceitos para a realização de qualquer leitura", or in a classroom environment it is up to the teacher to determine what meanings and perspectives are acceptable for a particular reading task (ARROJO 1992/2003: 90). This new approach should then equip students with a greater awareness of their autonomy first as readers, but also as citizens who play different roles within institutions and communities and have the power to change them – since they are human construes.

All in all, in this first paper there appears to be a call for action by Arrojo and Rajagopalan, particularly as far as school teachers are concerned – as the title indicates. As they embrace the deconstructionist perspective on reading, and thus largely discard the so-called traditional idea of neutrality and lack of perspective associated with reading, they demand that school teachers not only become aware of these perspectives and bias, but also render them as clearly and transparently as possible to their pupils. This way, pupils should be able to understand why certain interpretations tend to be privileged over others - not because these interpretations are *the* correct ones in absolute terms, but rather because they reflect the values of that teacher, that institution, that community at that moment. Consequently, pupils would also become aware of the active role that they play in society, since these values that inform correct interpretations, for example, are established by them as well - and not by some exterior force to which they must necessarily bow.

In a similar line of thought, Arrojo examines the history of translation theory as it is presented in Steiner's renowned *After Babel: Aspects of Language and Translation* (STEINER 1975/1998). The paper is called "As Questões Teóricas da Tradução e a Desconstrução do Logocentrismo: Algumas Reflexões", or Reflections on Translation Theory and the Deconstruction of Logocentrism[64] (ARROJO 1992/2003: 71-79). Analysing the four-period division of the "literature on the theory, practice and history of translation" proposed by Steiner (STEINER 1975/1998: 248), Arrojo agrees with his conclusion that "over some two thousand years of argument and precept, the beliefs and disagreements voiced about the nature of translation have been almost the same" (STEINER 1975: 251 - commenting on Knox's *On English Translation*). For Arrojo, these recurring "beliefs and disagreements" essentially surround two issues, namely that of fidelity, and the relationships between original and translation (ARROJO 1992/2003: 73).

In her view, the problem does not lie in the solutions for the question of fidelity in translation, as proposed by countless translation scholars over these two millennia; rather, she asserts, the problem lies first and foremost in the very formulation of the question of fidelity (ARROJO 1992/2003: 73). In other words, within the limits of what she calls logocentrism, it would be impossible to find answers to the questions derived from it. These logocentric limits are largely responsible for the emergence of questions such as "is this translation faithful to its original?", but have not yet produced unanimous, objective answers to a single translation example. And in order for logocentrism to be able

64 This paper featured in Wolf's anthology *Übersetzungswissenschaft in Brasilien* under the title "Gedanken zur Translationstheorie und zur Dekonstruktion des Logozentrismus" - translated by Hans Vermeer (WOLF 1997: 65-70).

to produce adequate answers for the questions it sets forth, Arrojo believes that both man and reality would have to be centred on some sort of suprahuman, unchanging rationalism and logic in the same way that logocentrism is (ARROJO 1992/2003: 77). In this light, what Arrojo believes needs reformulating is not the answers to the questions, but rather *the questions*.

Therefore, in her opinion, instead of four periods, translation theory can be said to have undergone just two periods, marked by two key tendencies. The first tendency is preponderant and enjoys millennia of tradition; it asks time and again the same unresolvable questions, relentlessly. The second tendency, in contrast, is far from mainstream, but embraces the human, contextual, historical and local aspects of translation, and therefore is apt to reformulate the age-old questions repeated incessantly within the first tendency (ARROJO 1992/2003: 79). In other words, one could speak of an essentialist trend opposed to an anti-essentialist one marked by poststructuralist thought.

Similarly, in another essay, named "A Desconstrução do Logocentrismo e a Origem do Significado" (The Deconstruction of Logocentrism and the Origin of Meaning), Arrojo examines again how so-called essentialist theories perceive the issue of the origin of meaning, as opposed to how anti-essentialist perspectives deal with the same issue. In addition to that, she analyses how the question of literariness is handled within both theoretical standpoints.

According to Rosemary Arrojo, faced with the question "where is meaning?", the theories labelled logocentric would reply *outside* the subject or reader (ARROJO 1992/2003: 35). In this view, meaning is unquestionably embedded in the signifier (i.e. the *word* itself) and in the author's intentions, hence the possibility both of literal meanings (i.e. meanings that exist before interpretations and outside contexts) and of literariness as a textual feature. In this light, readers are "receivers" – as Jakobson's term[65] very aptly illustrates – whose main task is to *find* meanings in the text, including the literary, the ironic, the poetic or the literal ones (ARROJO 1992/2003: 35).

If one places more emphasis on the role of the author's intentions, the receiver's task is now to find out what the author meant to say. The author, in turn, is perceived as a patriarchal, authoritarian figure who holds the "direito indiscutível de determinar os destinos e contornos da sua 'prole'", or the unquestionable right to determine the fate and shapes of his "offspring" (ARROJO

65 For more on Jakobson's use of the term "receiver", cf. JAKOBSON & HALLE (1956/2002).

1992/2003: 36). Nevertheless, whether the emphasis lies on the signifier or on the author's intentions, the role played by the *word* remains the same:

(...) o significante (a palavra, o texto) funciona novamente como o invólucro duradouro e resistente capaz de aprisionar através dos tempos e em qualquer circunstância o significado autoral conscientemente pretendido. Ao leitor/receptor cabe apenas, nesse enredo, um papel filial e passivo (ARROJO 1992/2003: 36).

[(...) the signifier (the word, the text) works once again as the long-lasting and sturdy wrappings that imprison the author's conscious meanings through time and under all possible circumstances. In this context, what is left to the reader/receptor is merely a passive, filial role.]

The main weakness of these theories, Arrojo contends, lies in the fact that even though they place so much emphasis on the word and on the notion of transcendental meanings, they cannot clearly point out intrinsic textual distinctions between literal and figurative language, ironic and non-ironic passages, literary and non-literary texts. Indeed, the work of the North-American critic Eric Hirsch is emblematic of this weakness. In his 1976 *The Aims of Interpretation*, when faced with the task to determine clear distinctions between literary and non-literary texts, he admits that "the distinction can never be successfully formulated". Despite the impossibility, he argues that this distinction, as well as any other relevant "classifications of written speech", is in any case "ethically governed by the intentions of the author" (apud ARROJO 1992/2003: 36). However, as Arrojo very aptly points out, what Hirsch does not clarify is how one should proceed to find these intentions.

Relying mainly on the works of the French-speaking philosophers Jacques Derrida, Michel Foucault and Roland Barthes, Arrojo finds an alternative to the dilemma of the origin of meaning in poststructuralist reflection. As Saussure had already pointed out, signs are simultaneously arbitrary and conventional (see I, 5 and 6), the products of tacit agreements made amongst people within communities. Therefore, to think about the origin of meaning is inevitably to think about this pact, this tacit agreement from which the *production* of meaning derives. In this sense, one need not pay any vital respect to the text, as Barthes explains (BARTHES 1979: 78 – translated by Josué Harari).[66] Similarly, one need not pay any vital respect to the author, as Foucault argues:

The author is not an indefinite source of significations which fill a work; the author does not precede the works, he is a certain functional principle by which, in our culture, one limits, excludes and chooses; in short by which one impedes the free circulation, the free manipulation, the free composition, decomposition and recomposition of fiction (FOUCAULT 1979: 159 – translated by Josué Harari).

Indeed, Foucault stresses that the role that authors have traditionally played has been changing drastically in recent times (FOUCAULT 1979: 159):

Although, since the eighteenth century, the author has played the role of the regulator of the fictive, a role quite characteristic of our era of industrial and bourgeois society, of individualism and private property, still, given the historical modifications that are taking place, it does not seem necessary that the author-function remain constant in form, complexity and even in existence.[67]

In this light, Arrojo sketches a new answer to the question of the origin and location of meaning. Rejecting both the thesis of the *text* as the receptacle of stable meanings and the idea of the *author's intentions* as the protective cotton

66 In this same text, Barthes reminds us of the etymology of the word "text" to claim that the level of polysemy of a given text does not depend on how ambiguous its contents are: "The Text's plurality does not depend on the ambiguity of its contents, but rather on what could be called the *stereographic plurality* of the signifiers that weave it (etymologically the text is a cloth; *textus*, from which text derives, means 'woven')" (BARTHES 1979: 76 – his emphasis).

67 Perhaps the Pirate Parties of various EU member-states (and a few countries outside the EU as well) are already a sign of Foucault's and Barthes' "prophecy" in Europe regarding the role of authorship and ownership in our culture. Having obtained more than 7% of the votes in the 2009 European elections, thus beating a number of well-established parties, the Swedish Pirate Party was granted a seat in the EU Parliament, which later increased to two seats after the Lisbon Treaty came into force. Two of their chief proposals are that copyright term should be reduced to only five years, and that patents should be largely abolished. They claim that the current copyright laws are unbalanced and jeopardise the access to and exchange of information (for more on the Pirate Party International, see http://www.pp-international.net/). So perhaps in the same way as the Enlightenment saw the birth of the author, according to Foucault, our contemporary times will witness his gradual fading. This phenomenon is already to be seen in online platforms such as Wikipedia, for example, in which the entire notion of scholarship, authorship and power has been subverted. Various contemporary thinkers have been dedicating their works to this issue, linking the aims of libertarianism to the World Wide Web precisely because of its structure that seems to resist authority. For more on this, refer to works by Clay SHIRKY (1995) and Aleks KROTOSKI (2012 and 2013), for instance.

wool wrapping meanings, she then turns to *individual readers* and explains that it is *not* up to them either to determine the meanings of a particular text. Instead, she defends the thesis of the conventional sign and thus attributes the production of meaning to *social conventions*, very much in line with Stanley Fish's notion of "interpretive communities" (see III, 3.1). For her, in addition to shaping meanings, this intricate network of social conventions also determines the reader's own profile, wishes, circumstances and limitations (ARROJO 1992/2003: 39).

To come back to the question she implicitly asks in the title of the paper (where is meaning?), she would probably respond that it is not in the word, as words change across times and their meanings shift according to contexts; neither is it in the authors' intentions, as one hardly has any access to them, and even authors themselves may be doubtful about them;[68] neither is it in readers, as they alone cannot be responsible for these tacit agreements and conventions that underlie language (see footnote 122 on page 224); instead, meaning is located precisely in these tacit agreements and conventions, hence the conventionality of signs advocated by Saussure.

In these first three papers taken from *O Signo Desconstruído*, Rosemary Arrojo's strategy to oppose a traditional, dominant tendency – usually labelled essentialist, structuralist and/or logocentric – to a novel, fresh trend – commonly referred to as nonessentialism, poststructuralism and/or deconstruction – may incur difficulties. For those readers less acquainted with poststructuralist thought, her remarks may sound like an attempt to propose yet new sets of dichotomies, of absolute oppositions, of antagonistic and self-excluding extremes. It is without doubt understandable that a certain level of generalisation and categorisation are necessary for one to be able to write "academic" papers. Furthermore, it is expected that Arrojo's attitude should be perceived as radical or exaggerated at times, seeing as the views that she defends are far from prevailing. These generalisations and polarisations might nonetheless require more caution, as they are precisely what poststructuralist thought strives to relativise.

Besides the issues of the origin of meaning and its impact on the processes of reading, interpretation and translation, Arrojo examines some of the other main issues that make up the core of postmodern thought, as she announces in the Introduction to the aforementioned book (see above). In "A Noção do

68 For a curious example of an author who was unaware of her own intentions, cf. LEAL (2006) and IV, 1.

Inconsciente e a Desconstrução do Sujeito Cartesiano", or The Unconscious and the Deconstruction of the Cartesian Subject, she explains that Nietzsche and Freud are the most important precursors of poststructuralist thought particularly because they were the first to deconstruct the notions of "truth", "reason" and "reality".

As Arrojo points out, Nietzsche's most paramount contribution was to question the autonomy of the intellect and the subsequent illusion of the truth of things in themselves, as well as all other illusions that derive from this main one. Mostly taking into account Nietzsche's 1873 article "Über die Wahrheit und Lüge im außermoralischen Sinne", Arrojo analyses how the philosopher first deconstructs the intellect:

Denn es gibt für jenen Intellekt keine weitere Mission, die über das Menschenleben hinausführte. Sondern menschlich ist er, und nur sein Besitzer und Erzeuger nimmt ihn so pathetisch, als ob die Angeln der Welt sich in ihm drehten (NIETZSCHE 1873: 1).

Similarly, it is Freud who demonstrates how illusory the notion of the consciousness is, given the fact that it is always subject to the unconscious and its countless processes. Arrojo remembers the well-known metaphor of the horse and rider, whereby the horse is the id – i.e. the animal-like force – and the rider represents the ego, which in turn can be strong or weak, skilful or clumsy. This way, the rider can either control the horse and take it where he wants to go, or be controlled by the horse and be taken despite his own will (ARROJO 1992/2003: 14-15). Furthermore, Arrojo stresses the power of the superego, which the rider has to control in addition to the untamed horse (ARROJO 1992/2003: 15). She summarises the question of the power of both the intellect and of the unconscious as follows:

Apesar de dividido entre o senso moral imposto pela sociedade e a força do inconsciente, o homem ocidental, forjado no culto ao racionalismo, ilude-se com a sua suposta autonomia "consciente" – que não passa de uma instância derivada de processos inconscientes – e crê poder separar-se do "real", ou seja, crê poder olhar o real e o outro com olhos neutros;

crê, em suma, poder "descobrir" "verdades" que não sejam construídas por ele mesmo (ARROJO 1992/2003: 15).[69]

[Though divided between the moral sense imposed by society and the power of the unconscious, Western man, who was forged in the worship of rationalism, eludes himself with the alleged autonomy of "consciousness", which in turn is nothing but a derived instance of unconscious processes. He believes, thus, he can detach himself from what is "real", he can look at "reality" and other men with neutral eyes; he believes, in summary, he can "find" "truths" that he himself did not invent.]

Moreover, she emphasises the human need to live in society and to form communities as Nietzsche defends it, i.e. as a sort of urge both to survive against others and to live socially in the herd. And in order to live socially, a sort of "peace treaty" is required, as Nietzsche explains:

Dieser Friedensschluss bringt etwas mit sich, was wie der erste Schritt zur Erlangung jenes rätselhaften Wahrheitstriebes aussieht. Jetzt wird nämlich das fixiert, was von nun an „Wahrheit" sein soll, das heißt, es wird eine gleichmäßig gültige und verbindliche Bezeichnung der Dinge erfunden, und die Gesetzgebung der Sprache gibt auch die ersten Gesetze der Wahrheit: denn es entsteht hier zum ersten Male der Kontrast von Wahrheit und Lüge (NIETZSCHE 1873: 2).

To summarise the point, Arrojo turns again to Nietzsche and his conclusions in his "Über die Wahrheit und Lüge im außermoralischen Sinne":

Was ist also Wahrheit? Ein bewegliches Heer von Metaphern, Metonymien, Anthropomorphismen, kurz eine Summe von menschlichen Relationen, die, poetisch und rhetorisch gesteigert, übertragen, geschmückt wurden und die nach langem Gebrauch einem Volke fest, kanonisch und verbindlich dünken: die Wahrheiten sind Illusionen, von denen man vergessen hat, dass sie welche sind, Metaphern, die abgenutzt und sinnlich kraftlos geworden sind (…) (NIETZSCHE 1873: 2).

69 This quotation is the epigraph of the present section.

In other words, even though so much attention appears to be paid to the notion of truth and its implications, one must acknowledge its construed, illusory character. As a *human* product, the truth is highly changeable, easily manipulable and finite. In this sense, the gesture of accepting "the" truth should be accompanied by a movement of questioning this very truth.

In this paper, though very briefly, Arrojo demonstrates how Freud and Nietzsche are the greatest precursors of postmodern thought, and how their ideas can help us to rethink the premises and assumptions upon which we build our sciences and theories. After all, if one acknowledges that all our sciences and knowledge derive from unconscious impulses, and that they are nothing but linguistic construes – which in turn is all our human condition allows us – one must then be more critical of one's "truths".

In the paradoxical "A Noção de Literalidade: A Metáfora Primordial" (Literalness: The Primordial Metaphor), Kanavillil Rajagopalan and Rosemary Arrojo argue that the vast majority of institutionalised disciplines dedicated to language share a common premise, namely the possibility of an objective and decontextualised opposition between literal meanings and metaphorical meanings (ARROJO 1992/2003: 47). Indeed, the authors first examine how the concept of metaphor is defined within synchronic linguistics by analysing the work of the American linguist Jerrold Sadock. Taking his 1979 paper "Figurative Speech and Linguistics" into account, they demonstrate how Sadock defines figurative speech as a phenomenon outside linguistics: "All nonliteral speech (...), including metaphor, falls outside the domain of synchronic linguistics" (SADOCK 1979/1993: 42). According to Sadock, the reason why "the study of metaphor (...) would not be a proper subject for synchronic linguistics" is the fact that "the basis of metaphor is a kind of indirection that is shared with nonlanguage behavior" (SADOCK 1979/1993: 42 – cf. ARROJO 1992/2003: 47). For him "the underlying principles governing metaphor are of a general psychological sort and are thus not specifically linguistic" (SADOCK 1979/1993: 42; ARROJO 1992/2003: 47).

Based on Sadock's reflections, taken here as exemplary of synchronic linguistics, Rajagopalan and Arrojo derive three main underlying presuppositions, as follows: (i) there are clear distinctions between literal and non-literal, linguistic and psychological, synchronic and diachronic, as well as between language and thought; (ii) literal meanings, which are the object of synchronic linguistics, are evident in themselves and are thus immune to context or interpretation shifts; and (iii) there are neutral and transparent subjects who can carry out literal acts without revealing their time and circumstances. We can add here the division at which Sadock hints in the quotation above, namely metaphor or "indirect speech" is psychological whereas "literal" or "direct speech" is not psychological – hence its affiliation to linguistics.

Having made these initial remarks on linguistics, Rosemary Arrojo and Kanavillil Rajagopalan set out to demonstrate how other institutionalised disciplines dedicated to language appear to share these very same presuppositions. The first work they take into consideration is David Rumelhart's 1979 paper "Some Problems with the Notion of Literal Meanings". Although the cognitive scientist's reflection is somewhat similar to Sadock's, he accounts for elements largely neglected by Sadock, as Arrojo and Rajagopalan stress. They also emphasise the fact that Rumelhart, despite coming from a completely different area (he worked as a psychology professor), endorses the idea of a clear opposition between literal and metaphorical meanings, and assigns the task of investigating figurative speech to psychology (cf. RUMELHART 1979/1993: 78-79; ARROJO 1992/2003: 49). Nevertheless, in spite of his evidently essentialist matrix (he does place great emphasis on the "transmitted meaning" as something stable and intrinsic to the utterance), Rumelhart calls attention to other elements that may play a role in interpretation as a whole:

Linguistic utterances are always interpreted in some context. The context of utterance, along with any knowledge available to the listener, may potentially be employed in the process of constructing an interpretation of the utterance. Moreover, I suspect that this knowledge is not employed in any ad hoc way, say simply as a filter in choosing among the various possible readings a sentence might have. Rather, these elements play a central role in determining what interpretations are possible for a given utterance (RUMELHART 1979/1993: 76).

In addition to that, Rumelhart investigates the use of metaphors by children and, as Arrojo and Rajagopalan point out, comes to conclusions that briefly border on postmodernism. Rumelhart turns to the process of language acquisition to justify the importance of figurative language, not as a "special aspect of linguistic or pragmatic competence", but as the "very basis for this competence" (RUMELHART 1979/1993: 73). He explains that the use of figurative language is very much present during the process of language acquisition, thus discarding the linguistic proposition of literal meanings being the main ones or the first ones to be learnt, and non-literal meanings being secondary and derivative (cf. RUMELHART 1979/1993: 73; ARROJO 1992/2003: 51-52).

To take another discipline dedicated to language into account, Arrojo and Rajagopalan scrutinise John Searle's view on the matter, particularly in his 1979 book *Expression and Meaning*. Differently from Sadock, Searle does not propose an opposition between literal and metaphorical meanings: "Many writers on the subject (...) think there are two kinds of sentence meaning, literal and metaphorical. However, sentences and words have only the meanings that they have" (SEARLE 1979: 77). For Searle, the secret lies in the speaker's attitude or intention:

> *Strictly speaking, whenever we talk about the metaphorical meaning of a word, expression, or sentence, we are talking about what a speaker might utter it to be, in a way that departs from what a word, expression or sentence actually means. We are, therefore, talking about possible speaker's intentions* (SEARLE 1979: 77).

Therefore, Searle proposes a distinction between what speakers *mean* by uttering certain words or sentences – "speaker's utterance meaning" – and what words or expressions *actually* mean – "word" or "sentence meaning" (SEARLE 1979: 77). For Arrojo and Rajagopalan then, Searle's new approach based on speech acts is not exactly that "new" when one compares it to Sadock's or even Rumelhart's (ARROJO 1992/2003: 50). The difference is that for Searle "o poder da intenção do autor é (...) tão absoluto que resiste ao tempo ou a qualquer outra mudança", or the author's intention is so powerful that it endures time as well as any other change (ARROJO 1992/2003: 52).

From the point of view of literary theory, Arrojo and Rajagopalan assert that even though the distinctions and dichotomies applied to "literal" and "metaphoric meaning" have different names, they are nothing but obvious masks for the same linguistic categories. Kenneth Burke, for instance, distinguishes between "poetic meaning" and "semantic meaning" (ARROJO 1992/2003: 50). Not surprisingly, the former refers to the speaker's or author's emotions and subjective attitude, whereas the latter concerns neutral, emotionless utterances (ARROJO 1992/2003: 50). Therefore, based on the categorisation proposed by Burke, one could easily draw very similar underlying presuppositions to the ones suggested previously, based on Sadock's categorisation.

Another element that appears to be common to all the theorists analysed in the paper in question is the fact that, when looking at specific examples, they all have great difficulty keeping to the categories that they themselves propose (ARROJO 1992/2003: 50-52). Moreover, both traditional linguistics and literary theory seem to share the hope (the "theory hope") that one day a clear distinction between literal and non-literal speech will be devised: "Como a lingüística tradicional que, em vão, persegue o algoritmo da literalidade, a teoria da literatura se empenha em encontrar a fórmula infalível que pudesse esclarecer para sempre a especificidade da linguagem poética", or in the same way as traditional linguistics perceives the algorithm of literalness in vain, literary theory strives to clarify the specificity of poetic language once and for all (ARROJO 1992/2003: 53).

At the end of their paper, Rajagopalan and Arrojo turn once again to Nietzsche, Freud and Derrida. From the aforementioned essay "Über die Wahrheit und Lüge im außermoralischen Sinne", they take a very interesting quotation in which the German philosopher defends that the meaning taken as literal today was once a metaphor *created* – and not *discovered* – by man (ARROJO 1992/2003: 53):

Erste Metapher. Das Bild wird nachgeformt in einem Laut! Zweite Metapher. Und jedesmal vollständiges Überspringen der Sphäre, mitten hinein in eine ganz andere und neue. Man kann sich einen Menschen denken, der ganz taub ist und nie eine Empfindung des Tones und der Musik gehabt hat: wie dieser etwa die chladnischen Klangfiguren im Sande anstaunt, ihre Ursachen im Erzittern des Saite findet und nun darauf schwören wird, jetzt müsse es wissen, was die Menschen den »Ton« nennen, so geht es uns allen mit der Sprache. Wir glauben etwas von den Dingen selbst zu wissen, wenn wir von Bäumen, Farben, Schnee und Blumen reden, und besitzen doch nichts als Metaphern der Dinge, die den ursprünglichen Wesenheiten ganz und gar nicht entsprechen. Wie der Ton als Sandfigur, so nimmt sich das rätselhafte X des Dings an sich einmal als Nervenreiz, dann als Bild, endlich als Laut aus (NIETZSCHE 1873: 2).

To complement Nietzsche's thought-provoking proposition, the authors mention Freud and the unconscious. After Freud, one cannot speak of sciences, truths and "literal" meanings that are not products of interpretations, mediations between man and the world. In this light, propositions such as Sadock's,

Rumelhart's, Searle's and Burke's would be unacceptable. To finish their paper, Arrojo and Rajagopalan comment on Derrida's 1982 essay "White Mythology: Metaphor in the Text of Philosophy" (translated by Alan Bass), which first came out in 1972 under the title "La Mythologie Blanche (la Métaphore dans le Texte Philosophique)":

(...) esse "homem branco" de que fala Derrida empreende uma busca quixotesca daquilo que nunca encontrará fora de si mesmo. Portanto, a literalidade – a neutralidade, a razão, o puramente objetivo – é a grande metáfora, a metáfora primordial criada pelo homem que, entretanto, precisa se esquecer de que a inventou para não se lembrar de sua finitude e de suas limitações humanas (ARROJO 1992/2003: 55).

[(...) this "white man" about whom Derrida speaks has embarked on a quixotic search for something he will never find outside himself. Therefore, literalness – neutrality, reason or pure objectivity – is the great metaphor, the primordial metaphor created by man who must forget he created it in order not to remember his own finitude and human limitations.]

Therefore, it is in this sense that Arrojo and Rajagopalan defend the notion of literalness as the "primordial metaphor" – let us not forget the title of the paper: Literalness: The Primordial Metaphor. As Nietzsche asserts (1873: 2), strictly speaking the entire use of language can be perceived as metaphorical, as there is no intrinsic relationship between the world and language – signs are indeed arbitrary, as Saussure would say. These "metaphors", which were created by man, gradually become canonical and widespread to the extent that man seems to forget that he himself created them and promoted their establishment; to the extent that man feels the need to look for them *outside* himself, as something that exists prior to and beyond his *own* existence. In this light, it seems that the main aim of this paper is to ask readers to take the pillars of our society – neutrality, reason and pure objectivity – with a large pinch of postmodern salt.

And perhaps this aim, or rather the way it is presented in this text and not only in this text, could give rise to criticism both from those who defend poststructuralist views and (particularly) from those who oppose them.[70] For in

70 I will come back to the issue of the multifaceted character of poststructuralist thought in Part IV,

Arrojo's and Rajagopalan's effort to deconstruct the works of those perceived as essentialist, they appear to fully dismiss these works. In the paper in question, it seems as though their aim were to fully discard the distinction between "literal" speech and "figurative" speech – though they never actually say it with these words. In the hands of readers with little knowledge of poststructuralist thought, their paper could well be interpreted as an attempt to completely abolish this distinction as it cannot be formulated in absolute terms. But rather than the promotion of full invalidation, the point appears to be to simply call the readers' attention to the illusory, fragile character of allegedly objective distinctions such as literal meaning versus metaphorical meaning, to demand a more critical attitude, on the part of the reader, towards such distinctions. As we will see in different papers below as well, Arrojo's (and Rajagopalan's) words seem to imply a radicalism which, in the hands of those who oppose poststructuralist tendencies, becomes a powerful "counterattack" weapon. But let us not dismiss here the possibility of an intentional radicalism, typical of scholars who want to establish new tendencies in a given field, who want to oppose mainstream, canonical ideas. Indeed, a radical, militant attitude is very much in line with Arrojo's standpoint, as already argued in II, 1. I will come back to this question later in this chapter.

But let us look back for a moment at the discussion Arrojo and Rajagopalan carry out on Searle's propositions. The debate between Searle and Derrida on speech act theory (see III, 3.1) stretched from the 70s to the 90s (see footnote 118 on page 216) and indeed raised many of the questions discussed in this last essay by Arrojo and Rajagopalan. In yet another paper, entitled "Searle e a Noção de Literalidade" (Searle and the Notion of Literalness), Arrojo and Rajagopalan address Searle's language theory specifically. They begin the essay by announcing that, as far as Searle's work is concerned, their impression is that the American philosopher has strived to develop first and foremost a theory of literalness (ARROJO 1992/2003: 113):

Há diversos momentos críticos na evolução do pensamento filosófico de Searle que revelam um desejo subliminar de encontrar um valor constante, um elemento qualquer que pudesse ser estável e irredutível, através do qual se tornasse possível a compreensão desse fenômeno tão volúvel que chamamos de linguagem (ARROJO 1992/2003: 113).

mentioning examples of thinkers who would probably go as far as to argue that Arrojo's and Rajagopalan's attitude reveals a misreading of poststructuralist ideas.

> [There is a series of critical moments in the development of Searle's thinking that unveil a subliminal desire to find a constant value, an element of stable and irreducible nature that would allow for the understanding of this ever so changeable phenomenon we call language.]

They then list three of these moments perceived as critical, namely (i) Searle's disagreement with Austin as far as illocutionary acts are concerned, (ii) his disagreement with Keith Donnellan's propositions about the issue of reference, and (iii) the combination of Grice's pragmatics and the speech act theory, which in turn led to the notion of indirect speech act (ARROJO 1992/2003: 113).

As for (i) his disagreement with Austin, Searle was the first to suggest, back in the late 60s, that the distinction between locutionary acts and illocutionary acts was not sound because they were not self-excluding; in other words, there were many cases in which illocutionary acts comprised locutionary acts. For Arrojo and Rajagopalan, the problem lies in the evidence he brings forth to prove his point that signification itself sometimes suffices to determine the illocutionary force of a sentence. When analysing an utterance and showing that its illocutionary force is already clear on the locutionary level, Searle warns that this utterance must be serious and literal, as opposed to metaphorical or sarcastic. For Arrojo and Rajagopalan, Searle reached an impasse, closed a vicious circle: for in order for his proposition to be true, in order for a locutionary act to be able to reveal, by itself, its entire signification, it depends on its literalness – but its literalness also depends on the ability of the locutionary act to unveil its full meaning, and on it goes (ARROJO 1992/2003: 114-115). However, Searle appears to find solace under his principle of expressibility, whereby "whatever can be meant, can be said" (SEARLE 1979: 68). This principle, claim Arrojo and Rajagopalan, is more of a linguistic dogma as it is largely taken for granted within Searle's theory. For them, this principle is an apologia to literalness, a metaphysical commitment to the notion of literalness (ARROJO 1992/2003: 115). In this sense, this first instance taken from Searle's work reveals indeed a "subliminal desire" (cf. ARROJO 1992/2003: 113) to define and establish literalness as a constant, universal value.

Similarly, (ii) Searle's disagreement with Donnellan, years later, appears to point towards this wish for a theory of language centred on a notion of literalness, of fully objective meanings. In 1966, Donnellan claimed to have solved the problem of reference by devising two categories, two uses of definite descriptions, namely attributive and referential (DONNELLAN 1966: 364):

A speaker who uses a definite description attributively in an assertion states something about whoever or whatever is the so-and-so. A speaker who uses a definite description referentially in an assertion, on the other hand, uses the description to enable his audience to pick out whom or what he is talking about and states something about that person or thing. In the first case the definite description might be said to occur essentially, for the speaker wishes to assert something about whatever or whoever fits that description; but in the referential use the definite description is merely one tool for doing a certain job (…).

Searle's objection is expressed in his 1979 paper "Referential and Attributive", in which he proposes yet another set of categories, this time more comprehensive than Donnellan's: the pure, objective meaning of an utterance versus the meaning intended by the speaker. Rajagopalan and Arrojo demonstrate how Searle departs from this basic distinction to fully discard Donnellan's views on reference, once again centring the discussion on these constant, literal meanings – which, in their view, Searle does not really demonstrate, but simply takes for granted as some sort of dogma.

As for (iii) Searle's indirect speech acts (cf. SEARLE 1975), Rajagopalan and Arrojo explain that his very defence of a first, objective speech act that leads to the performance of this same speech act, as opposed to a speech act that performs a further speech act indirectly, unveils once again his strong belief in literal meanings as a constant value in linguistics. Indeed, this expectation is confirmed not only by these three instances taken from Searle's work, but also by two papers dedicated specifically to the question of literal meanings – "Literal Meaning" (1978) and "Metaphor" (1979). For Arrojo and Rajagopalan, even though Searle announces in both papers that he intends to challenge the mainstream ideas associated with the question of literalness, he ends up endorsing these ideas as he believes first and foremost in the possibility of solely objective, self-explanatory and ambiguity-free meanings. For them, Searle disregards Freud's propositions about the unconscious, also neglecting the fact that utterances are invariably interpreted by listeners who, in turn, are not capable of recovering the speaker's intentions in their totality (ARROJO 1992/2003: 120).

In conclusion, it seems that however much Rajagopalan and Arrojo agree that Searle's contribution may have had a great impact on contemporary linguistics, they remain very critical of these contributions. Their criticism stems from Searle's obstinate defence of literal meanings without much consideration or questioning, a concept – however ill-defined –around which he structures most of his propositions. They quite rightly assert that Searle's project is an instance of a logocentric project, a project based on words that can function out of context and whose meaning can be defined regardless of the receiver and of the situation.

In another joint paper entitled "A Crise da Metalinguagem: Uma Perspectiva Interdisciplinar" (The Crisis of Metalanguage: An Interdisciplinary Perspective) Kanavillil Rajagopalan and Rosemary Arrojo analyse the way in which linguistics and literary theory describe the same phenomenon, which in this case is language in a broader sense (or more specifically metalanguage). According to them, institutionally speaking linguistics and literary theory have language as their main object of study, but despite this apparent confluence of aims there is hardly any conflict between the two disciplines because "o tácito acordo mútuo faz com que cada lado se preocupe com justamente aquilo que os seus vizinhos tendem a descartar como menos relevantes", or their tacit, mutual agreement allows them to worry about the issues that their neighbours discard as less relevant (ARROJO 1992/2003: 57). In the case of linguists, they claim to have the key to the language enigma, whereas those in literary studies take pride in its allegedly exclusive access to the aesthetic values of language (ARROJO 1992/2003: 57). Departing from this tacit agreement, Arrojo and Rajagopalan explain that their objective is to show how the two disciplines are not that far away from each other as they claim to be, particularly because their metalanguages cannot be as hermetically isolated and impermeable as they seem (ARROJO 1992/2003: 58).

By looking at the issue of metalanguage within linguistics, Rajagopalan and Arrojo assert that the main belief that permeates the entire discipline is the possibility of a neutral metalanguage – neutral in the sense that it has not been contaminated by everyday language. For them, Bloomfield's position on this matter is a corollary of this belief. Indeed, Bloomfield defends, in his 1944 paper "Secondary and Tertiary Responses to Language", that one must distinguish between what laymen say *in* a certain language and what they say *about* a certain language (ARROJO 1992/2003: 58). In other words, the metalanguage of linguistics cannot be contaminated by what Bloomfield calls "folk linguistics" (apud ARROJO 1992/2003: 58); rather, it must be purely and rigorously scientific. As Arrojo and Rajagopalan very clearly demonstrate, this anxiety to find a metalanguage "livre dos males que afligem seu próprio objeto de estudo", or free from the evils that afflict its own object of study (ARROJO 1992/2003: 58), has its roots in logic tradition. Their thesis is that today linguists still seem to be trying to accomplish, in natural languages, what can only be done in formal languages, as the Polish logician Alfred Tarski proved in the 1930s. Tarski's conclusions may be roughly summarised as follows: in order for one to formulate the truth of a given system, one must resort to a different system, more complex and complete than the first one. From this point of view, in order to be fully scientific, linguistics would need to come up with a system more complex than natural language, which would then work as its metalanguage. However, one such task is impossible because natural languages work as their own metalanguage. As Arrojo and Rajagopalan summarise (ARROJO 1992/2003: 59):

(...) para Tarski, estava definitivamente descartada a possibilidade de se definir, na língua natural, a noção de verdade sem cair em paradoxos, pois a língua natural continha sua própria metalinguagem (o que, em última análise, equivale a reconhecer o caráter auto-referencial da língua natural).

[(...) for Tarski, the possibility of defining the notion of truth in natural languages without meeting paradoxes was fully discarded because natural languages contained their own metalanguage (which in other words is the same as acknowledging the self-referential character of natural languages).]

Yet, despite Tarski's evident pessimism, linguistics appears to have found consolation in the works of the British anti-formalist John L. Austin, particularly in his distinction between the "serious" (speech-acts intended to work literally) and the "non serious" (speech-acts of actors and poets); not surprisingly, with the former being his object of study (ARROJO 1992/2003: 60). I will come back to Austin's categories further in this section.

The way literary theory approaches its metalanguage is, according to Arrojo and Rajagopalan, strikingly similar, particularly within North-American new criticism. Critics as William K. Wimsatt, for example, advocate a clear distinction between the literary or poetic language and the language the critics use to describe it (ARROJO 1992/2003: 61). It is as Christopher Norris summarises it:

New Critical method was rational enough in its mode of argumentation but kept a firm distance between its own methodology (...) and poetic language. (...) The poem in short was a sacrosanct object whose autonomy demanded a proper respect for the difference between it and the language that the critics used to describe it (NORRIS 1982/2002: 8).

From this point of view, as Arrojo and Rajagopalan assert, these apparently completely different disciplines seem to have embarked on the same logocentric quest for scientific distinctions between language and metalanguage, poetic language and common language; they seem to nurture the same illusion whereby man would be able to look at the world and not become mixed with it (ARROJO 1992/2003: 61), which Freud and Nietzsche have already proven to be impossible.

Arrojo and Rajagopalan conclude their paper by mentioning Derrida's deconstruction of key texts on philosophy, linguistics, literary theory and psychoanalysis, as well as its subsequent impact on the Humanities since the early 1970s. Besides, they mention the contributions of three other theorists whose works cannot be labelled as deconstructionist but as poststructuralist, namely Foucault, Barthes and Fish. Thanks to the works of these theorists, interdisciplinarity is no longer a mere effort of good will, but rather a necessity, an inevitable condition (ARROJO 1992/2003: 62). As their arguments in this short essay illustrate, linguistics and literary theory (to mention only two areas within the Humanities) appear to share the same logocentric matrix, some of the same dilemmas and illusions. In this sense, given its more overtly interdisciplinary character, translation studies is an area where the "crisis of metalanguage", as Arrojo and Rajagopalan put it, together with the logocentric dilemmas and illusions, is made more evident.

In view of their conclusions in this particular paper, one wonders whether the institutional and academic separation of linguistics, literary studies and translation studies – as it is observed at numerous European universities, for instance – is indeed mutually beneficial (as already addressed in I, 2). If interdisciplinarity is no longer a mere effort of good will, but rather an inevitable condition, would it not make more sense to have these areas actively collaborate? I will come back to this issue in II, 2.3 and 2.4, as well as in IV, 3.

For now let us look at another short paper, entitled "Compreender X Interpretar e a Questão da Tradução" [Understanding vs. Interpreting and the Question of Translation]. Arrojo begins the paper by listing the countless dichotomies produced by "our logocentric tradition", amongst which understanding versus interpreting is one of the most influential in translation studies (ARROJO 1992/2003: 67). As she explains, from an essentialist point of view, "understating" and "interpreting" would be different activities, the former being the objective, neutral decoding of words and texts, and the latter the subjective, derivative of the first. In other words, one would first have to *understand* a certain text correctly, without interfering with its original meanings, in order to be able to *interpret* it afterwards, thereby revealing one's own ideology.

Arrojo stresses that this recovery of stable meanings is largely taken for granted in the Western world, permeating most of our notions of knowledge and language (ARROJO 1992/2003: 68). Nevertheless, it is in translation theory that the opposition between understanding and interpreting becomes particularly problematic. This issue has already come up in Part I, where Arrojo's paper "The Revision of the Traditional Gap between Theory and Practice and the Empowerment of Translation in Postmodern Times" (ARROJO 1998b) was the basis of the discussion about translation theory. In this 1998 paper, Arrojo mentions the works of theorists belonging to different areas (Mounin, Newmark, Baker, Lefevere, amongst others) and how all of them appear to partake in this very notion of stable meanings and neutral readings of texts (*understanding* rather than *interpreting*). Here, in the paper in question (ARROJO 1992/2003: 67-70), Arrojo mentions a Brazilian example of a translation model devised in 1984 and whose objectives and limitations resemble the ones labelled "essentialist" in her 1998 paper.

In 1984, Francis Aubert (see I, 3) published a paper called "Descrição e Quantificação de Dados em Tradutologia" (Data Description and Quantification in Translatology) in the Brazilian journal *Tradução e Comunicação: Revista de Tradutores* (no. 4). According to Arrojo, although Aubert acknowledges the limitations of linguistics, as well as its difficulty in dealing with subjectivity, he believes that in the future these constraints will be unravelled thanks to the advances made in the so-called "linguistic science" (ARROJO 1992/2003: 69). Therefore, relying on his underlying belief in the dichotomy understanding versus interpreting, Aubert sets out to execute his project of translation systematisation, which in turn should lead "at least" to the following consequences:

(1) o estabelecimento de correlações significativas entre as modalidades de tradução e a tipologia das línguas envolvidas;

(2) o estabelecimento de correlações significativas entre as modalidades de tradução e a tipologia textual; e

(3) a delimitação do escopo de liberdade do tradutor dentro dos limites impostos pelas restrições estruturais, culturais e textuais, ou seja, determinar a margem de variação inter- e intra-individual (apud ARROJO 1992/2003: 69).

[(1) the establishment of significant correlations between translation types and the typology of the languages involved;

(2) the establishment of significant correlations between translation types and text typologies; and

(3) the scope delimitation of the translator's freedom within the constraints posed by structure, culture and text; in other words, the establishment of the inter- and intra-individual margins of variation.]

Not surprisingly, by the end of the essay Aubert concedes that the current version of his model still contains a number of limitations particularly because it cannot deal with the subject of creativity in translation. As Arrojo points out, he admits that the data he gathered does not allow for the generalisations he intended to make; hence he himself perceives his model as not *yet* ready to contribute to translation teaching and practice (ARROJO 1992/2003: 70).

Projects such as Aubert's[71] – or Newmark's, Graham's, Mounin's, amongst others, as Arrojo explains (1998b – see I, 5) – are exemplary of the logocentric tradition, and the enormous gulf between their objectives and their achievements illustrates how illusory the possibility of neutral understanding as opposed to subjective interpreting is. Indeed, Arrojo summarises the question as follows:

O projeto logocêntrico está fadado à frustração e ao insucesso pois ignora a temporalidade, a finitude e a mortalidade de todos os empreendimentos humanos e trata categorias inevitavelmente marcadas pelo tempo e

71 Let us bear in mind here that the paper by Aubert that Arrojo analyses was published in 1984, and that since then he has written numerous different papers. It is beyond the scope of this book to examine to what extent these essentialist tendencies of which Arrojo speaks remain in Aubert's work until today.

produzidas por sujeitos sempre situados em algum contexto sócio-cultural como instâncias "divinas", acima de qualquer perspectiva ou interesse subjetivo (ARROJO 1992/2003: 70).

[The logocentric project is doomed to failure and frustration because it takes temporality for granted, as well as the finite and mortal condition of every human enterprise; it perceives as "divine" and above individual perspectives and subjective interests categories that are inevitably marked by time and produced by people inserted in a given socio-cultural context.]

In other words, research in the field of translation studies – as well as any kind of research for that matter – should always have this finite, human perspective in its horizons. From this point of view, translation studies is actually an extremely fruitful field largely unexplored. Indeed, this is what Arrojo discusses in another paper entitled "A Pesquisa em Teoria da Tradução ou o que Poderia Haver de Novo no *Front*" (Research in Translation Theory: What is New on the Front). She begins the penultimate essay of her book *O Signo Desconstruído* by scrutinising the dichotomy theory versus practice. This basic opposition, which is taken for granted in most fields of knowledge, is perceived as particularly problematic in translation studies (see Part I) precisely because of the uniqueness of the translational act (ARROJO 1992/2003: 107).

Arrojo suggests that the etymology of the words theory and practice alludes to the image of the observer and the object that is observed. "Theory", from Greek and Latin, would be closely related to the verbs "observe", "examine", "investigate", whereas "practice", from Greek, is somewhat similar to the idea of "action" (cf. HOUAISS 2001). As already mentioned in Part I, one tends to associate different ideas to the notions of theory and practice precisely because, as Arrojo very aptly explains, one cannot establish an objective and clear division between theory and practice. In other words, different theories will actually *create* different objects (or practices), and not simply and neutrally *describe* the same objects (ARROJO 1992/2003: 108).

The dichotomy theory versus practice goes straight to the heart of one of Derrida's most central deconstructionist efforts, namely the dichotomy subject versus object which, together with the question of the signifier and the signified, makes up the core of his work (see the beginning of II, 2.1). As Gayatri C. Spivak explains in the translator's preface to the English version of Derrida's *De la Grammatologie* [*Of Grammatology*],

The opposition of the subject and the object, upon which the possibility of objective descriptions rests, is also questioned by the grammatological approach. The description of the object is as contaminated by the patterns of the subject's desire as is the subject constituted by that never-fulfilled desire. We can go yet further and repeat that the structure of binary oppositions in general is questioned by grammatology (SPIVAK 1976/1997: lix).

Therefore, as subject and object, theory and practice mutually and constantly define and shape each other, with ideology playing a major role in the process. Back to Arrojo's paper, after a brief analysis of the conflict theory versus practice, she returns to the question she asked in the title of her essay, i.e. what are the new research possibilities in translation studies? For her, what seemed particularly auspicious at the time (i.e. the early 1990s) was, indeed, ideology, as the set of values and convictions that permeates the production of meanings in a particular place and moment in time, and makes up the perspectives through which one theorises and classifies the world. She is interested, in other words, in the way every theory (or science) redefines its objects of study and creates its own truths (ARROJO 1992/2003: 111-112). She closes this essay by mentioning Thomas Kuhn and his 1970 book *The Structure of Scientific Revolutions*, in which he advocates that new theories and sciences do not arise and become a new paradigm because they "find" new "truths", but rather because of their persuasiveness and ability to identify with the needs and interests of the community to which it is targeted (ARROJO 1992/2003: 112 – see I, 6). In this sense, Arrojo appears to be interested precisely in these underlying motivations and interests that make translation theories canonical or not.

The last paper we will analyse in the present section is actually the antepenultimate paper of the book *O Signo Desconstruído*, entitled "O Ensino da Tradução e seus Limites: Por uma Abordagem Menos Ilusória" (or The Limits of Translator Training: Towards a Less Illusory Approach). Arrojo opens the essay with a classroom experience she went through in a translation course within the BA in English (Translation Studies) at PUC-SP (Pontifícia Universidade Católica de São Paulo). In short, the anecdote goes as follows: one of her students was translating a computer manual and, not having found a number of words and expressions she needed to translate, the student turned to the teacher (Arrojo) with a list of words, requesting her to provide correct translations. Perplexed, Arrojo asked the student why she believed that Arrojo should have the answers to her questions, to which the student replied that, as a translation specialist, Arrojo should know the Portuguese words that matched those English terms (ARROJO 1992/2003: 99).

Arrojo then explains how her student's attitude is very much in line with "our logocentric tradition", whereby the expectation towards translator training is precisely to enable students to translate *anything* – as long as from and into the languages they have mastered – *anywhere, anyhow* – even a list of words detached from a context, as was the case of the manual in question (ARROJO 1992/2003: 100). She speaks of the term "logocentric" as defended by Derrida, i.e. as a strong underlying belief in stable meanings, which in turn allows for the perfect and clear-cut distinction between subject and object; in other words, it allows for the idea that meanings can function regardless of the context in which they are embedded and regardless of the subject that makes use of them. Another consequence of this belief is the idea of literal meanings, meanings inherent to words and words only, once again completely apart from contexts and subjects, as pure, detachable objects.

For her, this "Cartesian" notion of language is strongly predominant in translation studies, as Nida's famous train metaphor very aptly illustrates (ARROJO 1992/2003: 101). Seen from this perspective, translating (or interpreting) should first and foremost be free from the translator's interference, from his/her interpretation as such. In this light, the main objective of translator and interpreter training should be to have students acquire as many translation pairs, as many equivalent words and expressions as possible. Devising glossaries and word lists would thus be an efficient method to assist students in this task, and so would translation exercises that focus primarily on these transferences – rather than on the translation process itself (ARROJO 1992/2003: 103).

As Arrojo quite rightly asserts, one such notion of translator and interpreter training implies that all one needs in order to translate is the knowledge of the languages involved and a suitable glossary or dictionary. Further, this notion of translator and interpreter training may imply

> *profissionais em formação (…) são (…) colecionadores de significados congelados e intercambiáveis de uma língua para outra e de um texto para outro e, o que é ainda mais grave, a alienação, o não-reconhecimento de seu papel essencial de autor e construtor de significados that* (ARROJO 1992/2003: 103).
>
> [professionals in training (…) are (…) collectors of frozen and interchangeable meanings from one language or text to another; but what is

even worse is the alienation, the nonacknowledgement of their essential authorial role as meaning builders.]

For her, these implications are very much in tune with her student's expectations. Because Arrojo was an experienced translator and translation professor, her student thought it was only natural for Arrojo to know a far larger number of equivalent expressions (in their case in English and Portuguese) than she did.

Arrojo then describes what, to her, should be the aim of translator and interpreter training. The starting point should always be the questioning of stable meanings and activities or theories that imply the existence of stable meanings. In Arrojo's view, pupils must first and foremost learn that readings, interpretations and translations inevitably reflect not only the subject-reader or translator, but also the time and cultural community in question (ARROJO 1992/2003: 103). In this light, more than mastering one-to-one correspondence between languages, translators should know their cultural community well, as well as the tendencies and characteristics that are perceived as adequate in that community at that moment (ARROJO 1992/2003: 103).

As for her student, Arrojo believes that the starting point of her translation task should have been, rather than the search for exact matches, the familiarisation with that text type within that community – the acceptable format, the jargon, the style. Surely at some point she would have to look for glossaries and specialised dictionaries, but this step should by no means constitute the out-and-out translation process.

As regards the limitations of translator and interpreter training, Arrojo concedes that no course can embrace each and every career possibility offered within the area of translation and interpreting. In contrast, nevertheless, one *can* build *awareness* within a translation or interpreting course – awareness of the role that translators and interpreters play in society and in history and awareness of the attitude expected of these professionals (ARROJO 1992/2003: 105). As an example of an awareness-raising activity, she suggests the critical comparison of canonical translations and translations done by students, aiming at showing how each text seems to be bound to a certain ideology, a certain moment in history, a certain interpretation (ARROJO 1992/2003: 105). Finally, Arrojo admits that this awareness-raising process can be rather painful and laborious as one must constantly question and challenge "common sense", well-established precepts produced by our age-old "logocentric tradition" (ARROJO 1992/2003: 105).

The issue of translator and interpreter training has already been addressed in Part I, and will be discussed in detail in Part IV as well. Arrojo's reasoning in this last paper seems very much in tune with poststructuralist thought in general, particularly in terms of the idea both of a structure of higher education that questions its "utilitarian purposes" and of a theory that is suspicious of its "utilitarian purposes" as well (cf. DERRIDA 1983: 13 – see I, 2). Theory and higher education are more awareness-raising than practice or market-oriented.

It is interesting to notice that Arrojo's propositions in this last essay are not at all radical. Thinking about it, much of what she says is undoubtedly a reality, at least to a certain extent, even in the classroom of the most essentialist professor. Surely every glossary, for example, is a product of circumstances and time and must be updated from time to time – and both teachers and pupils *do* realise it. Dictionaries are not immortal either, and most students are certainly aware of the need to have new, up-to-date dictionaries. Furthermore, pupils probably know that they should be suspicious of dictionaries, that language is far more dynamic than a dictionary allows it to be and that the actual use of language may be a far cry from dictionary definitions. Also, many translation teachers certainly use Arrojo's translation comparison exercise and probably even arrive at similar conclusions.

The difference, however, lies in the perspective – is the glass half full or half empty? Arrojo's intention seems to be to stress this finiteness, this instability, to put it in the spotlight instead of simply brushing the issue in passing. For even if the most essentialist person may be aware of the limitations of glossaries and dictionaries, many certainly do not take these limitations beyond that level and apply it to language as a whole, to translation, to science, to human undertakings in general. As for the translation exercise proposed by Arrojo, even though numerous translation teachers (whose affinities are not necessarily poststructuralist) may use it in a similar fashion and even come to similar conclusions, their objective might not be to show how translations reflect personal choices, cultural communities and ideology; rather, the purpose of one such exercise might probably be to determine what translation is *better*, without much consideration of the whys and wherefores of these conclusions (for a specific example of this attitude, see Part III, 3).

As already mentioned above, although many of Arrojo's works may often (sometimes unfairly) be criticised for their radicalism, this last essay could be criticised for not bringing many new contributions. But once again, let us not forget that the relationship between structuralism and poststructuralism (see I, 6) is one of complementariness as well, and therefore full radicalism is not in question. The point is more often than not to call attention to a shift in existing perspectives – as is the case of this paper by Arrojo – instead of proposing entirely new perspectives.

The question of the radicalism often expected of so-called poststructuralist thinkers will repeatedly come up again in Part III and will be addressed in Part IV as well – and so will the question of the relationship between structuralism and poststructuralism. Let us now look at the second book by Arrojo, originally published six years before *O Signo Desconstruído*, in which some of the issues discussed here will be brought up again.

2.2 Oficina de Tradução

Nossa tradução de qualquer texto, poético ou não, será fiel não ao texto "original", mas àquilo que consideramos ser o texto original, àquilo que consideramos constituí-lo.

Rosemary Arrojo[72]

Oficina de Tradução (or Translation Workshop) is perhaps one of the best-known books on translation in Brazil. It is part of the syllabus of a number of translation courses both at BA and MA level. Even those professors who do not fully agree with Arrojo's allegedly radical propositions seem to find the book an excellent classroom resource (I will come back to this question later). This book also makes up a section of the anthology organised by Michaela Wolf in 1997, entitled *Übersetzungswissenschaft in Brasilien: Beiträge zum Status von „Original" und Übersetzung*, as some of its chapters are German versions of chapters in Arrojo's book.

Oficina de Tradução first came out in 1986 and was in its 5th edition in 2010. Unlike *O Signo Desconstruído* (see previous section), which might be perceived as a more theoretical[73] book, *Oficina de Tradução* is meant to be

72 This excerpt is quoted and translated later in this section (see page 148).

73 "Theoretical" in the sense of a discourse that stimulates reflection rather than a discourse that

more practice-oriented. As the author explains in the first chapter (ARROJO 1986/2002: 7-10), the name of the book was taken from a course she had attended at the Johns Hopkins University in 1981, which was indeed called "Translation Workshop". This course was mostly practice-oriented, and the main task of its six participants was to compare and discuss their own translations, along with canonical translations of literary works (ARROJO 1986/2002: 9). Therefore, it is in this "hands-on" mood that Arrojo sets out to write this very short book – its nine chapters comprise just over 80 pages.

Notwithstanding its practical character, the first few chapters of the book – particularly chapters 2, 3 and 4 – are more theoretical, very much in line with *O Signo Desconstruído* (see Section 2.1). The abundant use of examples, however, makes these chapters sound less abstract. In Chapter 2, named "A Questão do Texto Original" (or On Original Texts[74]), the author departs from traditional views on translation, namely Catford's, Nida's and Tytler's, to build the notions of "original" and "translation" (ARROJO 1986/2002: 11-13). Given the target public that Arrojo has in mind (i.e. mainly translation students), she explains these authors' views by using allegories and examples. Then she turns to a short-story by the Argentinean writer Jorge Luis Borges, called "Pierre Menard, Autor del Quijote" (BORGES 1986: 17-22 – see II, 1 and 2.3 and 2.4), in order to relativise these traditional concepts of translation and originality proposed by the abovementioned translation scholars.

In Borges' story, a first-person narrator tells the reader about the works of Pierre Menard, a fictitious Frenchman of letters. The analysis of a recently published catalogue of Menard's works reveals his evident affinity with authors such as Descartes, Leibniz, John Wilkins, Ramón Lull and George Boole, who in turn had dreamt of and devised projects for a non-arbitrary, universal language. Menard's works point towards his ambitious dream of reaching truth as an absolute, objective concept found in language; in other words, they point towards his vision of a language capable of neutralising ambiguities and immune to different interpretations, contexts and timeframes. Nevertheless, his very bibliography is enough to prove the impossibility of his aspirations since he himself appears to have produced different translations of the same text, as well as different versions of the same text, thus showing just how non-objective language can be (cf. ARROJO 1986/2002: 18).

openly refers to a particular practice (see Part I).

74 In Wolf's anthology this chapter was translated as "Pierre Menard und eine neue Definition des 'Originals'" by Johanna Klemm (WOLF 1997: 25-34).

Yet it is not Menard's bibliography (his "obra visible", as Borges' narrator puts it – BORGES 1986: 17) that makes the impossibility of his dream evident. Instead, it is his unique work, the one beneath the surface, the endlessly heroic one ("la subterránea, la interminablemente heroica, la impar" – BORGES 1986: 19). This invisible work, as Arrojo calls it, consists of the reproduction of Cervantes' *Don Quijote* and is invisible because it was neither published nor concluded, in spite of its utmost importance (cf. ARROJO 1986/2002: 19). As the narrator explains, Menard

> *no quería componer otro Quijote – lo cual es fácil – sino **el Quijote**. Inútil agregar que no encaró nunca una transcripción mecánica del original; no se proponía copiarlo. Su admirable ambición era producir unas páginas que coincidieran – palabra por palabra y línea por línea – con las de Miguel de Cervantes* (BORGES 1986: 19 – his emphasis).
>
> [did not want to compose another Quixote – which is easy – but *the Quixote* itself. Needless to say, he never contemplated a mechanical transcription of the original; he did not propose to copy it. His admirable intention was to produce a few pages which would coincide – word for word and line for line – with those of Miguel de Cervantes] (BORGES 1964: 39 – translated by James E. Irby).

In other words, all Menard had to do in order to accomplish his mission was to know Spanish well, to restore Catholic faith, to fight against Moors or Turks, to completely forget the history of Europe between 1602 (when *Don Quijote* first came out) and 1918 (when Menard lived) and, finally, to be Miguel de Cervantes (cf. ARROJO 1986/2002: 19). Nonetheless, despite his efforts and optimism, all he managed to do was to copy Cervantes' words, one by one. Yet, this reproduction, however exact and accurate, still sounded different when compared to Cervantes' original precisely because the times were different, hence the impossibility of perfect equality, perfect equivalence. Menard's undertaking was a total failure, as the narrator explains, because Cervantes' words did not have the same impact as they had had three centuries before, giving rise to different interpretations by different, "contemporary" readers (ARROJO 1986/2002: 21-22).

Menard's wish to achieve perfect totality, free from subjectivity and ambiguity, and then his subsequent confrontation with reality reminds me of a one-paragraph story also written by Borges, called "Del Rigor en la Ciencia" (BOR-

GES 1972: 143). In this short-story a certain empire developed its cartography to such an advanced level of perfection that the map of the entire empire, for instance, coincided with the empire itself, both being exactly the same size. Given the uselessness of one such map, future generations "lo entregaron a las inclemencias del Sol y de los Inviernos. En los desiertos del Oeste perduran despedazadas Ruinas del Mapa, habitadas por Animales y por Mendigos" (BORGES 1972: 143) – "abandoned it to the Inclemencies of the Sun and of the Winters. In the deserts of the West some mangled Ruins of the Map lasted on, inhabited by Animals and Beggars" (translated by Harold Morland and Mildred Boyer – BORGES 1985: 90).

In Arrojo's book, Borges' short-story "Pierre Menard, Autor del Quijote" works as an allegory of the traditional views on translation represented, in this case, by Catford, Nida and Tytler. According to her, Borges' story teaches us first to question traditional outlooks on language and then to perceive translation as far more than the mere transfer or transport of stable meanings through languages because, as our fictitious Pierre Menard showed us, even within a single language meanings can only be determined temporarily, within a given context and timeframe and according to a particular interpretation (ARROJO 1986/2002: 22-23). In this sense, the image of the text as a receptacle of meanings is replaced by the image of the text as a palimpsest. The definition of the noun "palimpsest" reads as follows: "a parchment or the like from which writing has been partially or completely erased to make room for another text" (cf. *Random House Unabridged Dictionary*). As Arrojo puts it, "o 'palimpsesto' passa a ser o texto que se apaga, em cada comunidade cultural e em cada época, para dar lugar a outra escritura (ou interpretação, ou leitura, ou tradução) do 'mesmo' texto", or the "palimpsest" is the text that is erased according to time or place so as to give rise to a new writing (or reading, or interpretation, or translation) of the "same" text (ARROJO 1986/2002: 23-24). She closes this chapter on original texts by stressing both the production of meanings inherent to any reading (or translation) task and the impossibility of protecting so-called "original" meanings, as Borges' Menard illustrates so well (ARROJO 1986/2002: 24).

Chapter 3, entitled "A Questão do Texto Literário" (or On Literary Texts[75]), opens with a justification of the importance of literary texts in translation studies. For Arrojo, literary translation appears to be the Achilles' heel of any

75 This chapter featured in Wolf's anthology under the title "Eine neue Definition des Literarischen", translated by Annette Wußler (WOLF 1997: 35-42).

translation theory, neglected both by translation studies and literary theory – hence its significance for her work. Perhaps the initial and crucial hindrance lies in the difficulty in clearly distinguishing between literary and non-literary texts, but she will come back to this issue later in the chapter. At first, she presents the opinions of different writers on translation, all of which are negative – despite their completely different backgrounds. As she clarifies, these authors (Robert Frost, Paul Valéry, Heinrich Heine, Vladimir Nabokov and Marion Sorescu) perceive translation as an inferior undertaking because it is incapable of capturing the "soul" of literary texts (ARROJO 1986/2002: 25-27).

As Arrojo did in the previous chapter, she places these poets and writers side by side with Pierre Menard and the so-called "traditional" translation theorists (Catford, Nida and Tytler), for whom translations do not (or should not) alter original texts in any way. Catford, for example, will go as far as to say that "yes" is not a translation of "oui" because in French one can say "si" as well, thus indicating that it would be impossible to find a "perfect equivalent" in this case (CATFORD 1965: 41). Indeed, this view seems in agreement both with Menard's line of thought and with the abovementioned writers' expectations when it comes to literary translation.

Rosemary Arrojo then tries to sketch an alternative to this traditional view by rethinking the very notion of "literary". She asks herself how two texts as distinct as Milton's *Paradise Lost* and Pound's "In a Station in the Metro", for example, can both be acknowledged as poems. By drawing inspiration from Stanley Fish, particularly from his paper "How to Recognize a Poem When You See One?" (see III, 3.1), she suggests that the answer lies in the reading strategy adopted by readers rather than in the texts *themselves* – seeing as Pound's and Milton's aforementioned examples hardly bear any textual resemblance.

In order to clarify her point, Arrojo proposes a small experiment which I have used with my translation students repeatedly with great success.[76] The experiment, as I normally carry it out, consists of dividing the class into two groups and asking them to translate a short text in a way that works well in the target culture (in my case, I usually hand them Portuguese translations of the texts below and ask them to translate them into German, as my students are

76 I first read Arrojo's *Oficina* during my BA at Universidade Federal do Paraná. In fact, this experiment she proposes was carried out by one of my professors, Maurício Cardozo, within a translation course in 2004.

mostly Austrians learning Portuguese as a foreign language). The first group receives the following text, which is a note written by a guest to his host:

This is just to say I have eaten the plums that were in the icebox and which you were probably saving for breakfast. Forgive me, they were delicious: so sweet and so cold (ARROJO 1986/2002: 32).

As regards the second group, they must work with the following poem by William Carlos Williams (1934):[77]

This is just to say

I have eaten
the plums
that were in
the icebox

and which
you were probably
saving
for breakfast

Forgive me
they were delicious
so sweet
and so cold

(ARROJO 1986/2002: 32-33)

The results are simply impressive: even though the two excerpts contain the same words, the reading strategy adopted by the students, which in turn is strongly influenced by the way each text is presented to them, shapes the way the entire task is carried out.[78] Whereas the first group tends to finish rather quickly and have no further questions about the text, the second group needs a considerably longer amount of time and often asks various questions about the possible connotations certain words may have. Indeed, their translations reflect their reading strategy, i.e. the ones who believe to be faced with a note

77 This poem by Williams has been used by countless thinkers both to defend and oppose poststructuralist perspectives in literary theory – for more on this, refer to the debate between Altieri and Fischer (ALTIERI 1979; FISCHER 1979), as well as to Neil Easterbrook's 1994 paper (EASTERBROOK 1994).

78 Arrojo's experiment is strikingly similar to an experiment carried out by Stanley Fish (whom she quotes a few times in her *Oficina de Tradução*), which he describes in "How to Recognize a Poem When You See One" (FISH 1980: 322-337 – see III, 3.1).

translate it in a prosaic fashion, without any further concerns. As for the ones who believe that they have a poem in front of them, their translation usually strives to create sonorities (as the sibilants that permeate the English text) and rhythm (marked by the division into verses and stanzas). In addition to that, their attitude reveals their concern about the English (or Portuguese) words "plums" and "cold", especially, and their possible double meanings and allusions.

As Arrojo very aptly explains in this chapter, our reading expectations and strategies tend to adapt to the reading context. When faced with a note, we know what kind of structures and what kind of style to expect, so we read the note with these expectations in mind, which in turn derive from our vast note-reading experience. In contrast, when we are faced with a poem, the "same" elements may strike us as poetic rather than prosaic precisely because we read poems with "poetry-seeing eyes" (cf. FISH 1980: 326). Needless to say, these reading strategies will then play a fundamental role in the interpretation and hence the translation we make of a particular text. Arrojo emphasises the impact of the word "plums" in both cases, for instance, claiming that in the note they are taken literally, matter-of-factly, while in the poem they acquire symbolic, metaphoric meanings. In fact, most English dictionaries will mention the fact that plums may refer to "an excellent or desirable thing, as a fine position", or even to "an unanticipated large increase in money or property, as an unexpected legacy; a windfall" (cf. *Random House Unabridged Dictionary*). Moreover, Arrojo calls the readers' attention to the fact that when one thinks of "plums" in Brazil, what first springs to mind is "prunes", which are dark and dry, hardly bearing any resemblance to "plums". To make matters worse, most Brazilians would probably think of "prunes" as an effective home remedy for intestinal disorders, an interpretation far from traditionally poetic. In this sense, a functional translation into Portuguese would have to take these elements into account and perhaps replace "plums" with another fruit – like "apples", "cherries" or "peaches".

Therefore, this little experiment shows that literariness is not produced by means of textual elements, it is not necessarily intrinsic to texts, but rather a product of our interpretation, of the reading circumstances and strategies we adopt. Indeed, the experiment shows how neither the author's intention nor the text itself retain literary features as such; after all, if they did, the group who thought they had a silly little note before their eyes should have "detected" these literary textual features, these products of the "author's intentions", and yet they did not – and do not every time this experiment is repeated. So to conclude this chapter on literary texts, Arrojo advocates the importance of interpretation, particularly when it comes to translation. Every translation is inevitably the result of an interpretation, which is always conditioned by its general circumstances.

What Arrojo nonetheless fails to acknowledge is the paramount role played by the translation brief, the *skopos* of a given translation. Indeed, when it comes to the everyday life of translators, the translation brief (or "Übersetzungsauftrag", in Vermeer's words – cf. REIß & VERMEER 1984, for instance) will help to shape most of the reading and translation strategies employed, which does not mean to say that the translator's ideology and circumstances can be forgotten. Nevertheless, the translation brief will certainly help the translator to choose what interpretation to privilege in his or her translation. Arrojo maintains that "qualquer tradução de 'This is just to say' seria necessariamente o reflexo da interpretação que, *por alguma razão*, decidíssemos privilegiar", or any translation of "This is just to say" would inevitably reflect the interpretation, *for whatever reason*, one decided to privilege (ARROJO 1986/2002: 36 – my emphasis). Well, in most cases of professional translation this "reason" is provided by the initiator of the translation process and should be accommodated as well as possible by the translator. In any case, overlooking the contributions of the German functional approach is by no means exclusive to Rosemary Arrojo; unfortunately, the lack of translations into Portuguese (or even into English) of the works by the functionalists has led to the scarce knowledge many Brazilian theorists appear to have of it – and here one could go as far as to say that this scarce knowledge seems to be predominant outside the German-speaking world as a whole (cf. LEAL 2010b).

The last chapter in Arrojo's *Oficina de Tradução* to deal with more abstract issues is Chapter 4, entitled "A Questão da Fidelidade" (or On Fidelity)[79]. She returns to Borges' Pierre Menard to emphasise that readers (as well as translators) cannot prevent their contact with texts from being mediated by their own socio-historical context, their circumstances and their ideology (ARROJO 1986/2002: 38). In order to make this point clearer and to link it to the notion of fidelity in translation, Arrojo presents a rather unexpected example, namely a "Cleopatra Costume Contest" taking place in São Paulo in the 1920s. Although the efforts of the contestants would definitely be invested in reproducing the most faithful version of Cleopatra possible, these "Cleopatras" would

79 In Wolf's anthology, this chapter was translated as "Eine neue Auffassung von Treue" by Annette Wußler (WOLF 1997: 43-48).

inevitably be marked by the style and trends dominant in São Paulo in the 1920s. In other words, the winner of one such contest would probably bear little resemblance to the winner of a similar contest taking place in the 1920s in New York, or in São Paulo in the 1990s. As Arrojo stresses, if this contest were to take place today and one tried, through historical research, to recreate the "one and only" Cleopatra, it would only be one possible version, permeated by one's own idiosyncrasies and beliefs. Not to mention the dominant trends and style of one's culture and time, in addition to other technical issues, such as the types of fabric available, the kind of film in which the images of the contestants would be captured, the kind of jewellery found in the market, amongst others.

In spite of its triviality, the example illustrates the condition of any translator when faced with a translation task: even if one's sole purpose is to reproduce the so-called author's intentions or the so-called "original text", these intentions or this text is nothing but *our construction of it*, which is strongly influenced by our personal history and context. Going back to the example of the costume contest, the Cleopatra perceived as the most "original" in a particular place and time will be the product of the predominant assumptions surrounding Cleopatra in that particular interpretive community. In this light, fidelity is strongly relativised and no longer refers exclusively to an "original" concept or intention. As Arrojo explains, "nossa tradução de qualquer texto, poético ou não, será fiel não ao texto 'original', mas àquilo que consideramos ser o texto original, àquilo que consideramos constituí-lo", or our translation of any text, poetic or not, will not be faithful to the "original" text, but rather to what we think the original text is, what we think constitutes it (Arrojo 1986/2002: 44).[80]

The title of Chapter 5 indicates the shift from theoretical to practice-oriented issues: "A Teoria na Prática" (Arrojo 1986/2002: 46), or Theory in Practice. In this chapter, Arrojo thoroughly analyses a poem by Brazilian poet Carlos Drummond de Andrade (arguably the best Brazilian poet) and an English translation by John Nist. At the end of the chapter, she proposes her own English version of the poem.

Unlike her theoretical reflections, her analysis of the poem does not contain any allegedly "radical" elements, quite the contrary.[81] She begins by care-

80 This quotation was used as epigraph to the present section.

81 As already indicated, the issue of "radicalness" will be addressed at length in IV, 2. As it is a manifold issue, let us leave it in suspense for the time being.

fully examining the title of the poem, "Áporo", studying its Greek etymology and its vocabulary entries in Brazilian dictionaries. As Arrojo remarks, her interpretation reflects her willingness to read the poem "poetically" and to attribute poetic meanings to the words and phrases so as to build a sort of puzzle. For her, the first and perhaps most important piece of this puzzle is indeed the title, which guides her interpretation throughout the process (ARROJO 1986/2002: 47-48). One by one she analyses the verses bearing in mind the tone and the motif set by the title. Without breaking the poem into the traditional linguistic levels (such as form and content, meaning and form, extra and intratextual elements, etc.), Arrojo examines all poetic features at once – including the rhyme scheme, rhythm and sonorities – along with their impact on the poem as a whole.

During the course of her analysis, she often refers to the poem as a meaning machine ("máquina de significados" and "máquina de significação" – ARROJO 1986/2002: 52), and by perceiving it as such she quite rightly asserts that

> *O jogo de leitura poética não deve descartar nenhum fragmento que possa ser empregado na construção de uma interpretação. Como nesse jogo não há lugar para acidentes ou casualidades, a máquina de significados, em que se transforma um poema no momento em que é lido, deve tentar incorporar aos seus mecanismos todos os elementos, mesmo aqueles que aparentemente nada significam* (ARROJO 1986/2002: 52).
>
> [The game of poetic reading should take into account each and every element that may help to build an interpretation. In this game there is no room for chance or accident; hence the meaning machine, represented by the poem the moment it is being read, must try and embody as many elements as possible, including those that may seem, at first sight, meaningless.]

This way, what she tries to do is incorporate as many elements into the analysis as possible, often remarking that such and such an element might enrich her interpretation even more ("tal interpretação pode ser ainda enriquecida" – ARROJO 1986/2002: 53).

When she gets to the translation by John Nist, Arrojo claims that the translator privileged elements that she did not privilege in her interpretation, hinting not only at the fact that his choices were different than hers, but also that his translation flattens the ambiguities and polysemy ("achata as duplicidades e ambigüidades" – ARROJO 1986/2002: 55) so crucial to her reading of the poem. Her reading of Nist's translation points towards a literal translation strategy, whereby most poetic features (including metre and rhyme) are not recreated and most ambiguities are eliminated. Yet, in spite of her overall disapproval of his translation, she does not use overly negative words, nor does she claim that Nist's translation is wrong or inadequate. Instead, Arrojo often stresses that it simply *differs* from *her* interpretation and *her* expectations. Not once does she say what would be *correct* or *incorrect*, or what Drummond might have *intended* to say.

Perhaps due to her dissatisfaction with John Nist's translation of Drummond's poem, she suggests her own translation, this time a version "'fiel' à leitura de 'Áporo' que apresentei", or "faithful" to the interpretation presented (ARROJO 1986/2002: 55). What is particularly striking about Arrojo's translation is that she keeps the source text title, i.e. "Áporo", and adds a footnote after the poem explaining the possible meanings of the word in that particular context. This way, she hopes to offer her reader the same "key" she was offered when she read the poem in Portuguese. The reasons why she opts for this strategy are as follows: firstly because the poem was published in an English anthology of Drummond's poems, thus indicating that the potential target public would probably be willing to read foreign poems and would even expect to encounter a few foreign words in the book; and secondly because even those readers whose mother tongue is Portuguese would need to check different dictionaries to be able to follow the reading she proposes of the poem (ARROJO 1986/2002: 56).

As regards the rest of the poem, she maintains more or less the same structure as Nist's version, altering certain words, expressions and verb tenses, and explaining why she thinks her particular choices seem "better" than his. Although she does not use the word "better", she does admit that when one has the interpretation that she proposed in mind, her translation offers more possibilities to the reader (ARROJO 1986/2002: 57).

Before we look into the remaining chapters, I must say that Chapter 5 points towards one of the most paramount questions of this book, a question that has already come up several times: do different translation theories necessarily imply different translation practices? (see Part I). While on the one hand Arrojo's theoretical considerations may be perceived as radical and overly complex (see footnote 81 on page 148), her work as a translator or translation critic is not at all radical. The most striking difference between what Arrojo does and what other theorists perceived as more "traditional" do can be summarised by the word *awareness*. Awareness about the notion of sign, meaning and reading, about the status of original texts and translations, and about the power of interpretation. Her translation critique does not consist of a list of mistakes, unlike most translation critiques found in the media, or unlike the works of other translation theorists, translators and critics – examples will follow in Part III. What Arrojo may see as a mistake in Nist's translation, though never accompanied by the word "mistake", usually consists of a *difference* between her personal reading of the poem and the reading Nist appears to suggest through his translation. A difference that is recognised as such and not necessarily as a problem. These differences are only considered problems when Arrojo tries to recreate, re-enact her personal reading of Drummond's poem and does not find the required means to do it in Nist's translation.

Not once does she say that her choices are "better" than Nist's, "better" in absolute terms. They may be considered "better" for her intended purposes. Nor does she speak of what the poem actually means or what Drummond actually must have meant. Rather than that, Rosemary Arrojo makes it clear from the beginning that reading a poem is, for her, a game of signification. In this light, any element can and should be taken as meaningful, as another essential piece to the poem-puzzle – see the last indented quotation above. Nevertheless, this meaningfulness is not necessarily attributed to the *text* or to the *original author*, but rather to *her interpretation*. This may appear trivial at first sight but gains increasing importance when one remembers that a lot of what is called translation criticism nowadays consists of – if anything – lists of alleged mistakes made by translators. Furthermore, there still are translation teachers and students speaking freely and naively about authors' intentions and *actual* textual meanings, and basing their work on these factors. In this sense, Arrojo's "method" of translation criticism is not at all radical, but rather coherent with the postmodern notion of ethics underlying her work as a translation theorist; a notion of ethics that still remains alien to most of our logocentric tradition.

Furthermore, the possible relationship between Arrojo's reflections and those of the German functional approach are highly interesting, as already pointed out above. In her theoretical texts, such as the ones that make up *O Signo Desconstruído*, she tends to disregard the role the translation brief plays in translation, privileging other key factors instead, like the translator's background, context and ideology. Despite that, in Chapter 5 of *Oficina de*

Tradução she does justify the insertion of footnotes and the use of foreign words by mentioning the type of edition in which the poem would come out and hence the most probable target audience and their expectations. It is these elements that determine and guide her translation strategies. These elements (type of publication and target reader) are far more palpable than the translator's background and ideology and play just as big a role. Of course one must take into account that this book first appeared shortly after the "cultural turn" (cf. SNELL-HORNBY 2006) took place. In any case, in view of the dialogue that was later established between Arrojo's and Vermeer's work (see I, 6), this unacknowledged affinity is worth mentioning – I will come back to this question later in this chapter.

Chapter 6, named "Exercícios de Tradução" (Translation Exercises), invites the reader to analyse two more poems and their translations – "Poema de Sete Faces", again by Carlos Drummond de Andrade, and "Seven-sided poem", by American poet Elizabeth Bishop; and Sylvia Plath's "The Rival" and Luis C. de Brito Rezende's Brazilian version "Rival". The objective is "tentar estabelecer as principais diferenças de significado entre nossa leitura do 'original' e a leitura sugerida pelo texto traduzido", or to try and establish the main differences in meaning between our interpretation of the "original" and the reading suggested by the translation (ARROJO 1986/2002: 58). Therefore, similarly to Chapter 5, Arrojo studies each text thoroughly and then carefully points out these "differences in meaning", later inviting readers to come up with their own translations.

"Recado ao Tradutor/Aprendiz" or "Message to the Translator/Translation Student" is the name of Chapter 7 (ARROJO 1986/2002: 76-78), which summarises the entire book. In this chapter, Arrojo stresses that when it comes to the complex activity we call translation, there are no magic rules or absolute models, as she hopes to have shown in her book. Side by side with language competence, she emphasises the importance of knowing how to read and knowing how to write. And here reading and writing must be understood as activities that entail constant meaning *production*, rather than meaning *protection* or *re*production. It is, therefore, a job that involves great responsibility and requires a thorough knowledge of the interpretive community in which one lives and its intrinsic rules. However, as Rosemary Arrojo very rightly stresses, knowing these rules does not mean blindly following them, but rather having privileged access to them in order to, at some point, change or influence them somehow – I will come back to this issue below. Finally, she closes the chapter by saying that translation theory is also a key element to which all so-called professional translators should pay attention in order to be fully professional and well-informed, aware of the nature of their work and their paramount role in society.[82]

Chapters 8 and 9 contain a very small glossary (with only four entries) and a short commentated bibliography, respectively. Given the popularity of the book, updating and expanding these last two chapters would be very useful for translation teachers and students – though I do not know whether the editorial arrangements in question would allow for that. All in all, in spite of its old age, the book remains a classic of Brazilian translation studies, perhaps the most popular in translation courses at all levels all over the country. It is certainly one of the very few Brazilian books on translation to have got five new editions over a 20-year period of time.[83]

Perhaps it is precisely its lack of radicalism that makes this book such a useful classroom resource for translation teachers. Those teachers who share Arrojo's postmodern views will emphasise the importance of interpretation and the notions of translation criticism and ethics that derive from it. On the other hand, the more "traditional" teachers will overlook the emphasis placed on the power of interpretation, stressing that Nist's translation, for instance, is inaccurate and poor, saying it requires corrections, etc. I myself know a number of translation teachers working at different Brazilian universities who either openly disagree with Arrojo's view (such as Paulo Henriques Britto – see Part III) or have little or no knowledge of poststructuralist thought and deconstruction. Yet these teachers use Arrojo's *Oficina de Tradução* and find it very handy.

In the next section, we will look into Arrojo's 1993 book on the interfaces between translation, deconstruction and psychoanalysis (*Tradução, Desconstrução e Psicanálise*). Before we do so, though, let us remember the questions raised in the present section, questions to which I will come back in Part IV: the idea of radicalism and poststructuralist thought (brought up in II, 2.1), the interface between the German functional approach and some of Arrojo's reflections, as well as the requirements to establish so-called professional translators (a question first raised in Part I).

82 This argument sheds light on the question of what professional translations should be like, an issue already raised several times in Part I and to which I shall come back in IV, 1 and 3.

83 Together with Paulo Rónai's *Escola de Tradutores* (see Part I), originally published in 1952 and in its sixth edition in 2010. Also, John Milton's *Tradução: Teoria e Prática* (MILTON 1998) was in its third edition in the same year.

2.3 Tradução, Desconstrução e Psicanálise

The only realistic approach to the teaching of translation should concentrate on efforts to provide future professionals with the critical apparatus that will allow them to discover what kind of strategy should be employed in each translation project that they decide to undertake. This critical apparatus can be better developed (...) with a group of students that are willing to analyze and debate their own methods of producing meanings and exercising power, and to give up the futile mission of finding the definitive dictionary or glossary, or the unanimously exact equivalent to a word or expression, or even absolute fidelity to an original, or, perhaps, the immortal translation of a text.

Rosemary Arrojo[84]

Similarly to *O Signo Desconstruído* (see II, 2.1), *Tradução, Desconstrução e Psicanálise* (1993) is a collection of papers dedicated to the interface between translation and deconstruction, but also between translation and psychoanalysis – and more often than not the interface amongst all three. Unlike the 1992 anthology of essays, however, all 10 papers that make up *Tradução, Desconstrução e Psicanálise* were written by Rosemary Arrojo. She justifies her choice for deconstruction and psychoanalysis as follows: "tanto a psicanálise quanto a desconstrução – ao praticarem uma reflexão que parte da inevitabilidade de uma teoria da interpretação que não se tece em torno de um enredo de perdas e ganhos – se encontram dentro dos limites generosos da pós-modernidade" (ARROJO 1993: 10), or both psychoanalysis and deconstruction can be placed within the flexible boundaries of postmodernism since both entertain reflections whose starting point is the inevitability of a theory of interpretation – a theory that is not formulated in terms of gains and losses. Precisely because the question of interpretation is so crucial for Arrojo, deconstruction and psychoanalysis seem to offer two similar channels, two cognate sets of lenses through which she can contemplate translation. In other words, deconstruction and

84 This excerpt is quoted on pages 173-174.

psychoanalysis constitute two powerful instruments to question the essentialist notions of stable meanings and pure objectivity because they both value interpretation so much – as something individual, subjective, strongly influenced by the unconscious and subject to numerous factors.

However theoretical or abstract the title of her book may sound, Arrojo stresses that her ideas are not destined to remain limited to the pretentious realms of theoretical reflections ("não são destinadas a ocupar apenas o espaço reduzido e pretensioso da reflexão teórica" – Arrojo 1993: 10); instead, the aim of her papers is also to analyse and question practical issues surrounding translation, its professionalisation and teaching ("têm como objetivo explícito também discutir e problematizar as questões práticas de tradução, de sua profissionalização e de seu ensino" – Arrojo 1993: 10).

In the first paper, entitled "A que são Fiéis Tradutores e Críticos de Tradução? Paulo Vizioli e Nelson Ascher Discutem John Donne" (What are Translators and Translation Critics Faithful to? Paulo Vizioli and Nelson Ascher on John Donne), Arrojo addresses the question of translation criticism from a deconstructionist point of view.[85] She takes a dispute that took place in Brazil in 1985 over two Brazilian anthologies of John Donne's poems. The quarrel started when Ascher published a not very flattering review of Vizioli's translated anthology of the abovementioned poet in the newspaper *Folha de São Paulo*, to which Vizioli replied a week later, and finally, another week later Ascher published his counterreply – both in the same newspaper (Arrojo 1993: 15). Arrojo proposes to investigate to what extent translator (in this case, Vizioli) and critic (here Ascher) regard the *same* text as the original text.

Initially, Arrojo briefly sketches the deconstructionist view on the issue of meaning. Drawing a concise overview of the so-called traditional outlook – that presupposes stable, objective and unchanging meanings – and then quoting Nietzsche as the first thinker to relativise these views and defend the fictitious character of the concepts of "truth" and "meaning", Arrojo comes to the following conclusion:

Assim, nenhuma tradução pode ser exatamente fiel ao "original" porque o "original" não existe como um objeto estável, guardião implacável das intenções originais de seu autor. (...) De maneira semelhante, ao avaliarmos uma tradução, ao compararmos o texto traduzido ao

85 This paper by Arrojo is taken as a starting point in Britto's "Fidelidade em Tradução Poética: O Caso Donne" (Britto 2006b – see III, 3.3).

"original", estaremos apenas e tão-somente comparando a tradução à nossa interpretação do "original" que, por sua vez, jamais poderá ser exatamente a "mesma" do tradutor (ARROJO 1993: 19-20).

[Therefore, no translation can be absolutely faithful to the "original" because the "original" does not exist as a stable object, as the undefeatable guardian of its author's "original" intentions. (…) This way, when one assesses the quality of a translation, when one compares the translated text to its "original", all one does is compare the translation to one's interpretation of the original which, in turn, can never be the exact same interpretation as the translator's.]

Quoting various passages from Ascher's review, Arrojo states that the source of Ascher's dissatisfaction with Vizioli's translation appears to stem from the critic's preference for yet another translation of the same poems done by the celebrated poet Augusto de Campos several years earlier. For Ascher, Vizioli's translation is too scholarly, literal and old-fashioned, whereas Campos' translation is creative, is the true work of a poet (ARROJO 1993: 20). Furthermore, in Ascher's opinion Vizioli's style prevents him from capturing the essence of Donne's poetry, which Campos, on the other hand, accomplishes masterly. In his reply to Ascher, Vizioli argues that his translations sound old-fashioned and scholarly *on purpose* – he was, after all, translating a 16th century poet. He also very rightly questions Ascher's criteria to determine that his translation is inferior to Campos', seeing as Campos seems to have taken far more liberties with the text than he did. What Vizioli finds particularly questionable in Campos' translation and Ascher's subsequent praise is the fact that Campos famously inserted in Donne's poem "The Apparition" a verse of a popular Brazilian song by Lupicínio Rodrigues (cf. LEAL 2010a – ARROJO 1993: 21).

Arrojo does not take Ascher's counterreply into account, but rather focuses on criticising Ascher's review by taking into consideration some of Vizioli's arguments in his reply. She firstly quotes T.S. Eliot, Ben Jonson, John Dryden and Samuel Johnson to show how the reception of Donne's work is rather diverse, and that there has never been a consensus as far as the so-called "essence" of his poetry is concerned (ARROJO 1993: 22). She then analyses a few examples taken from Campos' and Vizioli's translations and arrives at the conclusion that the translators appear to have privileged different factors in their translations. For her, Campos seems to favour Donne's wit, irony and paradoxes, as well as his use of puns and plays on words. Vizioli, in contrast, openly privileges the awful dualism of Donne's times, as he himself mentions in the preface to the anthology in question (ARROJO 1993: 23-24). And here Arrojo very rightly points out that both the wit and the dualism are features commonly attributed to Baroque literature (ARROJO 1993: 24).

Rosemary Arrojo also makes a few remarks on Campos' and Vizioli's standpoints as far as poetry and literary translation is concerned, trying to show that they speak from different standpoints – hence their different preferences and standards as well. Whereas Vizioli thinks that a 16th century poet must be presented to a 20th century reader as *a 16th century poet*, Campos, on the other hand, appears to believe that Donne's poetry should speak to the poets of the present – hence the newer vocabulary and insertion of references from popular music. Arrojo concludes that "tanto Paulo Vizioli quanto Augusto de Campos são 'fiéis' às suas concepções teóricas acerca da tradução e acerca da poesia de Donne e, (...) tanto as traduções de um, como de outro, são legítimas e competentes", or both Vizioli and Campos are "faithful" to their notions of translation and of Donne's poetry, and both translations are legitimate and competent (ARROJO 1993: 25).

Indeed, it is as José Garcez Ghirardi argues in his *John Donne e a Crítica Brasileira: Três Momentos, Três Olhares* [John Donne and Brazilian Criticism: Three Moments, Three Views] (GHIRARDI 2000: 85-106): Campos saw Donne as "the newest new" and proposed to update and renew his work through Ezra Pound's poetics (GHIRARDI 2000: 85). Vizioli, in contrast, strived to understand Donne's work as a product of his controversial times, for which "love" and "death" ("amor" and "morte") were the cardinal points – hence Vizioli's choice of title for his anthology: *John Donne: O Poeta do Amor e da Morte* (GHIRARDI 2000: 96).

And let us not forget that the types of publication in which both anthologies appeared are utterly different. Augusto de Campos certainly chose to translate Donne as a part of his poetical project. It is beyond the scope of this book to fully flesh out the context in which Campos' translations of Donne's poems were published; suffice to say that by 1978, when his anthology came out, Campos had already translated Pound, Mallarmé, Joyce, Cummings, Blake and Mayakovski, as well as published dozens of other volumes of his own poetry, along with essays on poetry and literary criticism. In this sense, and in light of his colossal literary production and impact in Brazilian literature, his translations of Donne were undoubtedly a key part of his oeuvre, arguably in an unprecedented way in Brazilian literature. Most readers of one such anthology were probably far more interested in Campos than in Donne; they probably wished to see how Campos incorporated the English poet into *his own* poetics. One could, therefore, speak of a strong sense of independence, or reinventing, or liberty by Campos.

Vizioli, conversely, probably received a translation assignment from J. C. Ismael, a renowned Brazilian critic who commissioned the translation and publication of a few English poets, including Blake and Donne. The anthology was bilingual, presupposing a target reader whose interests probably involve the comparison between translation and original; a cultivated reader, therefore, who would possibly be keen on historical factors surrounding Donne's life and work. In this sense, this translation project strikes one as far more dependent on the source text than Campos' translation project. Surely Vizioli kept in mind the idea of a possible contrast between his translation and Donne's original, hence his toil both to maintain the translation as close as possible to the source text and to grant the text an old-fashioned tone, very much in tune with the reception of Donne's original today, centuries later.

All in all, Arrojo's argumentation in this first chapter of *Tradução, Desconstrução e Psicanálise* is not only highly interesting, but also careful. She takes a rather touchy issue in translation studies, i.e. the assessment of translation quality, and demonstrates how a deconstructionist perspective can help to unveil the differences between two translations. These differences are perceived as such, and not as signs of inferiority or superiority. Instead of attempting to prove which translation is better, she strives to understand the possible nature of the differences between the translations – what motivated them and what interests lie beneath them. In this sense, one could even say that the strengths and weaknesses of both translations are rendered clear, but never in absolute terms; rather, these strengths and weaknesses are changeable, varying according to the context and the situation of reception. Let us take Campos' intertextuality between Brazilian popular music and Donne's poetry: depending on the reader's context, expectations and motivations (amongst various other factors), this liberty on the part of Campos may be perceived either as a pro or as a con – as Arrojo very aptly explains.

And here once again the argument of the German functional approach seems pertinent, even though Arrojo never mentions the translation brief or the *skopos* of both translations. As argued above, Campos and Vizioli certainly had completely different readers in mind, with utterly diverse interests and motivations. The translations were published in entirely different editions as well, which inevitably influences the translators' strategies. Of course the translators' personal history and style are key factors equally at stake here; nonetheless, one cannot ignore the impact of the potential target audience, for example.

The second paper of *Tradução, Desconstrução e Psicanálise* is dedicated to yet another sensitive issue in translation studies, namely its institutionalisation and the professionalisation of translators/interpreters. Since this issue has repeatedly come up here, and will arise again in II, 2.4, as well as in Parts III and IV, I will only address it briefly. The paper is entitled "As Relações Perigosas entre Teorias e Políticas de Tradução", or Translation Policies and Theories: A Dangerous Relationship (ARROJO 1993: 27-33). Her main argument in this essay is that from an essentialist point of view there is a dangerous confluence of the concepts of translation generally entertained by common sense, linguistics and translation theory. According to Arrojo, in all three there is a strong belief in a fully stable original text, completely independent from interpretations, contexts and circumstances – even though all three are confronted daily with the illusory character of one such conviction (ARROJO 1993: 28).

This basic premise deemed logocentric by Arrojo leads to two main consequences: firstly, there is a tendency to assume that all one needs in order to translate or interpret is the knowledge of at least two languages. If translating means recovering the stable, objective meanings present in the so-called original text, speaking two languages fluently is more than enough. Secondly, precisely because linguistic knowledge is all one needs, translation finds no exclusive niche at higher education, being relegated to a significantly inferior post within cognate areas (ARROJO 1993: 31). As for translation theory, Arrojo maintains that a similar attitude applies. Because most theories are essentialist, they too relegate the translator's activity to an inferior post, filled with frustration and unsuccess, and thus confirm the status attributed to translation both by linguistics and common sense (ARROJO 1993: 30).

For Arrojo, the way out of the dilemma is fairly simple: translators must first become aware of their power, their unavoidable interference with what they translate. This would be the first step towards a more honest relationship amongst the instances that inform translation – such as theory or even common sense (ARROJO 1993: 32). As already indicated, it remains unclear whether Arrojo would endorse a full process of independence for translation studies in the terms that it occurred at various European universities, for example (see I, 1 and 2).

More important than that, however, is whether her remarks remain relevant today, two decades later – and if so, in what terms? Surely even the most essentialist translation scholar would disagree with the "logocentric" views suggested by Arrojo in this paper. Nowadays it seems relatively "common sense" that translation amounts to more than transferring meanings – we can thank the "cultural turn" for that (see I, 6). Despite that, both the translator and translation remain fairly marginal, unacknowledged, relegated to inferior posts in the most varied realms of society. Numerous universities across the globe do offer a safe niche for translation studies, but this awareness of the translator's authorial role that Arrojo advocates seems far from achieved. In other words, the establishment of an exclusive niche to translation studies at university has not led to a significant elevation of the status of translation. Could it be perhaps because the vast majority of translation theory dominant today remains rather essentialist, and so does the attitude of most of those involved with translation? Let us keep these ideas in mind as we advance into the next papers and chapters, in which these same issues will be brought up again.

The first paper in *Tradução, Desconstrução e Psicanálise* to deal with the interface amongst these three cardinal points that guide her work is called "Laplanche Traduz o Pai da Psicanálise: As Principais Cenas de um Romance Familiar", or Laplanche Translating the Father of Psychoanalysis: Scenes of a Family Romance (ARROJO 1993: 35-50).[86] Arrojo puts Laplanche's eminent translation project of Freud's complete works in the spotlight, using psychoanalysis and deconstruction to guide her through Laplanche's endeavour, and eventually drawing important conclusions for translation studies – particularly as far as fidelity in translation is concerned (ARROJO 1993: 35-36).

In a nutshell, Arrojo takes Freud's notion of "family romance"[87] and projects it onto the relationship between Laplanche, Lacan and Freud. According to her, Laplanche openly wanted the translations of Freud's complete works to fully escape all possible "external" interferences – particularly Lacan's, which Laplanche fiercely opposed. Indeed, in the first volume of the *Œuvres Com-*

86 A version of this paper was published in English in 2004 in the book *Übersetzung, Translation, Traduction: Ein internationales Handbuch zur Übersetzungsforschung* vol. 1 (edited by Armin Frank, Harald Kittel and Norbert Greiner) under the title "Translation as an Object of Reflection in Psychoanalysis".

87 According to Freud, during a particular phase of their lives children are prone to gradually replacing their blind adoration for their parents with a feeling of rejection, revenge against them or even rivalry – especially with the parent of the same sex. This process would be the result of the child's developing intellect and newly-acquired abilities to compare and relativise. Freud argues that it is often a mild, temporary process that in fact hides the child's longing for the days when its parents seemed flawless and noble (cf. FREUD 1959: 74-78).

plètes de Freud/Psychanalyse, Laplanche does state that his aim is to restore Freud back to Freud, to render his words absolutely faithfully, using a Freudian French that should allow the French audience to relate to his writings in French in the exact same way as the German audience related to his original texts (ARROJO 1993: 36-37 – quoting Laplanche). In this context, Lacan is perceived by Laplanche as a key antagonist not only because of their theoretical disagreements, which are made clear in Arrojo's paper, but also – and more importantly – because of their *similar* ambitions. After all, one could say, as Arrojo very aptly points out, that both wanted to be Freud's sole legitimate heir in French-speaking territory (ARROJO 1993: 44). This noble right to heritage, nevertheless, is not without its perversities:

> *Seu [de Laplanche] desejo explícito de ser o verdadeiro porta-voz de Freud em francês não implica somente a eliminação de seus rivais e mestres próximos; esconde também uma fantasia mais poderosa, a fantasia de se colocar no lugar privilegiado de Freud (...), deixando de ser apenas um dos muitos recipientes da teoria psicanalítica* (ARROJO 1993: 44).
>
> [[Laplanche's] explicit wish to become Freud's true spokesperson in French does not solely entail the elimination of his rivals and close masters, but it also hides an even more powerful fantasy, namely the fantasy of seizing Freud's privileged place (…), and no longer being a mere recipient of psychoanalytical theory.]

It is, therefore, in this light that Arrojo proposes the motif of the family romance. Indeed, she understands Laplanche's translation project as an oedipal gesture towards Freud's text and goes as far as to claim that one such oedipal gesture is common to all translations, since all translations attempt to simultaneously replace a given original text and lengthen its life by introducing it in another language, culture and time (ARROJO 1993: 47). And here she draws a witty parallel with Freud's famous misunderstanding of an autobiographical text by Leonardo da Vinci, thoroughly discussed by Alan Bass in his 1985 paper "On the History of a Mistranslation and the Psychoanalytic Movement". The details of this misunderstanding are beside the point here; the idea that is crucial for Arrojo is Bass's argument that Freud's mistake was actually motivated by his strong will to develop his theories on infantile sexuality. Therefore, it is as though Freud had found an excuse in da Vinci's text to develop his theories further the way he wished to do it – we see what we want to see, do we not? Similarly, both Lacan and Laplanche use the pretext of absolute fidelity to Freud to rewrite psychoanalysis through their own point of view (ARROJO 1993: 47).

On account of this rather thought-provoking reflection proposed by Arrojo, the consequences for translation seem evident. Accepting the impact of these ideas on translation means accepting that any interpretation is, to a certain extent, fetishist as it is the product of a relation of transference – and transference here should be taken in a Lacanian way (ARROJO 1993: 47 – see below). So from a psychoanalytical and a deconstructionist perspective, original texts do not contain stable meanings, but rather are always subject to the reading and interpretation of an individual who, in turn, is strongly marked by both his/her unconscious and his/her circumstances, perspectives and motivations (ARROJO 1993: 48). The desire for absolute faithfulness in translation masks all sorts of other conscious and unconscious desires that cannot be ignored any longer, particularly not in translation studies.

In this paper, Arrojo also comments on the outcome of Laplanche's translation project of Freud, claiming that countless critics deemed it *unfaithful* – notwithstanding his declared intentions of being "absolutely" faithful (ARROJO 1993: 41-42). She quotes Catherine David, for instance, for whom Laplanche's translation exaggerates the use of neologisms, making the text sound artificial (ARROJO 1993: 42). Other critics quoted by Arrojo, such as Rubens Volich, argue that Laplanche inserts far too many changes to the text than actually necessary (ARROJO 1993: 42). Arrojo then concedes that many of these critiques might be influenced by "sibling rivalry" (ARROJO 1993: 42), but still they show how relative the notion of faithfulness is.

Another interesting element in Arrojo's paper is François Roustang's remarks on the attempt to establish Freud's *true* words *unanimously* within a culture – in this case, the French culture. For Roustang, one such attempt can only be a misunderstanding (ROUSTANG 1976/1986: 14 – translated by Ned Lukacher – cf. ARROJO 1993: 45-46):

If one looks for the effects of psychoanalysis, one can see them in that any group of psychoanalysts carries within it the principle of its own disintegration. If the group is stable and functions well, however, it is a proof of the contrary: it has definitely abandoned the Freudian discovery. In this sense, psychoanalysis is basically asocial (...).

In this light, one can perceive Laplanche's translation project and particularly its results as an example of Freud's "savage horde" (apud ROUSTANG 1976/1986: 14), for whom unanimity is an impossibility. Should that not be the case in translation studies as well, an area in which the disagreements appear to escalate at the same rate as the wish for unity and global theories? I will come back to this question in IV, 4.[88]

The fourth essay that makes up this anthology is entitled "A Tradução como Paradigma dos Intercâmbios Intralinguísticos" (or Translation as a Paradigm of Intralinguistic Exchange[89]) and opens with an epigraph by Octavio Paz in which he compares translating to the process of producing utterances or reading in one's mother tongue (ARROJO 1993: 51). Arrojo takes this initial idea to develop a stimulating reflection on the allegedly intermediary role played by translators. For her, the key question is how these mediators can keep the meanings they inevitably manipulate intact (ARROJO 1993: 55). And here she remembers Mounin's famous conclusion about translation being the scandal of contemporary linguistics (see I, 5) and asks, very much in tune with Paz's epigraph, whether any reading, any paraphrase, any linguistic exchange and any process of oral and written communication would not be equally doomed to this "scandal" ("Não estariam condenadas a esse 'escândalo' toda leitura, toda paráfrase, todo intercâmbio linguístico e todo processo de comunicação oral e escrito?" – ARROJO 1993: 55). Indeed, it is as Steiner reminds us (STEINER 1975/1998: 263 – his emphasis):

No two speakers mean exactly the same thing when they use the same terms; or if they do, there is no conceivable way of demonstrating perfect homology. No complete, verifiable act of communication is, therefore, possible. All discourse is fundamentally monadic or idiolectic. This was a shopworn paradox long before Schleiermacher investigated the meaning of meaning in his ***Hermeneutik***

Departing from these preliminary ideas, Arrojo then looks into the deconstructionist perspective on language – taking mainly Derrida's "Des Tours de Babel" into consideration. For Derrida, as for Arrojo, the "multiplicité ir-

88 For more on this wish for global theories, cf. LEAL (2010b).

89 This paper featured in Michaela Wolf's *Übersetzungswissenschaft in Brasilien* (WOLF 1997: 71-88) under the title "Die Übersetzung als Paradigma der intralingualen Kommunikation" (translated by Helga Ahrens).

réductible des langues" is a reality within *one* language too, as no meaning can be fully determined, not even inside a "single" language (DERRIDA 1985 – cf. ARROJO 1993: 56). Because of this "confusion babélienne" (DERRIDA 1985), translation makes up a paradigm in deconstruction. In other words, the logocentrism that Derrida does his utmost to deconstruct is nothing but the dream of a universal language, nostalgia for the origin, i.e. the opposite of what the myth of Babel teaches us. In this light, Arrojo maintains that within the deconstructionist reflection on translation, instead of a secondary operation coming after its original, translations even precede the so-called originals as their (the original's) very possibility (ARROJO 1993: 58 – quoting Peggy Kamuf, Derrida's translator).[90]

Along with deconstruction, Arrojo draws inspiration from pragmatic philosophers like W. V. Quine, for whom a linguistic theory should not contemplate language as a faithful reproduction of reality – as both common sense and numerous philosophical tendencies do (ARROJO 1993: 59). Based mostly on Quine's 1969 *Ontological Relativity and Other Essays*, Arrojo comments on the different linguistic experiments carried out by the philosopher that led to his well-known thesis of the indeterminacy of meaning and, hence, the indeterminacy of translation (ARROJO 1993: 63). In her view, "a compreensão, a recepção, a apreensão (...) do discurso do outro envolve um ato de 'interpretação radical' já que não encontra outra âncora a não ser a observação, o julgamento, a crença do intérprete-receptor", or speech comprehension, reception or apprehension (...) entails a process of "radical interpretation" since it finds no other anchor than the interpreter-recipient's observation, judgement and belief (ARROJO 1993: 63 – quoting Quine).[91] For this reason Quine's disciple Donald Davidson defends the interdependence between meaning and belief (ARROJO 1993: 59).

90 For more on the notion of translation as the possibility for the original to survive, see Derrida's "Des Tours de Babel" (DERRIDA 1985): "Telle survie [la traduction] donne un plus de vie, plus qu'une survivance. L'œuvre ne vit pas seulement plus longtemps, elle vit *plus et mieux*, au-dessus des moyens de son auteur. Le traducteur serait-il alors un récepteur endetté, soumis au don et à la donnée d'un original? Nullement" (his emphasis).

91 This reminds me of Arthur Schnitzler's short-story "Ich", in which the protagonist is caught in bewilderment at the complexity of language, feeling suddenly compelled to attach little notes to everything and everyone stating their names. It is as though he were indeed trying to provide language with this "anchor" it does not possess, thus freeing it from mere belief and observation and bringing uncertainty and changeability to an end. The end of the story is particularly remarkable: fearing that the protagonist has gone mad, his wife calls a doctor and, upon his arrival, the patient is wearing a sign on his chest saying "ich" (SCHNITZLER 1961/2006: 304-311).

Fleshing out this bond between meaning and belief, Quine and Davidson develop Neil Wilson's principle of charity further, claiming that any kind of linguistic exchange (including translation) involves a certain disposition on the part of the interactants, and this disposition may be perceived as a sort of charity. In other words, in order for people to communicate, a tacit pact must precede the actual communication, a pact whereby all parties involved confirm their willingness to play the language game and assume that their interlocutors are telling the truth, are making valid arguments (ARROJO 1993: 64).[92] In Arrojo's opinion, this notion of charity or solidarity is an antiessentialist move by these new pragmatists. In comparison to so-called traditional philosophy, what thinkers like Davidson and Quine do is attack the notion of truth as a precise and suprahuman concept, and substitute the notion of objectivity with the notion of solidarity (ARROJO 1993: 65 – quoting Rorty's 1991 "Solidarity or Objectivity?"). And here Arrojo goes further and links this notion of solidarity to Nietzsche's idea of the "Allzumenschliches" inasmuch as both reflections defend the illusory character of the platonic notion of truth – whereby meanings would be immune to perspectives and circumstances (ARROJO 1993: 66-67 – see II, 2.1).

Arrojo ends this paper by asserting that it is this solidarity that defines textuality and allows for the very existence of language (ARROJO 1993: 67). So going back to the initial idea proposed by Paz of a parallel between translation and communication within a mother tongue, translators – like individuals engaged in communications of any sort – can no longer be perceived as neutral mediators, as they too make these tacit agreements that enable language to function. As any individual, translators have their own interests and motivations which, in turn, play a pivotal role in these silent pacts, in their willingness to charity and solidarity towards their interlocutors. If there is any truth, any meaning, then it lies precisely there, in these pacts, in this solidarity.

In "A Tradução Passada a Limpo e a Visibilidade do Tradutor" (Translation Reviewed and the Translator's Visibility[93]), Arrojo departs from the tricky relationship between sign and referent (from the point of view of classical semiotics) and proposes an interesting parallel concerning the relationship between translation and original (from a predominantly essentialist perspec-

92 For more on Davidson's and Quine's works and on how their perspectives differ from those of so-called traditional philosophy, see John Murphy's 1990 *Pragmatism: From Pierce to Davidson* (a book Arrojo quoted repeatedly in the paper in question).

93 In Michaela Wolf's 1997 anthology, this paper was translated as "Die Endfassung der Übersetzung und die Sichtbarkeit des Übersetzers" by Helga Ahrens (WOLF 1997: 117-134).

tive). This ultimately leads her to the review or reformulation of the concept of translation outside the boundaries of essentialism, which then brings her to the conclusion that the translator's visibility is not an aim to be achieved, but rather an inevitable element entailed by translation.

Largely based on Derrida's essay "Différance" (DERRIDA 1972/1982: 1-28), Arrojo discusses the "problem" of the referent in semiotics, asserting, like the philosopher, that for semiotics the sign works as a kind of simulacrum of the actual thing, of reality. Viewed in this light, signs would be secondary and provisory in relation to reality, being relegated to the position of mere mediator (DERRIDA 1972/1982: 9 – cf. ARROJO 1993: 72). Strikingly similar is the relationship between original and translation from a more essentialist point of view. Translation – like signs – would work as a sort of simulacrum of the original text, the mighty primary source – like reality. Arrojo maintains that, from this perspective, this origin, this thing in itself is always postponed, deferred in the name of a *promise of presence* that is never fulfilled (ARROJO 1993: 74). Indeed, in the abovementioned work, Derrida explains that there is no way out of this language labyrinth since signs refer solely to other signs in an endless process of postponement and deferment – hence his coinage of the word "différance", whereby one of the issues hinted at is precisely this idea of deferment, from "différer" in French.[94]

Applying this idea to translation, Arrojo suggests that "desconstruída e passada a limpo, a reflexão sobre tradução abre mão do sonho da transferência intacta do 'mesmo' de uma língua para a outra, tão passionalmente perseguido pela metafísica do logos, e abre-se para a presença ubíqua do outro na (e da) linguagem", or deconstructed and reformulated, reflection on translation gives up the dream of intact transference of the "same" from one language to the other – a dream so passionately pursued by the metaphysics of the logos – and opens up to the ubiquitous presence of the other in (and from) language (ARROJO 1993: 77).

In this newer light, Arrojo remembers the contributions made by Lawrence Venuti (1986), particularly in the sense of advocating the translator's visibility as a tool to invert power relations, as well as by Lori Chamberlain (1988), who draws an interesting parallel between the marginality of translation and the position of women in society. Arrojo's reservations about these contributions aside, she ends this paper by defending the translator's visibility as a sign of the

94 Derrida's "différance" may hint at numerous elements in addition to the idea of deferment. For more on the coinage of this term, cf. his homonymous paper (DERRIDA 1972/1982: 1-28).

acceptance of the "other" in the translated text. This acceptance means that translators no longer unrealistically pretend to be neutral and absent, innocent and faithful, but rather acknowledge their unavoidable interference. This movement, Arrojo predicts, would help (or even *is* helping) to establish a new translation tradition located outside the age-old essentialist boundaries that associate translation to invisibility and guilt. She goes further and argues that "quanto mais visível se tornar a presença do tradutor no texto traduzido, quanto maior sua visão acerca do processo do qual é agente e promotor, menores serão as chances de que seja ignorado, marginalizado e indignamente remunerado" – or the more visible the presence of translators in translated texts, or the greater their awareness of this process in which they are agents and promoters, the fewer the chances of their being ignored, marginalised and poorly paid (ARROJO 1993: 85).

So in conclusion, even though Arrojo stresses that the translator's visibility is but an inevitable mark of any translation, she insists that it is imperative that translators become aware of this visibility, that they acknowledge it not as something to be avoided, but as the reason for the very possibility of translation. For her, it is this awareness and acknowledgement that will help to improve translators' working conditions. And, as she argues in different papers, the chief aim of translator training should be precisely to raise and foster this kind of awareness.

Another paper that intertwines translation, poststructuralist thought and psychoanalysis is called "Sobre Interpretação e Asceticismo: Reflexões em Torno e a Partir da Transferência", or On Interpretations and Asceticism: Transference as a Source of Reflection and Point of Departure (ARROJO 1993: 91-114).[95] As in several other papers of hers, Rosemary Arrojo begins this one by opposing the so-called modern project, with its promises of objectivity, its scientific ambitions and its asceticism, to the deconstructionist project, which in turn openly recognises a will and a desire behind this alleged asceticism (ARROJO 1993: 92-93). Arrojo claims that thanks to this deconstructionist perspective we can question and rethink some of the implications and objectives of the modern project.

As an example, Arrojo takes Saussure's contributions to modern linguistics – addressed in I, 6 – adding Jean-Jacques Lecercle's views (mainly in his 1990

95 This paper featured in Wolf's anthology *Übersetzungswissenschaft in Brasilien* with the title "Interpretation und Askese: Weitergehende Gedanken zur Übertragung" (translated by Hans Vermeer – WOLF 1997: 141-164).

The Violence of Language) together with Derrida's observations (taken especially from his *Positions* – DERRIDA 1987). In Arrojo's opinion, both Lecercle and Derrida point towards a "residue" ("resíduo", in Arrojo's words) in Saussure's theory, whereby everything that does not concern the structure he proposes – i.e. the parole, diachrony, the social, the history and so on – is simply left outside language and linguistics. For both thinkers, this residue, this reject is precisely where language takes place and therefore cannot be ignored any longer.

Arrojo adds that a further issue that the modern tradition would rather keep ignored and outside its structures and sciences is the unconscious (ARROJO 1993: 98). She very opportunely quotes Lacan's brilliant neologism linguisterie, a term that conveys the interference of subjects and their desires in and together with language. She warns, however, that recognising interference, acknowledging the influence of human desire and opening up to the "residues" of language entails by no means the acceptance of chaos or the abandonment of science. For her, it entails instead coming out of our platonic refuge and giving up our ascetic notion of science (ARROJO 1993: 100-101).

Exploring the unconscious a bit further, Arrojo remembers Lacan once again and adverts to the fact that the unconscious should not be perceived as opposed to consciousness, but rather as the difference between consciousness and itself, the inherent and irreducible difference (ARROJO 1993: 102 – quoting Shoshana Felman). In this sense, the unconscious is not merely an object of study, but it *constitutes* the subject as well, the one who studies (ARROJO 1993: 101). And as Lacan famously remarked, "the unconscious is the discourse of the other", which in a way is crucial to Freudian psychoanalysis, as Arrojo defends (ARROJO 1993: 103). The patient's symptoms are to be translated by the patient, to which the analyst reacts and replies. These reactions and replies come to constitute the patient's unconscious – as the discourse of the other (see above) – through the process of transference. In the context of psychoanalysis, transference is key between analyst and patient because the latter assumes that the former knows all the answers, holds the keys that will lead to a cure – in the same fashion as children believe that their parents possess all knowledge. Arrojo explains that this is how transference takes place, and this is how the discourse of the other comes to form one's unconscious (ARROJO 1993: 105).

This relation of transference, Arrojo argues, can be found in our relationships with texts. Quoting Shoshana Felman again, Rosemary Arrojo claims that, in this context, critics or readers play the role of patients, whereas texts are like analysts – i.e. the ones who supposedly hold all truths and answers that patients or readers wish to attain. And here Arrojo turns to Harold Bloom in his 1973 *The Anxiety of Influence* and associates the idea of transference to love (as Lacan does, too) and to influence (ARROJO 1993: 107). Arrojo explains that in Bloom's view, poems, for example, are but the response, the reaction to other poems to the same extent that poets are the response to other poets, or people in general are the response to their parents (see epigraph in III, 3.2). In Bloom's reflection, in order to overcome this influence and write, poets have to indulge in misreading, in being critical, in antithetically completing their precursor's poems. In Arrojo's point of view, an interesting instance of this process is the relationship between Freud and Nietzsche, thoroughly discussed by Derrida in his "Spéculer – Sur Freud". In summary, though one would say that the influence of Nietzsche on Freud is without doubt patent, not only does the father of psychoanalysis not admit it, but he also vehemently denies it (ARROJO 1993: 109-110). For Bloom, Arrojo clarifies, "strong" readers like Freud must disguise the violence of their reading and the extent of their debt to their precursors – "dissimular (…) a violência de sua leitura e o tamanho do seu débito" (ARROJO 1993: 110).

At the end of the paper, Arrojo insists again – this time quoting Bloom in his *A Map of Misreading* – that after Freud's and Nietzsche's contributions, one can no longer speak of interpretation as a meaning recovery activity, of pure relationships between subjects and objects (texts), of relationships that do not leave residues (see above) behind (ARROJO 1993: 111-112). So in this paper Arrojo proposes a thorough revision of the idea of interpretation, discarding the wish for asceticism and embracing the undeniable effects of transference – as it is understood in psychoanalysis.

In "A Literatura como Fetichismo e Algumas Consequências para uma Teoria da Tradução" (Literature as Fetishism: Consequences for Translation Theory), Arrojo proposes a similar kind of revision and reformulation, this time taking the psychoanalytical concept of fetishism as a starting point to reformulate the notion of literariness. At the beginning of her paper, Arrojo draws an overview of the allegedly traditional outlook on literature and its relationship with translation. She quotes writers as disparate as Dante Alighieri and Benedetto Croce as examples of this traditional view, whereby literary translation would involve falsification and loss (ARROJO 1993: 115-116 – see II, 2.2). Arrojo holds that these views derive from a conviction that literary texts possess intrinsic, constant features which, in turn, make them literary. So she takes examples from literary theory to show how different theorists have tried – and failed – to define the nature of poetry and literature in general. And here she quotes Wellek and Warren in their classic *A Theory of Literature*, along with the American philosopher Monroe Beardsley, as instances of these failed attempts (ARROJO 1993: 117).

Another example of this view that Arrojo deems traditional is to be found in Edgar Allan Poe's acclaimed essay "Philosophy of Composition", a text which, in Arrojo's opinion, represents the dream to reach a "pre-babelic state" ["estado pré-babélico"], to find an area of linguistic and artistic stability that escapes the arbitrariness of signs ("área de estabilidade linguística e artística [que] escap[a] da arbitrariedade do signo" – ARROJO 1993: 118). For her, attitudes such as Poe's reveal the desire to create a fictitious universe and remain its lord forever, the one and only that can dictate rules and establish meanings (ARROJO 1993: 120).

In this light, this traditional view described by Arrojo has much in common with Freud's idea of the "Kastrationskomplex" and the "Penisneid" (FREUD 1905/2001: 96 – cf. ARROJO 1993: 120-121). According to Freud, male children tend to believe that both boys and girls possess a penis. Facing the fact that their mother, for example, has no penis, means facing a strong "Kastrationskomplex", whereby boys believe that their mothers have been castrated and that the same fate might befall them as well (ARROJO 1993 – both Freud and Arrojo). And here Arrojo adds, using Freud's words, that amongst children of both sexes the male genitalia appear to be the only ones that matter, the only ones that truly *exist*. But this primacy is not of the genitalia themselves, but rather of the phallus (ARROJO 1993: 121). For her, Lacan's interpretation of this particular element of Freud's theory of infantile sexuality is essential: for the French psychoanalyst perceives the phallus first as a simulacrum, as a signifier (ARROJO 1993: 121). In this sense, castration in psychoanalysis entails the loss of this imaginary plenitude, the perception of an essential incompleteness ("perda de uma plenitude imaginária, percepção de uma incompletude essencial" – ARROJO 1993: 122).

It is in these terms that Rosemary Arrojo proposes an interesting parallel between this notion of castration and the production of literature and translation. For her, writing implies embracing this castration, accepting the imaginary character of our hopes for plenitude and completeness in language. Denying this linguistic castration would be the same as denying that "one's language is doomed to never being more than a good translation", in Jane Gallop's words (GALLOP 1984: 98 – cf. ARROJO 1993: 122). This attitude of denial would lead to what Arrojo calls fetishism, i.e. the excessive adoration of something that is not present, that is not "there". In other words, the attempts, such as Poe's, to keep one's text untouched, or the attempts on the part of critics to protect a certain canonical reading of a particular text is nothing but fetishism, the adoration of constant meanings that, in fact, are not physically "there". Arrojo adds that this fetishism is one of the multiple expressions of narcissistic love reflected onto the objects that we decide to favour over others ("amor narcisístico refletido nos objetos que decidimos privilegiar sobre os outros" – ARROJO 1993: 124).

Towards the end of this essay, Arrojo briefly discusses the concept of fetishism in Marx's work. In short, Marx defends the idea that value is something collectively attributed to objects within a certain community, raising their status in this community. In other words, objects do not possess any intrinsic, inherent value. From this point of view, perceiving economy, for example, as something natural and purely objective is mere fetishism, as economy does not *possess* any essential value but only the value we attribute to it (ARROJO 1993: 125). In this light, Arrojo returns to Poe in his "Philosophy of Composition" and argues that

> *Como o fetichista de Freud, Poe se enreda nos domínios do simulacro à medida que constrói um substituto para um original que não pode existir fora de seu próprio desejo e de sua perspectiva e que, precisamente por isso, tem que ser protegido dos olhos reveladores do outro. Como o fetichista de Marx, Poe também se encontra enredado no campo ilusório das propriedades e dos valores inerentes (...)* (ARROJO 1993: 126).
>
> [Like Freud's fetishist, Poe is caught within the limits of simulacrum in that he builds a substitute for an original that cannot exist outside his own desire and perspective and, precisely for this reason, this substitute must be protected from the revealing looks of the others. Like Marx's fetishist, Poe is also caught in the illusory field of inherent properties and values (...).]

Arrojo closes the paper by stating that Poe's wish for a mathematical language simply transcends the limits of human language because, as the historian John Forrester advocates, the aim of mathematical languages is to achieve *equality*, whereas human language communicates what is *different* (ARROJO 1993: 128).[96] So for Arrojo reading and translating are perfectly possible activities as long as one acknowledges the undeniable impact of one's own desires (ARROJO 1993: 129), instead of defending inherent values and illusory meanings.

Arrojo demands a similar attitude from translation teachers and students in the paper "Desconstrução, Psicanálise e o Ensino de Tradução" (Deconstruction, Psychoanalysis and Translator Training[97]). Based upon Stephen Straight's 1981 essay "Knowledge, Purpose and Intuition: Three Dimensions in the Evaluation of Translation", Arrojo claims that the traditional knowledge categories ascribed to translator training – in this case Straight's six categories: ecology, material culture, technology, social organisation, mythic patterns and linguistic structures – imply that translating is a suprahuman activity (ARROJO 1993: 133-134). In her view, the traditional, essentialist way of thinking has imposed "embarrassing constraints" ["limites constrangedores"] to the activity of the translator, unsurprisingly associating it to failure and incompetence (ARROJO 1993: 135). Arrojo maintains that this attitude has brought about two main consequences. Firstly, translation has not found a niche at universities precisely because of its potentially subversive character, along with its stubborn resistance to systematisation and taxonomy. Secondly, what we call translation theory does nothing but inform students of how secondary their future job will be, as well as of how unattainable the demands of this job are (ARROJO 1993: 137).[98]

Within deconstructionist reflection, in contrast, translation is not only a feasible activity, but it also deserves a special place, or rather it is in the very *spotlight* of contemporary thought on language and culture. But this acknowledgement of the fundamental importance of translation is not without its difficulties, since poststructuralist thought proposes revolutionary changes in the way one *perceives* translation and translator training. To begin with, knowledge is by no means understood as a set of static information contained in books and dictionaries waiting to be objectively apprehended by pupils (ARROJO 1993: 140). As Arrojo remarks, after Nietzsche and Foucault, knowledge is no

96 We have previously seen another argument against the idea of an exact, mathematical natural language in II, 2.1, proposed by logician Alfred Tarski.

97 A version of this paper was published in English in the American journal *TranScribe* in 2005 (number 1). The title Arrojo chose for the paper is "Deconstruction, Psychoanalysis and the Teaching of Translation". Moreover, Hans Vermeer translated this paper into German in Wolf's 1997 anthology *Übersetzungswissenschaft in Brasilien*, where it was published under the title "Dekonstruktion, Psychoanalyse und Translationslehre" (WOLF 1997: 165-182).

98 For a similar discussion on translator training, see II, 2.1.

longer a property that can be rationally discovered, acquired or passed down without regard for history and ideology (ARROJO 1993: 140 – quoting Michael Ryan). Similarly, the university is no longer conceived as a body of universal knowledge that should be preserved and transmitted to future generations (ARROJO 1993: 141).

In fact, when it comes to "passing down" knowledge, Arrojo defends the psychoanalytical concept of transference again. In order for students to "acquire" knowledge, a mechanism of transference has to be set in motion in the same way as in the context of psychoanalysis. Only when pupils establish a relationship of transference with their teachers will they engage in learning, in having passion and eagerness for knowledge (ARROJO 1993: 142-144). But is transference not an inevitable aspect of the teacher-pupil or psychoanalyst-patient relationship? Yes, answers Arrojo, but the difference is that in post-modern times both pupils and teachers should become aware of transference, should openly analyse it and challenge it. It should become clear, for example, that transference is ambivalent – instead of blind adoration, it entails perverse feelings of dispute as well (ARROJO 1993: 144). Viewed in this light, learning and teaching become relentlessly self-critical undertakings – as already addressed in I, 2. Arrojo emphasises that rather than learning a set of truths that will guide their future work, pupils should become aware of the mechanisms and power struggles underneath these truths; they should become aware of the forces of transference that underlie their relationships with teachers and original authors and texts; they should become aware of the (tacit) rules that organise their profession so that they can change them and make them more human, realistic and decent (ARROJO 1993: 145-147). This is the tone of Arrojo's call for action at the end of the paper (quoted directly in English):

the only realistic approach to the teaching of translation should concentrate on efforts to provide future professionals with the critical apparatus that will allow them to discover what kind of strategy should be employed in each translation project that they decide to undertake. This critical apparatus can be better developed in the context of a classroom and with a group of students that are willing to analyze and debate their own methods of producing meanings and exercising power, and to give up the futile mission of finding the definitive dictionary or glossary, or the unanimously exact equivalent to a word or expression, or even absolute fidelity to an original, or, perhaps, the immortal translation of a text (ARROJO 2005, 33 – see ARROJO 1993: 147-148).[99]

In the penultimate paper that makes up *Tradução, Desconstrução e Psicanálise* Arrojo revisits recurring issues not only in this book, but also in her work as a whole. So the psychoanalytical notions of unconscious and transference, Borges' character Pierre Menard, along with Bloom's concept of "strong reader" are some of the main subjects of "A Tradução e o Flagrante da Transferência: Algumas Aventuras Textuais com Dom Quixote e Pierre Menard" (Translation and Quixote's and Pierre Menard's Textual Adventures: Transference Caught Red-Handed).[100]

Departing from Borges' story "Pierre Menard, Autor del Quijote", Arrojo briefly discusses how the protagonist, Pierre Menard, seems to embody the key features of the so-called modern project (see II, 2.1 or 2.4) in his wish to write *Don Quijote* again, as if it were the first time (ARROJO 1993: 151-154). However burlesque or caricatural this character may be, Arrojo finds that he has a lot in common with numerous contemporary critics, linguists, translation scholars and translators, particularly as far as their strong belief in stable meanings is concerned. As an example, she mentions the English Cervantes specialist Anthony Close who, Arrojo quotes, speaks of Cervantes' intentions in *Don Quijote* as if he had had direct access to them, going as far as to explain what purposes Cervantes had in mind when he wrote the book and openly condemning interpretations that do not fit his analysis (ARROJO 1993: 154-155 – quoting Close in his 1972 paper "*Don Quijote* and the 'Intentionality Fallacy'").

Faced with Menard and Close's remarks as typically modern, essentialist remarks, Arrojo builds a bridge between their general way of thinking and contemporary thought, beginning with Foucault's comments on Cervantes' *Don Quijote* from *Les Mots et les Choses*. As I will argue in detail in II, 2.4, the character Don Quixote is himself a convinced believer in stable meanings. No wonder Foucault calls him "le héros du Même", Arrojo remembers very opportunely (FOUCAULT 1966: 60 – cf. ARROJO 1993: 156). As Foucault maintains about Quijote (FOUCAULT 1966: 60),

99 A part of this quotation is the epigraph of the present section.

100 A Spanish version of this paper featured in the journal *Debats* (number 75, 2001/2002) under the title "Algunas Aventuras Textuales con Don Quijote y Pierre Menard: La Traducción y lo Flagrante de la Transferencia", translated by Javier Mallo Martínez.

en sa réalité de pauvre hidalgo, il ne peut devenir le chevalier qu'en écoutant de loin l'épopée séculaire qui formule la Loi. Le livre est moins son existence que son devoir. Sans cesse il doit le consulter afin de savoir que faire et que dire, et quels signes donner à lui-même et aux autres pour montrer qu'il est bien de même nature que le texte dont il est issu. Les romans de chevalerie ont écrit une fois pour toutes la prescription de son aventure. Et chaque épisode, chaque décision, chaque exploit seront signes que Don Quichotte est en effet semblable à tous ces signes qu'il a décalqués.

In short, Arrojo explains, both Quixote and Menard (and Close) rely on the strict, universal and unchanging correspondence between language and reality, and are both defeated by the arbitrariness of signs and the impossibility of full repetition (Arrojo 1993: 156). So much for our heroes "du même" (cf. FOUCAULT 1966: 60).

Rosemary Arrojo then moves on to deconstruction, claiming that one of its chief contributions to the debate of language and meaning is the acknowledgement of the desire inherent to any reading (ARROJO 1993: 157). And here Arrojo finds the contributions by psychoanalysis particularly fruitful as well, since this desire of which Derrida, for example, speaks, is intimately related to the psychoanalytical notions of unconscious and transference. As already indicated, if one understands the unconscious not as the other extreme of consciousness, but rather as an irreducible difference between consciousness and itself, one can no longer clearly distinguish *the one who reads* from *what is read* (ARROJO 1993: 158 – quoting Shoshana Felman). By the same token, one cannot exempt any interpretation, any reading, any writing or any translation from transference (ARROJO 1993: 158).

Very similarly to her reasoning in "Sobre Interpretação e Asceticismo: Reflexões em Torno e a Partir da Transferência" (ARROJO 1993: 91-114), Arrojo fleshes out the implications and mechanisms of transference in literary criticism, again turning to Harold Bloom's concept of "strong reader" to justify a certain violence inherent to any reading (ARROJO 1993: 162 – quoting Bloom's *A Map of Misreading*):

Como a escritura, a leitura é uma forma de violência (...) contra o próprio texto/autor com quem o leitor pode estabelecer uma relação, já que esse texto/autor jamais será repetido ou resgatado num processo impessoal ou desinteressado, e sim tomado, possuído e transformado pelo desejo e pelas circunstâncias do leitor que com ele se misturar. (...) Os escritores e leitores "fortes" (...) são aqueles que têm a coragem de supor que "apagaram" o rastro do pai/precursor e inventaram um lugar e um estilo próprios para si mesmos (...).

[Like writing, reading is a form of violence (...) against the text/author with whom the reader may establish a relationship, since this text/author shall never be repeated or recovered through an impersonal, disinterested process; instead, it can only be seized, possessed and transformed through its reader's desires and circumstances. (...) "Strong" readers and writers are those who have the courage to suppose that they have "erased" the father's/precursor's traces and have established their own place and style (...).]

Viewed in this light, the translators' claims to invisibility and neutrality make them an exemplary Oedipus, Arrojo claims, caught between guilt and desire, between the consummated crime against the father/author and the need to conceal this crime and remain incognito ("preso entre a culpa e o desejo, entre o ato consumado da eliminação do pai/autor e a necessidade de esconder esse ato e de se manter incógnito" – ARROJO 1993: 162). One such feeling may be illustrated through a very curious example Arrojo quotes in this text, namely Lacan's English translator, John Forrester, and his difficulty to tell his own ideas from Lacan's:

The problems of being Lacan's translator over the past few years have led to a singular intellectual phenomenon for me: an incapacity to remember whether certain ideas, when expressed in English, are 'mine', my version of Lacan's ideas, or my translation of Lacan's French. Perhaps it is not surprising that this has been my lot, given that the act of translation is specifically aimed at repressing the original and replacing it with an exact replica (FORRESTER 1990/1994: 373 – cf. ARROJO 1993: 164).

Finally, Arrojo goes back to Menard and carefully examines the chapters of *Don Quijote* that our fictitious Frenchman of letters wanted to "reproduce", following the lead left by Borges in the prologue to his *Ficciones*, according to which the allusion to certain texts was not arbitrary (ARROJO 1993: 163). Indeed, many of Menard's desires and much of his transference relationship with

Cervantes and his text are unveiled through Arrojo's analysis. She then ends her paper by claiming that one such analysis that strives to grasp the intricate network of desires and ambivalent feelings inherent to any human endeavour can only be carried out outside the boundaries of logocentrism (ARROJO 1993: 170).

The last paper of *Tradução, Desconstrução e Psicanálise*, "Maria Mutema, o Poder Autoral e a Resistência à Interpretação" (Maria Mutema, Authorial Power and the Resistance to Interpretation), is somewhat similar to the penultimate paper in that it explores representations of translation and language in fiction. Maria Mutema is a character in João Guimarães Rosa's *O Grande Sertão: Veredas*, or rather she is a character in a subnarrative in Rosa's novel. But in her paper, which is indeed the longest by far, Arrojo does not merely keep to this subnarrative. Instead, she presents a thorough analysis of the entire novel from the point of view of psychoanalysis and deconstruction, investigating issues such as the struggle and impact of power both in the novel and as metaphors for language, reading, writing and translation.

Because Arrojo's analysis of the novel is rather lengthy and detailed, and because most readers are probably not well-acquainted with Rosa's *Grande Sertão*, I will limit myself to a brief commentary on some of the ideas she investigates. The leitmotif of her analysis appears to be the power of speech, the power of words. Rosemary Arrojo finds numerous examples in the novel of characters who exert a powerful influence over others through their use of language – Maria Mutema's being perhaps the most fierce, as her words literally lead to a man's death. Furthermore, she provides various instances of how the protagonist, Riobaldo, strives to attain power through his use of language.

The psychoanalytical vein of her analysis is also very prominent, as Arrojo comments on countless instances of the oedipal desire to take the leader's or father's place and authority in the characters' relentless struggle for power. Moreover, the idea of the family romance (mentioned previously in this section) is also very much present, as a few characters (especially Riobaldo) seem to be caught in a twofold relationship of love and rivalry with some of their leaders and role-models. In this sense, Arrojo argues that the novel can be understood as Riobaldo's arduous attempt to leave these precursors, these father figures behind, thus establishing his own identity, establishing himself as a leader.

Another issue that permeates both the novel and Arrojo's analysis is that of absolute opposites. A common version of the main subject of the novel points precisely towards this direction: good versus evil, god versus the devil, and the truth versus lies and falsifications. What Riobaldo realises in his long and tortuous journey is that such clear oppositions cannot be sustained in real life and, even more importantly than that, that they only exist and can only be found *inside* ourselves, and not as given, *external* facts. In this sense, Arrojo asserts that the novel is about the subversion of these absolute opposites so vital to the essentialist tradition (ARROJO 1993: 182).

All in all, Arrojo finds in *O Grande Sertão: Veredas* many thought-provoking examples of feelings and attitudes that are intimately related to the activities of the translator – reading, interpreting, writing. The ambivalent and manifold relationship between translator and author and the desire to simultaneously faithfully repeat and violently replace the translated text; the difficulty in sticking to extreme oppositions such as subject-object and signifier-signified, which in translation insist on remaining changeable and inconstant; the power of words and discourse that incessantly overwhelms those who work with texts; in summary, the issues that arise as soon as one perceives language and translation outside the belittling and claustrophobic limits of essentialism make up the core of her analysis of *O Grande Sertão: Veredas* through the lenses of contemporary thought.

* * *

As this analysis of Arrojo's *Tradução, Desconstrução e Psicanálise* has shown, the interface between translation, deconstruction and psychoanalysis leads to various thought-provoking issues, such as the power of interpretation; the influence of the unconscious on the human mind; the role played by transference and fetishism in human relationships; the instability of the notions of truth, objectivity, meaning and science; the hierarchies and power relationships typical of human organisations; the role of higher education in postmodern times; the place of translation studies at university; amongst numerous others. Many of these issues will come up again in what follows; some will be addressed at length in Part IV – such as the role of higher education, the place of translation studies at university, the instability of certain concepts associated with translation, and the power relationships typical of human organisations.

Similarly to II, 2.1 and 2.2, in 2.3 Arrojo's work was the source of stimulating reflections, leaving little room for disagreement. The radicalism of which the texts in Section 2.1 might be accused appears somewhat softened here, but it certainly remains. However, unlike *O Signo Desconstruído*, *Tradução, Desconstrução e Psicanálise* seems less likely to fall into the hands of less experienced readers, considering that the level of difficulty of the book seems considerably higher. In *O Signo Desconstruído*, Arrojo's papers are short and usually less dense in terms of references; in *Tradução, Desconstrução e Psicanálise*, in contrast, references are abundant and various concepts are largely taken for granted. In this sense, even though some of her propositions may still strike one as relatively radical, this appears to be less potentially "harmful" than in II, 2.1

Last but not least, Arrojo's position as for the institutionalisation of translation remains unclear. While on the one hand she appears to defend it, claiming an exclusive place for translation quite apart from linguistics, on the other hand, one such exclusivity would go against poststructuralist thought, which in turn favours interdisciplinarity over exclusive, isolated disciplines (see I, 2). This question will be brought up again in what follows and will be addressed in detail in IV, 3. Let us now move on to the last section of the present part, where a few of Arrojo's more recent papers will be examined.

2.4 Selected Papers

¡Más bien por imposible! dirá el lector. De acuerdo, pero la empresa era de antemano imposible y de todos los medios imposibles para llevarla a término, éste era el menos interesante. Ser en el siglo veinte un novelista popular del siglo diecisiete le pareció una disminución. Ser, de alguna manera, Cervantes y llegar al Quijote le pareció menos arduo por – consiguiente, menos interesante – que seguir siendo Pierre Menard y llegar al Quijote, a través de las experiencias de Pierre Menard.

Jorge Luis Borges[101]

In this final section of Part II we will examine four papers by Rosemary Arrojo published in Portuguese and English-speaking journals on translation studies. As all of her works analysed up until now stemmed from the 1980s and 1990s, I have decided to include four more recent papers (2000s) in the present chapter, all on relatively different issues. The first paper we will look into, entitled "Modernidade e o Desprezo pela Tradução como Objeto de Pesquisa" (Modernity and the Contempt for Translation as a Research Object), was published in the Brazilian linguistics journal *Alfa* in 2000, and is dedicated to the issue of research, or rather the object of research of translation studies – an issue already raised in Part I. The second paper, named "Shared Ground in Translation Studies", was written together with Andrew Chesterman (see I, 3) and came out in issue 12 vol. 1 of the journal *Target* in 2000 as well. This paper is somewhat similar to the first one, but focuses more on the nature of translation studies.

101 This excerpt was taken from Borges' "Pierre Menard, Autor del Quijote". James E. Irby's translation reads as follows: "Rather as impossible! my reader will say. Granted, but the undertaking was impossible from the very beginning and of all the impossible ways of carrying it out, this was the least interesting. To be, in the twentieth century, a popular novelist of the seventeenth seemed to him a diminution. To be, in some way, Cervantes and reach the Quixote seemed less arduous to him – and, consequently, less interesting – than to go on being Pierre Menard and reach the Quixote through the experiences of Pierre Menard" (Borges 1964: 40).

Next, the third paper, "O Tradutor Invisível por ele Mesmo: Paulo Henriques Britto entre a Humildade e a Onipotência" [The Invisible Translator on his Own: Paulo Henriques Britto between Humility and Omnipotence], is dedicated to a different issue altogether. Having first appeared in the year 2000 in the Brazilian journal *Trabalhos de Linguística Aplicada*, this paper is a response to Paulo Henriques Britto's fierce criticism of both Arrojo's work and poststructuralist thought as a whole – see III, 3.1. Although Britto's work will first be introduced in Part III below, when the motivations behind his criticism will become clearer, many of his objections to poststructuralist tendencies can be easily anticipated.

Finally, the fourth paper that makes up this section is called "Translation, Transference and the Attraction to Otherness: Borges, Menard, Whitman" and featured in the journal *Diacritics* in 2004. Not only is this the most recent work by Arrojo analysed in the present book, but it also exemplifies the latest developments in her research as the paper explores representations of translation in literary texts.

In her 2000 paper "Modernidade e o Desprezo pela Tradução como Objeto de Pesquisa" (ARROJO 2000a), Arrojo addresses the issue of research in translation studies firstly by drawing a general overview of the discipline from a postmodern point of view. She maintains that even though there may have been various tendencies and theoretical standpoints within translation studies in the past four decades or so, their main concern has always been the same: to establish appropriate methods applicable to the widest possible range of text types, as well as strict rules and principles for translating and assessing the quality of translations (ARROJO 2000a: 71 – quoting NEWMARK 1988[102]). For her, this concern is very much in line with the modern aspiration to systematise knowledge with a view to increasing efficiency – an aspiration derived from the enlightenment philosophy of the 18th century. In the case of translation studies, this would mean the establishment of a set of universal precepts fully immune to human circumstances and the limitations of science; the establishment, in other words, of a *true* translation theory (ARROJO 2000a: 72 – quoting AUBERT 1993[103]).

Arrojo then mentions two pioneer works in translation studies that epitomise this wish to bring to an end the "it-depends-on-the-case" philosophy commonly associated with translation (which I will take for granted here):

102 Newmark's *Approaches to Translation*.

103 Aubert's *As (In)fidelidades da Tradução: Servidões e Autonomia do Tradutor*.

Andrei Fedorov's 1953 *Vvdenie v Teorju Perevoda* and Vinay and Darbelnet's 1958 *Stylistique Comparée du Français et de l'Anglais* (ARROJO 2000a: 73). She then goes on to explain that

> *A modernidade nos estudos da tradução se associa não apenas à possibilidade de transformar tanto a teoria quanto a prática em objetos de uma ciência (...) puramente objetiva e, portanto, universalmente aplicável, mas também à suposta superioridade da teoria e dos teóricos – sobretudo da linguística e dos linguistas – em relação ao empirismo de tradutores e comentaristas comprometidos com outras áreas do conhecimento* (ARROJO 2000a: 74).
>
> [Modernity in translation studies is associated not only with the possibility to turn both theory and practice into the objects of a (…) purely objective and hence universally applicable science, but also with the alleged superiority of theory and theorists – particularly of linguistics and linguists – over the empiricism of translators and thinkers from other areas.]

From her point of view, therefore, modernity has brought to translation studies an urge to scientism which is still very much present today (see Part I). She nonetheless believes that this scientistic impulse has not given rise to a single instance of universally applicable knowledge – neither in translation theory nor in translation practice (ARROJO 2000a: 75). Despite this, many of the theorists engaged in translation studies do not appear to have been affected by this failure and remain optimistic for a better future (their glass certainly seems half full):

> *(...) as tentativas de eleger algum ramo dessa disciplina [linguística] para servir de base para a elaboração de um conjunto de conhecimentos pretensamente "objetivos" e de aplicação universal, quaisquer que sejam as circunstâncias e as línguas e culturas envolvidas, continuam atraindo a grande maioria dos pesquisadores da área (...)* (ARROJO 2000a: 75).
>
> [(...) the attempts to elect a branch of this discipline [linguistics] to work as the basis of pretentiously "objective" and universally applicable knowledge, irrespective of the circumstances, languages and cultures involved, still appeal to the vast majority of the researchers in the area.]

As examples of this attitude, Arrojo mentions Wolfram Wilss, Julianne House and Joseph Graham, quoting some of their works from the early 1980s. She then mentions Mona Baker (1992) and Newmark again (1988)[104] as examples of translation scholars who defend (or at least defended in the works in question) linguistics as the discipline to which translation studies should ideally be affiliated because, for these researchers, translation is predominantly a linguistic phenomenon.

To quote a more recent thinker, who appears to have a tendency towards scientism and who defends linguistics as the "science" to which translation studies belongs, Arrojo remembers Peter Fawcett's 1997 *Translation and Language*, in which he claims that "since linguistics (...) has produced such powerful and productive theories about how *language* works, and since *translation* is a *language* activity, it would seem only common sense to think that the first had something to say about the second" (FAWCETT 1997: 1 – his emphasis – cf. ARROJO 2000a: 76). Rosemary Arrojo then compares Fawcett's theoretical undertakings and his subsequent humble accomplishments – at least as far as an absolute and universally applicable theory of translation is concerned – to those of Mounin, as already thoroughly discussed in her 1998 paper (see I, 5). Indeed some of Fawcett's conclusions are very similar to those of Mounin thirty years earlier, pointing to the allegedly chaotic, almost rebellious nature of translation and how it seems to resist systematisation (ARROJO 2000a : 77-78).

Arrojo then briefly takes the discussion in a more political direction, asserting that this obsession with scientism in translation studies (and not solely translation studies) ends up transcending the avowed objective to study translation mechanisms and becomes the representation of the autonomy and dominance interests of the scientists (ARROJO 2000a: 78). She finds in LYOTARD (mostly 1988/1997) a source of clarification for this phenomenon, as he explains that modernism has in its heart the idea of both the project as a will and an objective, and of legitimation, of a need for universal truths – hence its conspicuous scientism. Legitimation can only take place through the universal, the permanent, the unchangeable. Lyotard (translated by Julian Pefanis, Morgan Thomas and Don Barry) goes further and asserts that

104 See footnote 102 on page 181.

Scientific reason is not examined according to the (cognitive) criterion of truth or falsity, on the message/referent axis, but according to the performativity of its utterances, on the (pragmatic) axis addressor/addressee. What I say has more truth than what you say, since I can "do more" (gain more time, go further) with what I say than you can with what you say (LYOTARD 1988/1997: 64 – cf. ARROJO 2000a: 78).[105]

In this light, Arrojo feels that countless scholars engaged in translation studies are far more concerned with establishing a science, and thereby strictly controlling the activity we call translation – which is rather utopic – than about observing the actual reality of translation (ARROJO 2000a: 79 – see I, 4).

She then draws a parallel between translation studies and literary theory from the point of view of this difficulty with its own research object. For her, new criticism is an example of a similar process whereby a discipline, in this case literary theory, tried to become scientific and strictly objective – indeed very much in line with the process through which translation studies has been going. Nonetheless, this project also seemed to be doomed to failure as very few developments were actually made in the sense of finding the essence of poetry, delimiting the functions and contours of literary theory, and establishing the distinguishing features of literary texts – and here she quotes I. A. Richards and Schorer et al. as examples of these unfulfilled scientistic ambitions (ARROJO 2000a: 80).

Next, she analyses I. A. Richards' 1948 "Science and Poetry" more closely, aiming at showing that although translation studies and literary theory appear to have gone through a similar process of allegedly scientific and objective systematisation, there is a difference of paramount importance between the two processes. For her, unlike in translation studies, in literary theory there is hardly a desire to control, to discipline, to chastise their research object, i.e. literary texts. Much as they may wish to formalise and systematise the *analysis* and *criticism* of their object of study, there is first and foremost a strong feeling of *acceptance* of literary texts as they are – rather than an obsession to control the *production* of these texts, as is the case in translation studies (ARROJO 2000a: 80-83).

Finally, Rosemary Arrojo briefly draws a parallel between translation studies and linguistics again, claiming that, ever since Mounin, there has been a

105 A part of this quotation was used in the epigraph of III, 3.

hope that one day linguistics will finally save translation studies from illegitimacy, that a branch of linguistics will uncover the key that will allow translation studies not only to become a fully independent discipline, but also to have the right instruments to control its research object adequately. She then mentions a few examples of intersections between literary theory and linguistics (amongst which is Michael Riffaterre's "Describing Poetic Structures: Two Approaches to Baudelaire's 'Les Chats'"), showing that the influence of the latter over the former does not compare with the influence of linguistics over translation studies. Riffaterre's well-known conclusion in the abovementioned paper is that structural linguistics is the one that would have to adapt in order to be able to accommodate the needs imposed by literary texts – an attitude scarcely to be expected of translation scholars (ARROJO 2000a: 83-85).

Arrojo concludes the paper by stating that both translation studies and literary theory have inherited a problematic relationship with their research objects from modernism, a relationship based upon the utopic notions of infinity and immortality. For her,

> *Tanto para os estudos da tradução, como para os estudos literários, de vocação essencialista, o grande apelo da modernidade e suas promessas de objetividade e universalização parece ser a possibilidade de validar seus próprios pressupostos e expectativas referentes à possibilidade de significados perfeitamente estáveis e características textuais intrínsecas* (ARROJO 2000a: 85).
>
> [Both for translation studies and literary theory – with an essentialist vein – the great appeal exerted by modernity and its promises of objectivity and universalisation seems to be the possibility to validate their own expectations as far as perfectly stable meanings and intrinsic textual features are concerned.]

In view of this obsession with universalisation and the flattening of differences, Arrojo believes that the myth of Babel remains the best narrative to illustrate the so-called modern ambitions. Having been condemned to difference (and translation) by an infuriated god, modern man is still trying, through his supposedly objective measures, to defeat and take the place of this god. In this light, it is up to postmodern man, she asserts, to prepare the epistemological ground that will allow us to recognise the ineluctably human nature of our divine aspirations – and hence their impossibility as well – accepting language as a space where *difference* is exercised (ARROJO 2000a: 86).

By now the reader is certainly well-acquainted with Arrojo's arguments and style, so this paper probably strikes one as typical of her. By dissecting the research object of translation studies and comparing it to those of linguistics and literary theory, all of which through a poststructuralist lens, she strives to expose their differences and similarities, weaknesses and strengths. Her attack on scientism is very much in tune with poststructuralist criticism as a whole, and so is her suspicion of attempts to embrace the universal and permanent.

However, what is not clear from her argumentation is her attitude towards the relationship amongst translation studies, literary theory and linguistics. She does make it crystal clear that she by no means believes in linguistics as the be-all and end-all of translation. In fact, she even seems to claim that the influence of linguistics on translation studies is more of a harmful one. As for literary theory, she appears to criticise it for its obstinate hope of one day finding the missing piece of the puzzle that will allow them to finally reach full objectivity – very much in line with translation studies. On the other hand, though, Arrojo seems to admire literary theory for at least accepting, embracing its object of study as it is – unlike in translation studies, as the aim of numerous scholars is precisely to control their object of study. But still a few issues remain unclear.

Surely it is only natural that translation studies should have a more passive attitude towards linguistics than, say, literary theory, because translation studies is a significantly younger field, having produced fewer canonical works and enjoying less acceptance or even awareness on the part of nonspecialists. The point here is not that translation studies *should not* question linguistics as, for example, literary theory does, but rather that the predominantly passive attitude of translation studies towards linguistics is not particularly noteworthy. More importantly than that, nevertheless, is the question of whether these cognate areas should not be attempting to work together. Arrojo is keen on defending the independence of translation studies, particularly from the evil influence of linguistics – and I do agree with her to a certain extent. But let us not forget that relativising something is not the same as abolishing it. Certainly linguistics is not the all-important part of translation studies, but would the study of linguistics actually harm translator and interpreter training? I will return to this question in IV, 3.

In another paper published two years earlier ("Os 'Estudos da Tradução' como Área de Pesquisa Independente: Dilemas e Ilusões de uma Disciplina em (Des)Construção" – "Translation Studies" as an Independent Research Area: Dilemmas and Illusions of a Discipline in (De)Construction – Arrojo 1998a), Arrojo addresses this question more directly, though without bringing forth suggestions as for how (or even whether) translation studies should interact with cognate areas. In this paper, largely based on Steiner in his well-known *After Babel*, she thoroughly explains (and mentions examples) that the wish to establish an independent discipline (called translation studies) and the belief in truly objective, ubiquitous translation theories go hand in hand (see I, 4). Further, she is convinced that until today numerous translation scholars still derive most of their reflections from linguistics – despite the alleged independence of translation studies – and essentialism. As examples she quotes Mona Baker, Susan Bassnett, Andre Lefevere, Kirsten Malmkjaer, Basil Hatin and Ian Mason, all in works from the 1990s (except for Lefevere, whose cited work is his 1978 "Translation Studies: The Goal of the Discipline"). Similarly to her 2000 paper (ARROJO 2000a), in this earlier essay she emphatically states that despite their ambitious objectives and declared scientism, these theories linked with linguistics and essentialism have not been able to formulate a single translation universal, a single objective, unquestionable and universally applicable translation rule. She does end the paper on a cheerful note, though, attributing the recent boom of translation journals, academic events and translation courses to reflections labeled poststructuralist. Let us bear these issues in mind for now as we advance to the next papers by Arrojo. Since they cannot be addressed them here at length, I will come back to them in Part IV.

The paper "Shared Ground in Translation Studies" (ARROJO & CHESTERMAN 2000) is an attempt towards a compromise between Rosemary Arrojo and Andrew Chesterman as far as translator and interpreter training is concerned. More than a compromise between these two scholars, nevertheless, this joint paper appears to point towards a compromise (or "reconciliation", in their words) between what the authors call "essentialism" and "non-essentialism", or "postmodern cultural studies" and "an empirical approach" (ARROJO & CHESTERMAN 2000: 151). After a brief introduction, what follows is a list of 30 theses on translation and translation studies that represent their shared ground.

The first half of these theses (ARROJO & CHESTERMAN 2000: 152-154) concerns the definition of translation and the aims of translation studies. Even though the authors agree that translation studies "seeks to understand the phenomenon of translation", they do not propose a definition of the term "translation"; instead, they favour an idea of multiplicity and changeability according to the scholar behind the definition (ARROJO & CHESTERMAN 2000: 152). Accordingly, they concede that the word "translation" may acquire different connotations in different languages, or even within the same language, and that investigating these differences constitutes a valid research topic. Similarly, they agree that examining what is perceived as translation in different cultures at different times is one of the aims of translation studies. As far as data are concerned, they defend their ineluctably subjective nature and their mandatory character simultaneously – since one cannot do research without data (ARROJO & CHESTERMAN 2000: 152).

In addition to studying what is labelled as translation, Arrojo and Chesterman assent that translation studies concerns what gets to be translated and what does not, as well as who gets to do translations in different cultures across times (ARROJO & CHESTERMAN 2000: 153). Also, defining what kinds of translations are labelled as "more central" or "more peripheral" is another aim of translation studies. Likewise, the discourse on translation produced within different cultures at different moments in time constitutes an interesting research topic in the area, particularly in terms of the metaphors different thinkers use in order to convey their ideas (ARROJO & CHESTERMAN 2000: 153). As far as translation types are concerned, both scholars maintain that devising typologies is paramount, "both within a culture/period and more generally" (ARROJO & CHESTERMAN 2000: 153-154). Finally, Arrojo and Chesterman explain that the definition of translation is closely related to the notion of quality; therefore, another goal of translation studies is to analyse how diverse cultures perceive the issue of translation quality, seeking to determine whether there are common criteria amongst cultures across times.

As for the question "why is this (kind of) translation like this?", the two thinkers come up with seven more theses (ARROJO & CHESTERMAN 2000: 154-155). In summary, they claim that explaining the phenomenon of translation should be one of the objectives of translation scholars, but one such explanation need not follow a specific format, but rather can take into account as many aspects as possible (ARROJO & CHESTERMAN 2000: 154):

the translator's personality, gender, cognitive state, personal experience, decision processes; the influence of the client, of the skopos, of features of the translation situation itself, the source text, the source and target lan-

guages, the implied readers; socio-cultural and historical factors, ideological factors, values, ethical considerations…

Moreover, they concede that, as a human activity, every translation is unique, at least to a certain extent. Despite that, however, they claim that by observing numerous translations one *can* identify certain patterns and thus derive general statements about these regularities (ARROJO & CHESTERMAN 2000: 155).

The last nine theses address the question "what consequences do translations have?" (ARROJO & CHESTERMAN 2000: 155-156). In a nutshell, Chesterman and Arrojo call the readers' attention to the potential effects of translation, both in the sense of influencing cultures (as originals do) and in the sense of triggering value judgements from recipients and/or critics. Analysing these influences and value judgements is a part of translation studies. Furthermore, prescriptive statements about the "desirable effects" of translations may be useful for translation trainers and trainees, though the question "desirable to whom?" remains pertinent (ARROJO & CHESTERMAN 2000: 156). Finally, the last thesis states that the issue of translation raises countless questions surrounding the notion of ethics – examining those should also be an aim of translation studies (ARROJO & CHESTERMAN 2000: 156).

Having established their "shared ground", Chesterman and Arrojo move on to their own individual codas (156-159). Chesterman begins by defending the notion of "interpretive hypotheses" as crucial elements of descriptive research. And here he provides the following definition for "interpretive hypotheses": "they are hypotheses which claim that it is useful (in some way) to interpret translation in this way". For him, this is the starting point of "any conceptual analysis" (ARROJO & CHESTERMAN 2000: 156). Arrojo, in contrast, begins by stressing that if it is impossible, in translation research, to depart from a pre-established, universal definition of translation, research on translation will invariably be marked by the researcher's subjectivity and context. She adds that, for similar reasons, however descriptive one's intentions may be, it is impossible not only to fully distinguish between the descriptive and the interpretive, but also to formulate a universal translation theory (ARROJO & CHESTERMAN 2000: 158).

Chesterman proceeds by adding that unstable though meanings may be, some are more stable than others, more context-bound than others (ARROJO & CHESTERMAN 2000: 156). Arrojo, conversely, emphasises that meanings are *always* unstable and context-bound, and that this factor is of utmost importance for the understanding of the finite, local character of any translation (ARROJO & CHESTERMAN 2000: 158).

As for translation norms, Chesterman argues that "statements of norms" are not prescriptive but rather descriptive, as opposed to norms, whose connotation tends to be exclusively prescriptive (ARROJO & CHESTERMAN 2000: 156-157). Arrojo does agree with his remarks, but warns that statements of norms are not *solely* descriptive either, as they invariably reflect the perspective of the one who describes. Moreover, she stresses that norms can never be "universally valid or applicable" (ARROJO & CHESTERMAN 2000: 159). Still on the issue of norms, Chesterman admits that attempts to describe so-called translation "regularities or universals" may strike one as prescriptive. He nonetheless believes that these statements can "function as explanations". In other words, instead of being read as norms to be followed, these norms can be understood as justifications for the translator's behaviour (ARROJO & CHESTERMAN 2000: 157). Arrojo then replies that regardless of their prescriptive or descriptive character, these norms (or "regularities and universals") cannot be detached from the context that gave rise to them, from the interpretive influence of the researcher. In this sense, norms are construes marked by a specific viewpoint. Therefore, calling them "universals" is, in Arrojo's view, an authoritarian gesture, because it entails an imposition of values and a disregard for other perspectives (ARROJO & CHESTERMAN 2000: 159).

Finally, regarding translation theory, Chesterman defends the idea of a set of what he calls "predictive hypotheses" based on translation conditions. The better the hypotheses, the better its application to translation practice. And here Chesterman warns readers that these predictive hypotheses are not prescriptive since they "do not impinge on the translator's freedom to make choices". He ends his coda by defining translation theory: "a logically linked set of well-corroborated hypotheses (interpretive, descriptive, explanatory and predictive hypotheses)". In conclusion, he concedes that one such theory should always remain open to "new refinements" and will always be subject to replacement by a better theory (ARROJO & CHESTERMAN 2000: 157). In response, Arrojo argues that translation theory should not make predictions; instead, it should raise the students' awareness about their responsibility and active role "in the establishment of all sorts of relationships between cultures" (ARROJO & CHESTERMAN 2000: 159).

At the end of the paper, the authors invite readers to write responses to their debate. In fact, this discussion became a forum that stretched over three issues of *Target*, attracting all sorts of responses from different areas of translation studies – for more on this debate, see *Target* vol. 13:1, 13:2 and 14:1. Furthermore, various replies appeared in different journals across the globe as well, so the debate was not restricted to *Target*. But let us now focus a bit longer on the debate between Arrojo and Chesterman.

Arrojo's acceptance to establish "shared ground", to "reconcile" predominantly essentialist and predominantly anti-essentialist perspectives seems odd. Was it perhaps a good-neighbour gesture? In any case, by looking at their codas at the end of the paper, it becomes clear that they have hardly managed to establish any shared ground at all – in spite of the fact that the 30 theses were devised four-handedly. In fact, these theses appear to address neither Chesterman's nor Arrojo's interests in translation studies, as their codas make evident. Various examples could be drawn from their debate to illustrate this point, but let us focus, for the sake of brevity, on the one that speaks closest to the present book, namely the issue of translation theory. Chesterman may even concede that translation theories cannot be universal, but he still speaks of translation "universals and regularities" across times and cultures; similarly, Arrojo may even concede that translation theory may comprise so-called general translation typologies, but she still contends that the purpose of translation theory should be to raise awareness about the role of translation and translators in society.

Even if we take for granted that reconciliation is not possible here, nor is it desirable – as argued in Part I – Chesterman and Arrojo do not appear to be speaking of the same concept when they refer to, for example, translation theory. This certainly comes as no surprise to the reader, especially if we remember those two notions of translation theory repeatedly mentioned in Part I. If we accept that Arrojo's notion of theory is more awareness-raising, whereas Chesterman's is more practice-oriented, we will understand that they cannot – will not and need not – establish "shared ground". However enlightening their debate may be, it may strike one as misleading since all it does is reveal their irreconcilable differences – which is by no means a problem, but it is certainly no consensus either. The expression "shared ground" seems indeed to point to an idea of consensus, and their use of words such as "reconciliation" does confirm this impression. Their debate has shown that a space where both perspectives can live in relative harmony may even exist, but *reconciliation* seems out of the question.

Chesterman himself addresses these questions and expresses his doubts in a 2005 paper entitled "Consilience and Translation Studies" (CHESTERMAN 2005). As already mentioned above, different scholars have expressed their opinions and doubts as to how "shared" and how relevant their "shared ground" is – one example is Sonja Tirkkonen-Condit, Jukka Mäkisalo, Riitta Jääskeläinen, Mirja Kalasniemi and Pekka Kujamäki's paper "Do we Need a Shared Ground?" (published in *Target* 13:2, 2002). What is concerning is the fact that some of these discussions seem to end in a "theory hope" note, pointing towards a bright future in which there will be a universal consensus and translation studies will be "one". Chesterman, for instance, comments in the abovementioned paper: "So although we are *still* far from a shared philosophical paradigm, we can perhaps *see the beginnings of the basis* for one" (CHESTERMAN 2005 – my emphasis). If we look back at Roustang's indented quotation in II, 2.3, we will remember that *not* having shared ground (in the sense of consensus) is perhaps crucial for the existence of poststructuralist thought. So whether Arrojo was convinced of that, but decided, nevertheless, to formulate these shared theses anyway remains a mystery. Perhaps for her the idea of "shared ground" referred chiefly to a common space for these disparate views to cohabit rather than to formulate unanimous theses. Some of the issues in Arrojo's debate with Chesterman will reappear in her debate with Britto in Part III. Moreover, some of these questions will be specifically addressed in Part IV.

On a very different note now, let us examine the short paper "O Tradutor Invisível por ele Mesmo: Paulo Henriques Britto entre a Humildade e a Onipotência" (The Invisible Translator on his Own: Paulo Henriques Britto between Humility and Omnipotence), published in Brazil in 2000. Part III will be devoted to Britto and his work as a translation scholar and translator. Suffice to say that he is one of the best-known contemporary Brazilian translators, having published over 100 translated works. Arrojo bases this essay on a newspaper interview with Britto entitled "As Aspas da Tradução" [The Quotation Marks of Translation], published in the newspaper *Folha de São Paulo* on 27th February 2000.

As it will become clearer in Part III, by the year 2000 Britto and Arrojo had already had a few academic quarrels, mostly over the issue of the impact of poststructuralist thought on translation studies. So Arrojo's choice of this particular interview is by no means random. The title of the newspaper interview, "As Aspas da Tradução", is due to Britto's claim that translators are actually authors between quotations marks, and that in order to be a translator, one must be humble – hence the quotation marks (quoted in ARROJO 2000b: 161). Needless to say, this affirmation triggers Arrojo's entire response, whose work has been largely dedicated to the elevation of the status of the translator as a result of his/her ineluctably authorial role in translation. But Britto seems to perceive this issue differently, as he claims that he tries to interfere as little as possible with the text, thus remaining as invisible as he can (ARROJO 2000b: 161). When asked about whether he is affiliated to a particular translation movement, he says that he is first and foremost a translator and not really a scholar, but is very much aware of the theoretical discussions in the area.[106] He then declares that he strongly disagrees with Venuti's overall argumentation on the translator's visibility and much prefers Pym's view, whereby the place of the translator is restricted to prefaces and footnotes, and remaining invisible is an objective to be pursued (ARROJO 2000b: 161).[107]

Arrojo briefly comments on and quotes Venuti's *The Translator's Invisibility* (1995), adding insight from poststructuralist thought (similarly to her other papers) to clarify some of his views. For her, remaining invisible is not only undesirable, but also impossible, since the translators' choices constantly reveal their personal history, their ideology and their unconscious processes. In her view, the introductory paper written by the interviewer in question, Maurício Santana Dias, endorses her argument. Dias maintains that Britto is responsible for the Brazilian accent of authors such as Henry James and Salman Rushdie, and adds that this accent is not only Brazilian, but also clearly "carioca", or from Rio de Janeiro (the city where Britto comes from – ARROJO 2000b: 160).

So to return to the subtitle of her paper, "Paulo Henriques Britto between Humility and Omnipotence", Arrojo's final argument is that by claiming to be an author between quotation marks, Britto is not being humble at all. In fact, he is being quite authoritarian because he believes himself capable of "supra-human omnipotence" ("onipotência sobrehumana" – ARROJO 2000b: 164). For Arrojo, and for poststructuralist thought in general, not leaving one's marks, one's residues (see II, 2.3) in any reading or writing task is simply inhuman. In this light, Arrojo takes Britto's claim to humility as a fairly arrogant gesture – as though he were somehow able to erase himself when he translates. Being humble in translation means, as Arrojo explains, admitting that one can *but* interfere, that one can *but* leave traces behind, and that this is the very possi-

106 This affirmation is the epigraph to III, 2.

107 I will take for granted here that the reader is acquainted with Venuti's and Pym's reflections. For similar remarks on the issue of invisibility by another translator, see the introduction to Part I.

bility that our human condition allows us (Arrojo 2000b: 164). Being humble is, therefore, accepting this human finiteness whereby meanings are attributed to words only temporarily, in a particular context and under numerous personal and subjective influences. By calling himself an author between quotation marks, Britto is saying that he has found *the* correct and faithful meaning of the original, and has transparently and mechanically rendered it into Portuguese. In Arrojo's opinion, one such attitude is far more of omnipotence than it is of humility (Arrojo 2000b: 164). I will address Britto and Arrojo's disagreements at length in Part III. Let us bear in mind these initial divergences as we head towards the end of Part II.

Let us now move to the last paper we will analyse in Part II, entitled "Translation, Transference and the Attraction to Otherness: Borges, Menard, Whitman" (Arrojo 2004).[108] As already mentioned above, recently Arrojo's research has been dedicated chiefly to representations of translation in fiction, with the works of Borges, Kafka, Poe, Saramago, Guimarães Rosa, Calvino and Kosztolányi at its forefront. Indeed, Arrojo is currently preparing two books to be published on her recent undertakings. In this 2004 paper, she investigates not only how the idea of translation is conveyed in Borges' short-story "Pierre Menard, Autor del Quijote" (see II, 2.2), but also the "relationships that are generally established between translators and authors" (Arrojo 2004: 31). Her main aim is indeed to provide possible answers to the following questions (Arrojo 2004: 31-32):

Understood as an intrinsically performative textual activity, translation is generally viewed, in Borges's terms, as a form of rewriting which is not in any sense neutral or secondary to the original. If, in such terms, both the so-called original and the translated text seem to enjoy a similar status, what kind of exchange might there be between the two? And, at the same time, if translators cannot, in any sense, be "invisible" in their translations, and, like authors, at least on some level, do mean what they say, what might it represent, for a translator, at a certain point, to choose a certain text to translate?[109]

108 This paper was also published in 2006 in the proceedings of the Conferencia Internacional de Traducción e Intercambio Cultural en la Época de la Globalización (held at the Universitat de Barcelona), under the title "La Traducción Literaria en la Época Contemporánea".

109 This issue will be brought up again, this time by Paulo Henriques Britto, in III, 3.2.

Indeed for her, Borges' story is "a story about translation and transference, or translation as transference, or even translation as a response to influence" (ARROJO 2004: 35), and here the term "transference" is used in a Lacanian fashion (ARROJO 2004: 34 – see II, 2.3).

Initially, Rosemary Arrojo takes the relationship between the protagonist, Pierre Menard, and the author he wishes to "translate", Miguel de Cervantes, and compares it to Borges' relationship with Walt Whitman. In a few side notes, Arrojo does mention that the story brilliantly illustrates the absurdity of absolute faithfulness in translation, an issue far from resolved in contemporary translation studies. She even suggests that the story may be read as a sort of "ironical criticism of the call for faithfulness and invisibility typically associated with traditional translation theories and practices" (ARROJO 2004: 32). Nevertheless, her interest seems to lie closer to this analogous relationship between the pairs Menard and Cervantes, and Borges and Whitman.

In the story, Menard is obsessed with *becoming* Cervantes somehow – so much so that his "visible" work, his own published texts, did not free him from the dominance of the persistent idea of *repeating* Cervantes. And here let us not forget that Quixote, too, was obsessed with "becoming" what he had been reading, which places Menard into a quixotic mission as well. Occasionally turning to psychoanalysis, Arrojo sees in Menard's strong desire to become Cervantes a wish to escape his own self, geography, native language, age and, most of all, "his petty bourgeois life as a minor writer and obscure scholar" (ARROJO 2004: 35). More than becoming Cervantes, Arrojo asserts, perhaps Menard wished to replace him, to outwit him, and though he never used the words "translation" or "translator", he rather controversially tried incredibly hard to remain faithful to Cervantes' text (ARROJO 2004: 36).

According to Arrojo's research of Borges' biography and bibliography, this particular short-story brings forth "Borges's most profound meditation on the vicissitudes of writing and translating literary texts under the influence of a strong precursor" (ARROJO 2004: 38), a fact she does not take as accidental, since the story appears to have been written in a rather critical moment of the Argentine's life. Indeed, his father – "his first important precursor" – had died a few months earlier, and Borges himself had recently recovered from a life and death experience caused by an accident that had taken place in 1938 (ARROJO 2004: 39). For Arrojo, Menard represents much of Borges' own father, a mediocre writer, but also Borges' own fear of confining his literary career to such unexceptional bounds – as both his father and Menard had done. In this sense, the story can be interpreted as a milestone in Borges' life, a kind of liberation from these psychological "constraints".

As for the leitmotif of translation, Arrojo understands it as a way not only to imitate, but also to provisorily take over the authorial position of those we admire (ARROJO 2004: 41 – see II, 2.3). It is known that Borges, having been brought up in a bilingual (English-Spanish) environment, began translating at the age of nine, with the publication of the Spanish translation of Wilde's story "The Happy Prince" (ARROJO 2004: 31). He then went on to translate Kafka, Virginia Woolf, Joyce, Faulkner, Poe, Whitman, Hart Crane, Chesterton, Apollinaire, Browne, Papini, Novalis, and Hawthorne, amongst others (ARROJO 2004: 31). But his story with Walt Whitman is far more complicated than that, as he could not simply translate Whitman. In fact, he announced that he would translate his *Leaves of Grass* in 1927, but the translation only actually came out in 1969 (*Hojas de Hierba* – ARROJO 2004: 42).

At this point, Arrojo speculates about the reason for such a delay. In 1919 Borges' first poem, "Himno del Mar", was published in Europe. She carefully examines the poem to show how Whitman's work appears to be the "subterranean motivation" (ARROJO 2004: 42 – quoting "Pierre Menard, Autor del Quijote") behind it. His autobiographical accounts and criticism do confirm Arrojo's suspicions, as Whitman was indeed considered *the* great influence in Borges' entire career – even by Borges himself. In addition to that, Borges' first poem seems to have been received as an "obvious" instance of this strong influence exerted by Whitman in his works. In this light, Arrojo goes on to conclude that "if Cervantes was 'the subject presumed to know' in Menard's biography, Whitman represented for the younger Borges the desirable, idealized possibility of a life devoted to writing" (ARROJO 2004: 45). Perhaps this is the reason why, Arrojo conjectures, it took Borges 42 years to finally conclude his translation of *Leaves of Grass*. Perhaps Borges needed to wait until he felt confident enough about his own career, until he had lost the fear that the critics would forever compare him to Whitman and thus restrict his work to "an echo of other echoes", as Menard's was (ARROJO 2004: 45, 47 – quoting "Pierre Menard, Autor del Quijote"). For Arrojo, it was important for the young Borges to "emulate his own idealized version of Whitman" through his "Himno del Mar", whereas the older Borges seemed to be "interested in turning the American poet into a reflection of himself, as he seems to find comfort in the idea that Whitman's real life was perhaps not very different from his own" (ARROJO 2004: 47). Borges' publication of Whitman's Spanish translation may indicate that the Argentine had finally overcome his infatuation with Whitman seeing as, by 1969, Borges was unquestionably a major author (ARROJO 2004: 47).

To conclude her paper, Arrojo returns to the notion of translation as a means to deal with influence, with transference, and explains that

> *Even though his [Borges'] long-lasting love affair with Whitman's work and persona was obviously more productive – and had a happier end – than Menard's unacknowledged obsession with Cervantes, in both cases translation seems to have played a fundamental, even therapeutic, role in helping these two ambitious, Quixotic readers deal with the conflicts of influence and act out their desire to be someone else* (ARROJO 2004: 48).

Moreover, Arrojo sees translation as an instrument to allow one to escape oneself temporarily which, in Borges's case, "might have helped [him] escape Borges and the Borgesian, [allowing him] to pretend to be an-Other and, in the Other's name, to write, in his native Spanish, that which he was unable to address in his own texts and in his own name" (ARROJO 2004: 48).

Arrojo's reflection in this paper appears to be very much in line with what she herself – as well as other scholars whose affinities lie with poststructuralist thought, such as Cristina Rodrigues, for instance (see I, 6) – defends as interesting and somewhat new research in translation studies. Even though numerous avid Arrojo readers may feel disappointed that she appears to have abandoned her somewhat militant defence of deconstruction, or attack to essentialism, it seems understandable that she should want to shift her focus slightly after decades of almost belligerent struggle – whose outcome remains fairly negligible considering how much translation studies has grown in recent years (I will address this question in what follows and in IV, 4).

By analysing Borges' biography and oeuvre and drawing an interesting parallel between that and his own short-story "Pierre Menard, Autor del Quijote", Arrojo touched on several thought-provoking issues surrounding the notion of translation – such as what is the nature of the relationship between author and translator, what is the impact of translation on the translator, what role does translation play in the life of an author, how does one manage one's admiration for the work one is about to translate, what is the impact of transference in the work of a writer and/or translator, amongst various others. And by doing so, Arrojo very openly and freely spoke of speculation or personal opinions, and thus hinted in no way at absolute truths and verifiable theses. More importantly than that, she did not attempt to derive a set of rules from her analysis, allegedly objectively determining exact and universal answers to the questions suggested above. Instead, her reflections are avowedly limited, personal and subjective, and these characteristics by no means jeopardise her argumentation. Indeed, one finds in her paper a great source of inspiration, of awareness-raising insight, very much in tune with the idea of translation theory repeatedly discussed in Part I – as a practice in itself with no direct commitment to translation practice. Her paper is certainly *not useful* as far as translation practice is concerned, but nevertheless (and because of it) it remains a powerful source of reflection.

* * *

This chapter on Rosemary Arrojo's work has certainly confirmed the expectations drawn in II, 1, dedicated to Arrojo's academic life. This analysis of a selection of her texts has shown how multifaceted her work is, sometimes focussing on the interface between translation and psychoanalysis, sometimes emphasising a deconstructionist perspective on translation; sometimes dealing with translator training, sometimes addressing literary translation; sometimes speaking of the Humanities as a whole, sometimes focussing on translation studies as a discipline; sometimes studying the links between translation and linguistics, sometimes between translation and literary theory; sometimes drawing inspiration from representations of translation in fiction, sometimes finding allegories for translation in psychoanalysis.

Indeed, her work addresses various crucial issues in translation studies at the moment, including what the purpose of translator training should be, whether translation theory should be practice-oriented, how the subjectivity and unconscious of the translator may work, the (un)importance of linguistics in translation studies, the issue of the translator's (in)visibility, the issue of translation criticism and quality assessment, amongst many others. Radical though her work may appear to some, it is undeniably thought-provoking.

Speaking of radicalism, as this question has been raised several times in Part II, let us briefly close it here (and then raise it again in IV, 2). The entire atmosphere surrounding postmodern thought is clearly one of upheaval, hence the radicalism of which countless thinkers – not only Arrojo – are accused. Accordingly, opposing millennia of tradition, as is the case of Arrojo in translation studies, requires a certain militant attitude. And let me make it clear here that for some trends of poststructuralist thought her work is not at all radical and militant but, at the same time, radicalism – or perceived radicalism – is the only way for her to get her ideas across.

Nevertheless, her emphasis of certain extremes, for example, could be dangerous. By relentlessly opposing structuralism to poststructuralism, essentialism to anti-essentialism, logocentrism to deconstruction, as well as (more generically) old and traditional to new, Arrojo appears to endorse the very age-old tradition she criticises, a tradition that believes in extreme, independent opposites. And however often she reminds the reader that the relationship between structuralism and poststructuralism is one of complementariness as well, she certainly places them on seemingly antagonistic sides far more often. Similarly, even though she may sometimes take the context of the thinkers she so fiercely criticises into account – such as Mounin or Newmark – she usually appears to largely neglect their times and contexts and discard their contributions. Finally, it is true that she sometimes concedes that their contributions do have some relevance, but mostly she groups all these scholars labelled essentialist together and flattens their differences, their heterogeneity.

But then again, by defending these thinkers' contributions despite their essentialist tendencies, by addressing them in their variety despite their similarities and, finally, by emphasising the constant complementariness between structuralism and poststructuralist thought, Arrojo would probably not have attracted the amount of attention she did by being arguably radical. And this is to a certain extent one's lot when one chooses the non-mainstream path – as Arrojo did.

So all in all this question of radicalism is at least twofold. There is the radicalism potentially perceived especially by those less acquainted with poststructuralist thought, for whom some of Arrojo's propositions may sound like an outcry for the full abolishment of the notions of truth, reality, reason and science, along with the full disregard for the works of countless thinkers she labels essentialist. On the other hand, there is the radicalism that any thinker whose affinities lie with poststructuralist thought might perceive in her work. And "radicalism" is perhaps not the most adequate term to designate this, since it refers to a certain exaggeration, a certain emphasis on polarities that, in view of the first notion of radicalism presented previously, may result in the very defeat of her ultimate objectives. So here again the idea of a double gesture appears to be pertinent: a certain radicalism (in the first sense) is required to break with traditional views; excessive radicalism, in contrast, is not even in question since deconstruction has nothing to do with destruction. In other words, her work can but sound radical to some, even though radicalism is simply out of the question (see IV, 2).

Another expectation drawn in II, 1 concerned the recent shift in her research. This question is particularly relevant because for some of those who have been following her career this change might appear disappointing at first sight. This shift in Arrojo's research in recent years is nevertheless an interesting development. It is as though Arrojo no longer felt the need to relentlessly – sometimes "radically" – defend poststructuralist thought, to attack essentialist tendencies. Those issues that seem to have always been close to her heart – contemporary thought, literary translation, representations of fiction in translation (see II, 1) – remain very much present, but the balance amongst them has changed, making her work sound less "militant". In the meantime perhaps Arrojo has grown tired of so much opposing, or perhaps she now feels that poststructuralist thought is not as alien to academia as it used to be, say, two decades ago – though some claim to have no knowledge of it whatsoever, as I have argued elsewhere (LEAL 2010b).

Many of the questions listed in the first paragraphs of these final remarks will be brought up again in Part III, dedicated to Brazilian translator and translation scholar Paulo Henriques Britto. Britto's standpoint and theoretical views will offer a counterpoint to Arrojo's standpoint and theoretical views. Britto's work not only represents the resistance to poststructuralist tendencies in translation, but it also resembles the works of numerous other scholars (outside Brazil as well) who either oppose or remain listless to poststructuralist thought. Yet, even though Arrojo's and Britto's works may strike one as antagonistic at first sight, the reader should beware of antagonisms – after all, our glass should be half empty *and* half full.

Part III: Paulo Henriques Britto

Que precisão tem o amor de linhas retas
se paralelas afinal são nada mais
que a garantia do infinito desencontro?

Paulo Henriques Britto

Paulo Henriques Britto is one of the most prominent contemporary literary translators in Brazil. Indeed, he has translated over 100 volumes mostly of prose and poetry originally in English, including authors as diverse as Ian McEwan, Emily Dickinson and Lord Byron. He currently translates almost exclusively for Companhia das Letras, one of the largest and most prestigious publishers in Brazil. Additionally, he has been working as a translation professor at the Pontifícia Universidade Católica do Rio de Janeiro since the late 1970s. As an academic, he has written numerous papers in Portuguese and English mainly on literary translation. As a translator and writer, he has won nine different literary awards (more below) in the past fifteen years, most of which for his own poems and short-stories. In fact, Britto has been attracting increasingly more attention for his work as a writer, with six poetry volumes, one of short stories and one of nonfiction at the time of writing – in addition to an English translation of some of his poems, and a Portuguese edition of one of his poetry collections.

In this part, we will look into Britto's life and work so as to try and understand the standpoint from which he speaks as a translation theorist. Let us first examine, in Chapter 1, how he became involved with literature and translation, as well as his career as a writer and literary translator. After that, in Chapter 2, we will look into his academic career – his education, the courses he teaches at university, his research groups and the theses he has supervised over the years. These initial chapters will help to sketch the outlines of the standpoint from which he speaks as a translation scholar. Finally, in the last and longest chapter (3), we will listen to his voice as a translation scholar by analysing some of his papers on translation.

1. Standpoint as a Writer and Translator

O poeta é um fingidor.
Finge tão completamente
Que chega a fingir que é dor
A dor que deveras sente.

Fernando Pessoa

In an interview given to Bernardo Mello Franco (FRANCO 2010), Britto says that he fell in love with poetry at about the age of 10, when he first lived in the USA:

Na verdade, descobri a poesia em língua inglesa, quando morei nos Estados Unidos (...). Quando era menino, no Brasil, os modernos eram ignorados na escola. Só lia Olavo Bilac, Gonçalves Dias... Poesia era uma chatice, que a gente tinha que decorar para ler no Dia das Mães. (...) Nos Estados Unidos, me deram Shakespeare, Emily Dickinson, Walt Whitman. Foi um choque. Quando voltei ao Brasil é que descobri, com Fernando Pessoa, que também existia poesia boa em língua portuguesa.

[Actually, I first discovered poetry in English, when I lived in the United Sates (…). When I was a boy, modernist poets were ignored in Brazilian schools. All we read was Olavo Bilac, Gonçalves Dias... Poetry was boring, something we were forced to memorise for Mother's Day. (…) In the USA they gave me Shakespeare, Emily Dickinson, Walt Whitman. It was a shock for me. When I came back to Brazil I realised, thanks to Fernando Pessoa, that there was also good poetry in Portuguese.]

He then went on to read Hawthorn, Dickens, Poe, Conan Doyle and Chesterton. Back in Brazil, he started reading Machado de Assis and, as a teenager, came across Fernando Pessoa, which was a "true revelation" ("verdadeira revelação" – LEÃO 2010) to him, making him want to write poetry seriously, with literary aspirations – and not simply for fun, as he had been doing ever since he was six (LEÃO 2010). This serious interest in poetry, together with the discovery of poetry in Portuguese, led him to Carlos Drummond de Andrade and Manuel Bandeira, our most important modernist poets.

When he was about 17 he became interested in music, particularly in Caetano Veloso, Gilberto Gil (Tropicália was at its peak) and Chico Buarque, as well as in Bob Dylan and Jim Morrison. This interest led him to Augusto de Campos' *Balanço da Bossa*, a book "fundamental na minha formação, que me levou a me interessar por crítica e teoria", or essential to my development – it awakened my interest in literary criticism and theory (LEÃO 2010). Yet more writers would influence him greatly in the years to come: Freud, Kafka (his favourite), Joyce, Beckett, Clarice Lispector, Graciliano Ramos, Cortázar, Gombrowitz, Sartre, Mário de Andrade, and Campos de Carvalho. Still in his youth, in a period he calls "development phase" ("fase de formação" – LEÃO 2010), he read various linguists and thinkers, such as Chomsky, Popper and especially Wittgenstein, but also novelists like Dostoyevsky, Tolstoy, Melville, Flaubert and, above all, Proust. Four poets/critics were equally important to him, namely Eliot, Pound, Haroldo de Campos and Augusto de Campos. Finally, the poets Wallace Stevens and João Cabral de Melo Neto were the last to strongly impact and mark his work for decades to come (LEÃO 2010).

In the early 1970s Britto moved to the United States again to study cinema at the San Francisco Art Institute, but dropped out and went back to Brazil to do an English and Portuguese teaching degree at the Pontifícia Universidade Católica do Rio de Janeiro – between 1975 and 1978. After he obtained his degree, he was hired by the same university mostly to teach English, English literature and translation within the same major that he had just taken. At this point he had already been working as a translator and had a few published works. In 1979 he began his MA in linguistics, which he concluded in 1982 with a thesis on Portuguese semantics ("Conectivos Oracionais do Português: Uma Proposta de Análise Semântica" – Portuguese Conjunctions: A Semantic Analysis).

Also in 1982 he published his first collection of poetry, entitled *Liturgia da Matéria*. Unfortunately there have not been many reviews of this book so far and it is sold out. Between 1982 and 1989 he published more than 40 different translations – most of which of entire books – by authors such as Chomsky, Jack Kerouac, Wallace Stevens, Raymond Chandler, Thomas Brown, John Donne, Susan Sontag, Lord Byron and Naipaul, amongst many others. In 1989 his second volume of poetry came out: *Mínima Lírica*. By the way, in 1986 he began translating for Companhia das Letras, one of the most renowned and prestigious Brazilian publishing houses, and until today, more than 20 years later, Britto writes and translates almost exclusively for them.

However, it was not until 1997 – 50 more translations later, including Emily Dickinson, Henry James, Gertrud Stein, Virginia Woolf, John Updike, Salman Rushdie, Charles Dickens, Dylan Thomas, George Orwell, D. H. Lawrence and E. L. Doctorow (awarded the "Prêmio Paulo Rónai"), amongst various others – that his third collection of poetry came out, this time attracting more attention. *Trovar Claro* was awarded the "Prêmio Alphonsus de Guimaraens" and called the attention of critics for its metalinguistic vocation ("vocação metalinguística" – CAMPOS 2010). Indeed, Jorge Lúcio de Campos maintains that "Seus poemas remoem a si mesmos em interrogações e contra-interrogações que os mantêm sempre tensos e acesos. Ali o poeta e poema se interrogam e avaliam: para que sirvo, senão para ser nada?", or his poems ruminate themselves through questioning and counter-questioning, a process that keeps them constantly tense, alight. In them, poet and poem question each other, ponder: what am I for other than nothing? (CAMPOS 2010). Similarly, Izacyl Ferreira speaks of a "lírica indagadora" (FERREIRA 2010), or inquiring lyricism, and Pessôa adds that "Se o poeta testemunha o seu próprio ofício enquanto o executa, para além do como se faz, perguntamos: o que é desse ofício?", or if the poet witnesses his own craft while exercising it, we ask, in addition to how he does it, what is of this craft? (PESSÔA 2010).

Between the release of *Trovar Claro* and his next poetry volume, *Macau*, six years went by in which he published more than 20 different translations. A few new names he translated during this period include Thomas Pynchon, Philip Roth, Elizabeth Bishop and Ian McEwan, amongst others. With *Macau* he won the "Prêmio Portugal Telecom", one of the most esteemed and financially rewarding Brazilian literary awards. The title of the book, *Macau*, is an extraordinary metaphor that Ferreira understands as follows: "língua nada hegemônica é o nosso idioma, arte nada popular é a poesia, espécie nada privilegiada é o homem se não tem a proteção de um deus, uma utopia, qualquer esperança de redenção", or our language [Portuguese] is not at all hegemonic, poetry is not at all a popular kind of art, men are not at all privileged unless they have divine protection, a utopia or some kind of hope for redemption (FERREIRA 2010). Furthermore, Ferreira speaks of "Niilismo, humor e metapoesia", or nihilism, humour and metapoetry (FERREIRA 2010). In other words, his "metalinguistic vocation" seems to persist, and nihilism and humour go hand in hand setting the tone of most of the poems, without, nevertheless, unleashing neither misery nor bliss ("não chegam, um a desatar o riso, outro a desatar o pranto" – FERREIRA 2010). For Manuel da Costa Pinto, columnist of the newspaper *Folha de São Paulo, Macau* is "humor e rigor, subjetividade e racionalismo, lírica e antilírica. Não é fácil definir 'Macau'. Um livro, enfim, que acumula várias camadas de leitura, às quais devemos acrescentar ainda vestígios da atividade de Britto como tradutor (...)" (PINTO 2003) – or *Macau* is humour and rigour, subjectivity and rationalism, lyric and antilyric. It is not easy to define "Macau", a book with various reading layers to which one must also add traces of Britto's work as a translator.

Between 2003 and 2010 Britto published another three books: *Paraísos Artificiais* (2004 – short-stories), *Tarde* (2007b – poetry) and *Eu Quero é Botar meu Bloco na Rua* (2009 – nonfiction). In 2012, his latest volume of poetry, *Formas do Nada*, came out. As far as his work as a translator is concerned, in the last decade or so he has translated a few more books by Roth, Updike and Pynchon, along with many other authors – more than 20 books in total. His collection of short-stories, *Paraísos Artificiais*, received the prestigious "Prêmio Jabuti" (second place), and so did *Tarde* (third place), along with a second "Prêmio Alphonsus de Guimaraens". Moreover, his *Macau* was edited[110] in Portugal in 2010, and some of his poems were translated into English by Idra Novey and published in the United States (BOA Editions) in 2007 under the title *The Clean Shirt of it: Poems by Paulo Henriques Britto*. In 2008, a few of his poems were translated by Austrian poet Hans Raimund for the poetry festival VERSschmuggel.

So this very brief account of Britto's works as a writer and translator shows that his production is undoubtedly impressive. When questioned about the long intervals between his own books (1982, 1989, 1997, 2003, 2004, 2007, 2009 and 2012), he explains that

> *Escrevo nas horas vagas. No máximo, seis poemas por ano. É um trabalho muito esporádico, passo meses sem abrir o caderno. Gostaria de ter mais tempo para a poesia, mas ainda tenho que preparar aulas, corrigir provas...* (FRANCO 2010).
>
> [I write in my spare time. Six poems a year at the most. I do it sporadically – I spend months without touching the notebook. I would like to have more time for poetry, but have to prepare classes, mark exams…]

110 Brazilian books cannot be sold in Portugal. In order for the sale of Brazilian books to be allowed in Portugal, a Portuguese publisher must edit the book there, which was the case for Britto's book.

Britto is arguably one of our most prominent contemporary translators and writers. *Folha de São Paulo* uses the epithet "transformer": Britto [is a] poet, and one of the great transformers from English into Portuguese – "poeta e um dos grandes transformadores do inglês para o português" (MACHADO 2002). The newspaper *O Globo* speaks of a "poeta premiado, tradutor prestigiado", or award-winning poet, celebrated translator (O GLOBO 2009). Indeed, Britto was one of the four Brazilian writers to take part in the renowned Internationales Literaturfestival Berlin in September 2010, a festival that "present[s] a wide literary variety of contemporary prose and poetry from around the world". According to the organisers of the event, Britto is "one of Brazil's most distinguished translators" and "a cult poet in Brazil".[111]

Because the main focus of this book is translation theory, his vast production – more than 100 literary translated books, in addition to his own books – will not be analysed in detail. III, 2 will shed light on Paulo Henriques Britto as an academic, someone engaged at university for more than 30 years now, with scholarly production certainly not comparable to his colossal literary production.

111 The information about the festival was taken from their webpage – http://www.literaturfestival.com/ (last accessed in December 2010).

2. Standpoint as an Academic

Na verdade, não sou um estudioso da teoria da tradução, embora esteja a par das discussões mais recentes.

Paulo Henriques Britto[112]

As already briefly pointed out, Britto started working at university in 1978, already as a "professor" – though he would have been the equivalent of a "lecturer" in the UK when he first began. At the time, it was very common for Brazilian universities to hire students who had just finished their major, as was Britto's case. Of course nowadays there is hardly a field in which one can apply for a position at university without an MA and a PhD (perhaps only those fields in which specific education is still scarce), but until the late 1990s this was by no means rare in Brazil. He then went on to do his Master's already as a lecturer, between 1979 and 1982. In 2002 he was awarded a "Notório Saber" by the same university, which is equivalent to a doctorate and accredits knowledge and experience acquired outside the university.

Curiously, Britto chose to do his MA in linguistics – though the programme itself, which we call Letras in Brazilian Portuguese, involves literature and linguistics, and increasingly more translation studies as well (see I, 1). In those days, his academic production as such was far from vast: one article on Portuguese conjunctions published in a Brazilian journal in 1980, together with a few reviews and short essays published in newspapers and magazines. Having said that, one must also acknowledge the fact that university faculties were not nearly as pressured to publish back then as they are today. CAPES (Coordenação de Aperfeiçoamento de Pessoal de Nível Superior – Agency for Further Development of Staff in Higher Education) and Cnpq (Conselho Nacional de Desenvolvimento Científico e Tecnológico – National Agency for Scientific and Technological Development), our most important research funding agencies, now have complex evaluation systems whereby universities and individual professors receive acknowledgement (and funding) according to the number and nature of academic activities they carry out, amongst which scientific publications rank the highest.

112 Cf. Arrojo (2000b: 161) – see II, 2.4.

In any case, it was not until the mid-1990s that Britto started publishing papers in specialised journals, as well as book chapters – just under 50 works altogether.[113] Similarly, it was not until the mid-2000s that he began supervising BA and MA theses – perhaps because before then he did not have a PhD or equivalent degree. Therefore, one can say that by the time his academic activity became more intense, he had already translated dozens of books, published a few of his own books and been awarded a few times for his work as a literary translator and poet. In this light, the impact that his career as a translator and writer must have had on his work as an academic was probably strong.

Indeed, the vast majority of his papers published in specialised journals between 1995 and 2010 are about literary translation. Some approach his own endeavours as a literary translator, whereas others are about the craft of poetry translation (which, all together, delineate a model of poetry translation or criticism of poetry translation). Finally, some compare the work of the writer to the work of the literary translator. As for his papers on issues other than literary translation specifically, the most recurrent topic is his fierce criticism of poststructuralist "approaches" to translation, with a few papers having been written in direct response to Stanley Fish and Rosemary Arrojo (see Part II). As addressed in footnote 113 (on this page), many of the book chapters and newspaper articles published by Britto consist of his own poems, as well as of translation and literary criticism. In this sense, one can say that most of his publications are not specifically devoted to translation studies as such, and this is probably the reason why Britto claims not to be a translation scholar (see the end of II, 2.4). We will at any rate analyse many of his papers on translation more closely in III, 3.

But before we move on to the next, longer chapter, dedicated to the analysis of these papers, let us briefly look at his work as a professor, focusing mainly on the courses he has taught over the years, as well as on the theses he has supervised and the research projects in which he has taken part. According to his national CV,[114] he has been teaching more or less the same undergraduate

113 I last accessed his CV (see footnote 114 on this page) in January 2011. It is important to take into account that most of the approximately 25 book chapters published by Britto consist of anthologies of his own poems, as well as literary translation criticism and literary criticism. In addition to these nearly 50 papers and book chapters, Britto has published, since the early 1980s, nearly 30 newspaper and magazine articles and reviews, mostly on literature and literary translation, including some of his own poems.

114 Most of the information presented here about his publications, awards, education and academic work in general were taken from this source, this so-called "national CV". In Brazil, all scholars have their CV on a public platform called "Lattes", which in turn is monitored by the chief funding agencies we have in Brazil (CAPES and Cnpq – see above). The link to Britto's Lattes is the

courses since 1978, all within the BA and teaching degree in English and Portuguese. In addition to the English language (Syntax, Semantics, Contrastive Linguistics), he has been in charge of Poetry in English, Creative Writing, Linguistics, Semantics, Portuguese for Translators, as well as several translation courses – Technical Translation, Literary Translation, Translation into Foreign Languages and Introduction to Translation. Furthermore, he has been teaching the courses Literary Translation and Poetry Translation within the Specialisation in Translation offered by the same institution.

On the website of the Pontifícia Universidade Católica do Rio de Janeiro his name features amongst the supervisors of BA and Specialisation theses on translation. Between 2003 and 2004 he concluded the supervision of nine specialisation theses. Additionally, at the time of writing he has completed the supervision of 12 MA theses, along with two PhD theses. He is currently supervising three additional PhD theses. Taking all these different works into account, most are about literary translation, with a specific corpus – such as poems by Hilda Hilst, Campos de Carvalho and Elizabeth Bishop, a short-story by D. H. Lawrence, amongst others. Some are dedicated to translation in a broader sense, including specific linguistic phenomena in English and Portuguese and the use of dictionaries in translation. Finally, a few discuss poetry and music, with no avowed relation to translation.

As for his engagement in research projects, Britto has taken part in the following: Non-representational Approaches to Meaning: Aporias and Perspectives (from 2003 to 2005); Transformations in Contemporary Brazilian Poetry (from 2007 to 2009); Translation Studies: Linguistic and Cultural Aspects (from 2004 to date); Cultural and Linguistic Interfaces: Translation, Teaching and Bilingualism (from 2004 to date); Poetry Translation (from 2006 to date); and The Verse in Modern Brazilian Poetry (from 2014) – all originally in Portuguese. The main aim of these projects is to foster research within a single university, as well as between different universities.[115] The members of the group may organise events (such as conferences and colloquia) and publications together, engaging not only experienced scholars, but also students. The activities carried out by these groups also contribute to the abovementioned

following: http://buscatextual.cnpq.br/buscatextual/visualizacv.jsp?id=K4701673J9 (last accessed in February 2013).

115 The reader will have noticed that I have not mentioned any research projects for Arrojo, in Part II. These projects have become increasingly popular in the last decade or so, but were quite rare before the turn of the century, when Arrojo was still engaged at a Brazilian university.

evaluation system devised by CAPES and Cnpq to award recognition and funding to universities.

Furthermore, Britto worked as a member of the editorial board of the following Brazilian journals on literature and/or translation: *Fragmentos*, *PaLavra*, *Tradução em Revista*, *Eutomia* and *Cadernos de Tradução*. He is currently a member of the editorial board of the last three journals. Further, he has taken part in dozens of academic events on translation and literature, often as a guest or keynote speaker, both in Brazil and abroad. He presented his poems, along with his translator Hans Raimund, at the Vienna City Library in September 2011.

This first look into Britto's academic career reveals an atypical academic in a number of different ways. Firstly, it is quite striking that he has not got a PhD while most of his colleagues have at least a doctorate – many have a postdoctorate as well. Secondly, his academic production, not only in terms of theses supervised, but also as far as publications on translation studies are concerned, is not particularly numerous. Nevertheless, in light of his colossal literary production (see III, 1), one can hardly expect his academic work to achieve the same magnitude. In this sense, and taking into consideration the main subjects of his papers, together with the subjects on which his students have been working, Britto strikes one as somebody whose interests lie far closer to translation practice than to translation theory, in a more academic sense. Even though he is a professor in the full sense of the term, his performance as a translator and as a writer simply does not compare with his academic achievements. The intention here is by no means to belittle his academic work, which indeed is highly interesting and relevant – as the next chapter will reveal. However, these considerations are crucial to help to delineate the *standpoint* from which he speaks. As our discussion advances and we start looking into the way Britto perceives translation and translation studies, awareness about his standpoint will be of fundamental importance.

His engagement in research projects somehow confirms this idea of someone whose affinities lie with more "practical" or practice-oriented issues. Furthermore, it unveils his dedication to linguistics as well, and so do the courses he teaches both at undergraduate and postgraduate levels. Judging by his career turns, one could expect his interest to have drifted from linguistics, when he did his MA back in the early 1980s, to literature, after he had translated dozens of literary works and written a few of his own as well. His academic career as a whole appears nonetheless to indicate that linguistics remains one of his chief areas of concentration. Indeed, according to his national CV (see footnote 114 on pages 208-209), the main areas of research in which he is engaged are the following: Translation; English Language; Literature in English; Brazilian Literature; Semantics and Phonology. In light of this item of information only, one could hardly imagine that this researcher is also one of the most prominent literary translators of a huge translating country like Brazil. So in this sense Britto is also atypical – in terms of his academic interests (largely influenced by linguistics) and his work outside academia (as a celebrated literary translator whose specialty is poetry translation).

Yet being "atypical" is by no means a problem, but simply a trait we should take into account as we move on to the next chapter. However necessary, any attempt to define Britto would certainly run into generalisations and stereotypes, so as we try and carve the contours of his standpoint, let us be critical of this very attempt, in a Derridean double gesture (see I, 2). This "carving of contours" should of course be understood in a very limited way – i.e. the limits circumscribed by the present book.

From an academic point of view, Britto would probably be labelled as a linguist – on account of his education and academic career as a whole. And let us bear in mind that, in Brazil, most of the scholars engaged in translation studies would either be labelled as linguists or literary theorists since most of them do tend to have training in either one of these areas – seeing as there are only a few translation studies programmes in the country, most of which newly-founded. For the media, however, he is chiefly a superb translator and writer. His production as a literary translator and writer (particularly as a poet) unquestionably outshines his academic production. Indeed, his countless interviews fill search engines and platforms like "YouTube", for example. Therefore, as far as literary translation is concerned, these two elements – namely the fact that he can be said to be a linguist and that his experience as a translator surpasses his work as an academic – make him come across as a practitioner, and/or an academic whose main concerns are probably practice-oriented. As for the combination of linguistics and translation studies, my expectations are very much in line with the history of this "marriage" in previous years, as addressed both in Part I and Part II. In very general terms, there seems to be a tendency for linguists engaged in translation studies to have a rather "scientific", strict outlook on the discipline, seeking to devise models and precepts with a view to translation practice – very much in tune with that notion of translation theory as a set of precepts, as argued in Part I. Of course these conclusions concerning Britto are more expectations than assertions, and the next chapter will help to reveal to what extent they may or may not be pertinent.

Let us then study, in III, 3, some of Britto's papers on translation and translation studies. As already pointed out, bearing in mind the outlines of his standpoint briefly sketched in these initial chapters will help to understand his views.

3. Works

What I say has more truth than what you say,
since I can "do more" (gain more time, go further)
with what I say than you can with what you say.

Jean-François Lyotard

While it is true that amongst translation students Paulo Henriques Britto is best-known for his countless translations of authors such as Philip Roth, John Updike, William Faulkner, Ian McEwan and Elizabeth Bishop, it is also true that his academic disputes with Rosemary Arrojo (see Part II) have received a great deal of attention in academic circles. Furthermore, his model of analysis and criticism of poetry translation has divided the Brazilian academic community. In the present chapter, let us examine a selection of his papers published in specialised journals. For organisational reasons, they have been divided them into three categories, as follows: 3.1 Criticism of Poststructuralism (three papers), 3.2 The Craft and Art of the Poetry Translator (two papers), and 3.3 Translation Criticism (five papers). Although these three main issues tend to feature in most of his papers, there appears to be a focus towards one or the other, so this was my criterion in devising this structure.

Even though Britto's production as a translation theorist is solid and emblematic of Brazilian criticism of poststructuralism in translation, his works have not been compiled into a single volume yet. Instead, most of his papers have been published in different Brazilian journals, which is indeed the case of many translation scholars today given both the difficulty in publishing books and the profusion of (mostly electronic) translation journals in Brazil. Having said that, in view of this enormous literary production, translating and writing literature are probably his priority.

Similarly to Part II, dedicated to Rosemary Arrojo, I will describe and comment on his papers simultaneously, making clear what aspects seem conspicuous. Additionally, whenever possible I will try and refer back to his standpoint, aiming at anticipating and justifying some the motivations and interests lying beneath his theoretical undertakings. As in Parts I and II, a few issues raised here – the more lengthy or intricate ones – will be discussed in detail in Part IV, so that in this part the focus can remain on Britto's work.

3.1 Criticism of Poststructuralism

Diante da constatação de que Deus não existe, uns se tornam ateus; outros, porém, nostálgicos, preferem criar uma nova igreja, com teologia, liturgia, ritos, sacerdócio e tudo o mais, em torno de um altar onde está entronizado o grande deus Nada. Para estes, o fim dos absolutos torna-se o novo absoluto.

Paulo Henriques Britto[116]

In the present section we will examine three papers in which Paulo Henriques Britto presents a direct critique of poststructuralist thought on translation. For the sake of organisation, I will use reverse chronological order (2007, 2001 and 1995) as Britto's first and second papers (1995 and 2001) are direct responses to papers by Rosemary Arrojo and Stanley Fish, whereas the third one (2007a) is a general reflection on ideas perceived by Britto as poststructuralist. This way, this most recent paper will offer a preamble to the discussions carried out in the two earlier papers, in which more specific issues of poststructuralism and deconstruction are addressed.

"É Possível Avaliar Traduções" [Is it Possible to Evaluate Translations?] is largely based on a lecture given at the conference Etnografia e Tradução (Ethnography and Translation) at the Universidade Estadual do Rio de Janeiro in 2005. As we will see in III, 3.3, Britto has been working on a methodology for criticism of poetry translation for a number of years now. In the paper in question, as the title indicates, Britto asks himself whether one such criticism or evaluation is at all possible or even desirable, particularly in view of the ever growing dissemination of theoretical perspectives labelled poststructuralist.[117]

Britto starts his paper by describing what, to him, the goal of translation criticism is. Quoting Lefevere (1975: 3 – cf. Britto 2007a: 1), he claims that translation criticism is not up to the average reader, since s/he is probably making use of a translation precisely because s/he does not speak the source language. In this sense, it would be up for translation critics to

116 This excerpt is quoted and translated later in this section (see page 217).

117 For a discussion of literary translation criticism informed by poststructuralist values, see the first paper analysed in II, 2.3, entitled "A que são Fiéis Tradutores e Críticos de Tradução? Paulo Vizioli e Nelson Ascher Discutem John Donne" (Arrojo 1993: 15-26).

avaliar as traduções de poesia, com o fim de orientar o consumidor de traduções de poesia disponíveis no mercado, do mesmo modo que se espera dos especialistas em informática que orientem os consumidores de computadores e software, comparando os diferentes produtos e avaliando-os (BRITTO 2007a: 1).

[evaluate translated poetry so as to guide consumers through the translations available on the market – in the same way as one would expect computer specialists to guide software consumers, comparing and assessing the different products available.]

And for him this comparison and assessment should be carried out through rational and reasonably objective methods ("métodos racionais e razoavelmente objetivos" – BRITTO 2007a: 1), despite the so-called postmodern reservations to objectivity and rationality. Indeed, he summarises the allegedly poststructuralist argument against translation criticism as a strong will to abolish all kinds of value judgement due to their ineluctable, underlying subjectivity and lack of absoluteness (BRITTO 2007a: 4). Britto then goes on to explain that his aim as a translation critic is to

analisar o mérito de uma tradução de poesia com base nos recursos utilizados pelo tradutor em comparação com os usados pelo autor do original, ***sem a interferência de fatores subjetivos*** *— os pressupostos teóricos tomados como ponto de partida, as simpatias e antipatias pessoais, as preferências emocionais por certas palavras em detrimento de outras, etc.* (BRITTO 2007a: 1 – my emphasis).

[analyse the merits of a translated poem based on the comparison of the resources used by the original author to those employed by the translator *without the interference of subjective factors* – such as theoretical perspectives used as starting point, personal empathies and antipathies, emotional preferences for particular words instead of others, etc.]

Britto then concedes that full objectivity is certainly unattainable, which in turn does not by any means make him embrace the supposedly poststructuralist idea that, because full objectivity is impossible, it should not be strived for at all, it should be fully discarded. As an example of what, to him, is a hasty conclusion often drawn by poststructuralist thinkers on these grounds is the famous dispute between Derrida and Searle over the distinction between "normal" and "parasitic speech acts".[118] In Britto's view, just because one cannot distinguish fiction from reality through unanimously universal and absolute criteria, it does not mean that one should completely do away with this distinction. For him, abolishing such distinctions represents an outrage not only to common sense, but also – and perhaps more importantly – to science, to the basic parameters of scientific work ("parâmetros básicos do trabalho científico" – BRITTO 2007a: 6).

As an instance of the harm poststructuralist ideas could do to science, Britto mentions the problem of absolute zero in physics. As it happens, it is impossible to reach an absolute zero temperature (-273.15°C); nevertheless, because carrying out experiments at absolute zero is important in various different areas, scientists do their best to get as close as they can to it, and then make the necessary adjustments on the results obtained. However, Britto maintains, if scientists were influenced by poststructuralism, they might come to the conclusion that, since reaching absolute zero is impossible anyway, carrying out experiments that require this temperature is a waste of time – "se os físicos fossem pós-estruturalistas, talvez eles concluíssem que, como não é possível atingir o zero absoluto, seria perda de tempo formular experimentos que envolvam essa temperatura" (BRITTO 2007a: 6).

Britto then goes on to discuss the issue of scientism (see I, 1 and 2) in the Humanities. Firstly, he contends that ambiguity and incompleteness are normal parts of scientific work. But then he stops for a moment and questions whether the scientific method, the method of the so-called Natural Sciences, is indeed appropriate for the Humanities after all: "De fato, o fenômeno da linguagem natural humana levanta (...) questões de ordem muito diversa das que são encontradas no mundo natural, e é de se esperar que as ciências humanas recorram a princípios e métodos diferentes dos das ciências naturais", or indeed, the phenomenon of natural languages raises utterly different questions than the ones found in the natural world, and so it is only natural that the Humanities should make use of principles and methods that are different from those of the Natural Sciences (BRITTO 2007a: 7). Despite this, however, he believes that anti-scientism is not a solution either, particularly when it takes poststructuralist contours. For him, poststructuralist anti-scientism stands for

118 For more on the Derrida/Searle debate, cf., for example, MOATI (2009) and DERRIDA (1988/1992 – translated by Alan Bass, Samuel Weber and Jeffrey Mehlman).

the idea that because science cannot rely on an absolutely metaphysical foundation, it is not at all a suitable basis for reliable knowledge (BRITTO 2007a: 7).

To emphasise his point, he resorts to a generalisation and an allegory: "a inexistência de uma diferenciação entre duas categorias A e B que seja inteiramente livre de ambigüidades é razão suficiente para abolir todo o sistema de categorizações que leva à criação de A e B (?)", or the impossibility of a fully ambiguity-free distinction between two categories A and B is enough to abolish the entire categorisation system from which A and B stem (?) (BRITTO 2007a: 8). The allegory to illustrate his argument is the following: if trying to constantly stay healthy is impossible because we are all doomed to sickness and eventual death, should we arrive at the conclusion that taking care of our health is useless? – "Se a meta de se conservar sempre saudável é inatingível, já que todos nós estamos fadados a adoecer e morrer em algum momento, devemos concluir que cuidar da saúde é inútil?" (BRITTO 2007a: 8).

Towards the end of the paper, Britto, still focused mostly on the question of scientism, draws an overview of linguistics from the 1960s and 70s onwards. He explains that even though the desire to model linguistics after the Natural Sciences, including mathematical language and rigorous logics, dominated the field initially, it was gradually replaced by a feeling of suspicion of these same aspirations. And though in linguistics these scientific ambitions were not completely discarded, in literary theory they attracted their most radical opposition. In Britto's opinion, nevertheless, just because a mathematical formula or strictly logical postulate cannot comprehend a particular issue in its entirety, it does *not* mean that the scientific method, that reason cannot offer adequate methods to analyse this issue. In this light, he seems to perceive this anti-scientism in the Humanities – and especially in literary theory – with scepticism, as another instance of a hasty conclusion typically drawn by thinkers with poststructuralist affinities.

Britto then attacks deconstruction more specifically, claiming that

Diante da constatação de que Deus não existe, uns se tornam ateus; outros, porém, nostálgicos, preferem criar uma nova igreja, com teologia, liturgia, ritos, sacerdócio e tudo o mais, em torno de um altar onde está entronizado o grande deus Nada. Para estes, o fim dos absolutos torna-se o novo absoluto (BRITTO 2007a: 9-10).

[Faced with evidence that God does not exist, some become atheists; others, however, become nostalgic and set up a new church – with theology, liturgy, rites, priesthood and all the rest, around an altar where the great god Nothing is enthroned. For them, the end of the absolute becomes the new absolute.]

In other words, in Britto's view on deconstruction does away with numerous traditional dogmas only to elect new ones to replace them – nihilism and scepticism[119] being perhaps the most conspicuous elements of this "new dogma".

At the end of the paper, Britto proposes yet another analogy, this time featuring an experienced and diligent literary translator who, after years of hard and meticulous work with texts, comes across a paper on poststructuralist perspectives on translation. Suddenly our hypothetical translator finds out that, as it happens, the texts s/he has been working on for decades do not contain any meaning ("não contêm nenhum significado"); that, in fact, texts actually mean nothing ("afinal de contas, os textos nada significam"); that originals are nothing but a pretence derived from the Enlightenment ("o conceito de 'original' não passa de uma contrafação iluminista"); that the dichotomies original-translation and author-translator are mere ideological construes ("as oposições original/tradução e autor/tradutor são meros construtos ideológicos"); and that in a nonsexist and nonimperialist world there should be no distinction between the work of the translator and the work of the author ("num mundo não-sexista e não-imperialista não haveria qualquer distinção entre o trabalho de autoria e o trabalho de tradução" – BRITTO 2007a: 9-10). In Britto's opinion, this archetypical translator will at best listen to these assertions with a smirk on his/her face ("nosso hipotético tradutor só poderá encarar tais afirmativas com, na melhor das hipóteses, um sorriso nos lábios" – BRITTO 2007a: 10). He then explains, as already indicated in I, 3, that the suspicion and lack of interest with which practitioners tend to look at translation theory is due to allegedly poststructuralist positions as the ones quoted in this paragraph, and that exaggerated and hasty though their rejection of theory may be, it is not without justification.[120]

Finally, Britto goes back to the question he asks in the title of his paper (is it possible to evaluate translations?) and states that objectively evaluating and criticising translations is indeed possible and desirable, and that the starting

119 And here, inspired in Martha Nussbaum's paper "Skepticism about Practical Reason" (1994), Britto explains that this scepticism is not like Hume's – it is Sextus Empiricus scepticism, like Pyrrhonism.

120 See the long indented quotation by Britto in I, 3.

point should always be the very critique of these poststructuralist ideas – particularly of the idea that, without absolute values, there are no values at all ("na ausência de absolutos, não resta nenhuma valor"). He then ends the paper with a call for action: "Ousemos discordar: restam valores *relativos*, o que não é a mesma coisa que nada", or let us dare disagree: there are *relative* values, which is by no means the same as nothing (BRITTO 2007a: 11 – his emphasis).

In this paper, Britto's final allegory reminds us very vividly of his own standpoint. This experienced and diligent literary translator, who faces poststructuralist discourse on his work with a smirk on his face and feels suspicious of translation theory, could be interpreted as himself. As I will repeatedly argue in this chapter, Britto's criticism of poststructuralist thought on translation has two main components. Firstly, he speaks from the point of view of the practitioner; in other words, he appears to presuppose the marriage of theory and practice (see Part I) and finds so-called poststructuralist contributions to translation relatively useless, as they do not address, never mind systematise or simplify, translation practice. Secondly, he expects a radicalism of perspectives labeled poststructuralist which is simply not pertinent – though it is typical of those who oppose deconstruction and other poststructuralist tendencies. I will develop these two arguments below, as we look into two other papers by Britto on the same issue. As for his remarks on scientism and the adequacy of the methods of the Natural Sciences to the Humanities, I will address them at length in Part IV – though these issues will come up again in III, 3.2 and 3.3.

The second paper we will analyse in this section, entitled "Why deconstruct?" ("Desconstruir para quê?" – BRITTO 2001),[121] is a direct critique of one of Arrojo's papers on deconstruction, included in the abovementioned 1992 collection *O Signo Desconstruído* (see II, 2.1). The name of Arrojo's paper is "As Questões Teóricas da Tradução e a Desconstrução do Logocentrismo: Algumas Reflexões" (Reflections on the Theoretical Questions of Translation and the Deconstruction of Logocentrism).

Britto's main criticism is the fact that, in her paper, Arrojo very contradictorily censures so-called logocentric perspectives on translation but, at the same time, makes use of a number of these very same logocentric perspectives of which she disapproves to get her message across. In the paper in question,

121 Britto translated most of his published papers himself and made them available on his personal webpage (http://phbritto.org/ – last accessed in December 2010). Even though the references to his texts (including years of publication and page numbers) will take the Brazilian published versions into account, I will make use of his own translations here. Therefore, whenever I quote him directly in English, this means that the text in question has been translated by Britto himself.

Arrojo summarises the logocentric tradition into three assumptions, devised according to Ronald Knox's work. The first assumption refers to the possibility of (i) perfect equivalence between original and translation; the second maintains that original texts contain (ii) stable meanings that can be identified with the author's conscious intention; and the third assumption refers to the (iii) clear separation between a text's meaning and its style (BRITTO 2001: 42-43 and ARROJO 1992/2003: 74).

While Arrojo asserts that a translation can by no means be considered (i) equivalent to its original, Britto maintains that she acts as though she thought translations and originals were equivalent,

> *For the passages by Nietzsche and Mounin she quotes were not written by Nietzsche or Mounin: they are taken from texts written in Portuguese by Brazilian translators. Or are the translators Portuguese rather than Brazilian? They might be, for all we know; Arrojo does not name them, clearly because she thinks this detail is irrelevant to her purposes. It is the meaning of the passages that she is interested in, the ideas that Nietzsche and Mounin express in their texts; and she believes that these meanings, or ideas, have been transposed into Portuguese in the translations she quotes from in a reasonably reliable way. By treating translations as if they were originals and attributing them to the authors of the originals, Arrojo is clearly assuming the logocentric notion, [i.e.] translations are texts that are equivalent to their originals* (BRITTO 2001: 43).

As Britto quite rightly points out, by not mentioning the names of Mounin's and Nietzsche's translators – neither in the text nor in the bibliography – Arrojo does subscribe to the logocentric tradition she fiercely criticises, since she does not acknowledge the work done by these translators. From Britto's point of view, this is the same as saying that original and translation are equivalent, as she does not find it relevant to mention the names of the translators precisely because she seems to think that original and translation are the same, regardless of who translated it.

Nowadays, not mentioning translators' names is simply inexcusable in all sorts of publications, never mind in papers on translation studies – and even worse in those whose main subject is the importance and autonomy of the translator. Regardless of their theoretical standpoint, it is high time all involved with translation started paying more attention to who translated what. In the specific case of this book by Rosemary Arrojo, however, it is curious that only four of the papers (out of 17) do not include the translators' names in the bibliography; the other 13 do, some of which mentioning the very same books by Nietzsche and Mounin to which Britto refers. Therefore, it is not that the entire book neglects translators, but some chapters do, even though the book is in its second edition at the time of writing.

But let us look back at Britto's critique of Arrojo's paper. He then moves on to the second assumption Arrojo censures. His argument is that by using expressions such as "Mounin believes", for example, she appears to endorse the second logocentric assumption, too. In other words, if Rosemary Arrojo says that Mounin believes in something according to a text written by Mounin in French a long time ago and then translated into Brazilian Portuguese, it is because she believes that "meaning is a stable property of a text, which can be identified with the author's conscious intention in writing it, and which does not depend on the reader's actual circumstances" (BRITTO 2001: 44).

Accordingly, Britto claims that Arrojo seems to follow the third assumption as well, considering that in the paper in question she "assumes that meaning can be considered an object distinct from the style of the text in which it appears" (BRITTO 2001: 44). From his point of view, if she did not think so, she would have quoted Nietzsche and Mounin in German and French, respectively, and not in Portuguese. Indeed, he goes further and says that "by using a translation of the passage by Nietzsche (...) she makes it clear that the only thing she is interested in is the meaning, this object that can be conveniently detached from the text, and so necessarily from the author's style" (BRITTO 2001: 44).

These last two arguments go straight to the heart of the polarity structuralism "versus" poststructuralism. Let us once again look at the way Arrojo perceives the relationship between these two lines of thought:

> *(...) one can say that there is a paradoxical relationship between structuralism and poststructuralism, i.e. radical opposition and complementariness at the same time. In the transition (without clearly established boundaries) from structuralism to poststructuralism, a radicalisation of insights and presuppositions takes place, which in turn allows one to perceive poststructuralism as a* ***more attentive*** *and* ***less naive kind of structuralism*** (ARROJO 1996a: 60 – my emphasis).

Drawing her conclusions from Derrida's work, Arrojo makes it clear that the aim of poststructuralism is not to fully break with structuralism. Rather, one should complement the other. In I, 6, for example, we saw how the division between signifier and signified is necessary and yet the nature of this division must be questioned. Therefore, it is not that one should get rid of the division altogether; instead, one should perceive it more critically.

If poststructuralism were taken to an extreme (and structuralism too, for that matter) and completely broke with structuralism, communication would be far more chaotic than it actually is. In order for Arrojo to be able to write her paper, for instance, and be as radical and extremist as Britto appears to expect her to be, she would have to say "According to my reading of the translation of Mounin's text done by this particular translator". That, however, would probably not suffice, seeing as she would have to account for the background and ideology of the author, of the translator and of herself which, again, would be impossible, as everything would always be what she *perceives* them to be at a given moment in time. By advocating the instability of meaning or the importance of the translator's ideology, for example, deconstruction is not saying that communication is impossible and translations are completely different texts when compared to their originals. As the relationship between signifier and signified exemplifies, what seems to be the point is to question just how stable meanings are, or how negligible and neutral translators can actually be.

However potent Britto's arguments may be, they presuppose a radicalism which would completely break with structuralism and relegate language to the realms of impossible tasks. By not completely breaking with structuralism and simply looking at some of its core principles "more attentively" and "less naively" (cf. ARROJO 1996a: 60), communication is still possible – one would certainly not need a theory to show that communication is possible – and one can still say "Mounin believes" rather than "I believe, according to my background, ideology and intentions, that one can read Mounin's text in this or that way at this moment in this place". Nevertheless, in postmodern times, saying "Mounin believes" does not by any means imply that "I can tell exactly what Mounin was trying to say by reading his text and decoding the meanings he intentionally placed in it" (as Britto himself admits).

It seems that without the radicalism that Britto expects of translation theories labelled as poststructuralist, what he says is rather similar to what Arrojo says. His conclusion in the paper in question is an excellent example of this theoretical similarity (BRITTO 2001: 48 – my emphasis):

> *To conclude, I would like to return to the point that to relativize is not the same thing as to abolish. Deconstructive critique forces us to relativize a number of concepts – that is, to take them for what they are, fictions and not realities. But we cannot do without these fictions (…). Such concepts as "meaning," "original" and "equivalence" are indispensable assumptions of most textual practices, however imperfect they may be shown to be.* ***We must criticize them, we must be aware of their constructed nature, but we still need them.*** *The name of the logocentric game is language. To refuse to play it is to condemn oneself to silence.*

Indeed, Arrojo claims that relativising and criticising structuralism is *not* the same as abolishing it, whereas Britto claims that relativising and criticising *poststructuralism* is not the same as abolishing it. It is almost as though one of them saw the glass half empty and, the other, half full. Just how important a role these "fictions" play in Britto's work will become clearer in III, 3.3. It is nonetheless evident that Britto appears to expect Arrojo to "refuse to play" what he calls "the logocentric game", or the language game, and thus "condemn [her]self to silence", which she obviously does not do and does not intend to do. Likewise, taking Britto's theoretical standpoint into account, one could quite rightly expect him to argue that his translations are the neutral and transparent renderings of the original author's intentions, which he does not, either. I will develop these initial arguments in what follows.

Another of Britto's papers written in response to ideas labelled "poststructuralist" is called "Lícidas: Diálogo Mais ou Menos Platônico em Torno de 'Como Reconhecer um Poema ao Vê-lo', de Stanley Fish" (Lycidas: A More or Less Platonic Dialogue on Stanley Fish's "How to Recognize a Poem When You See One"), first published in 1995 in the journal *PaLavra* (no. 3), from the Pontifícia Universidade Católica do Rio de Janeiro. As its name suggests, this essay is a direct critique of Stanley Fish's paper "How to Recognize a Poem When You See One", which in turn was published two years earlier in Portuguese in the same journal (*PaLavra* no.1) under the title "Como Reconhecer um Poema ao Vê-lo" (translated by Sônia Moreira). The original English version had first appeared in Fish's 1980 *Is There a Text in this Class? The Authority of Interpretive Communities* (FISH 1980: 322-337).

In his paper, which was originally delivered as a lecture, Fish emphasises the importance of his "interpretive communities", this time not only in terms of the production of meaning, but also in terms of the "recognition" of different text types – in this particular case, poems. As Fish explains in the introduction to his 1980 book, "it is interpretive communities, rather than the text or the reader, that produce meanings and are responsible for the emergence of formal features" (FISH 1980: 14). Therefore, for him texts do not contain intrinsic, stable meanings, nor do individual readers have total freedom to interpret texts as they wish.[122] The secret would lie in these interpretive communities and the tacit agreements their members constantly make. According to Fish,

> *Interpretive communities are made up of those who share interpretive strategies not for reading but for writing texts, for constituting their properties. In other words these strategies exist prior to the act of reading and therefore determine the shape of what is read rather than, as is usually assumed, the other way around* (Fish 1980: 14).

From this point of view, Fish "dethrones" the text and favours readers and their interpretive strategies without, however, granting them full autonomy, which in turn would make them "independent agents" (Fish 1980: 14) and thus sabotage communication.[123] Indeed, these reading (or writing) strategies do not belong to individual readers, but rather (as pointed out in this last indented quotation above) to the interpretive community of which readers are members, since these communities "at once enable and limit the operations of [the readers'] consciousness" (FISH 1980: 14).

In his "How to Recongnize a Poem When You See One?", Fish shows the impact of these communities on the "reception" of text type. He bases his reflections on a more or less accidental classroom experiment that took place at

122 In II, 2.1 Arrojo came to a remarkably similar conclusion as for the location and origin of meaning – see page 119.

123 One such "sabotage" may be comparable to what happens to the protagonist of Peter Bichsel's short-story "Ein Tisch ist ein Tisch" (BICHSEL 1995). In the story, he gradually attributes new words to the world surrounding him – "Dem Bett sagte er Bild. Dem Tisch sagte er Teppich. Dem Stuhl sagte er Wecker. Der Zeitung sagte er Bett. Dem Spiegel sagte er Stuhl. Dem Wecker sagte er Fotoalbum. Dem Schrank sagte er Zeitung. Dem Teppich sagte er Schrank. Dem Bild sagte er Tisch. Und dem Fotoalbum sagte er Spiegel" (BICHSEL 1995: 22) – until he reaches a point in which he is fully isolated from other human beings, as he understands no one and no one understands him (for more on this, see the discussion on the origin of meaning mentioned in the previous footnote).

the University of New York in 1971. In short, what happened was that between the first and the second course he used to teach (the first one on stylistics and the other one on English religious poetry), he had left the names of five linguists on the board, the last of which followed by a question mark because he was not sure about its spelling (FISH 1980: 322-323). As the English religious poetry class began, Fish told his students that the list on the board was actually a poem, and so they immediately began to apply the interpretive strategies they had learnt to analyse what, to them, was the poem written on the board. It was not before long that they had "found" meanings to all the names on the board, in addition to a number of poetic devices.[124]

Going back to the question he very aptly asks in the title ("how to recognize a poem when you see one?"), he replies that the common sense answer, "to which many literary critics and linguists are committed", holds that "the act of recognition is triggered by the observable presence of distinguishing features" (FISH 1980: 325-326), i.e. intrinsic and clear-cut characteristics on the text surface. However, as his little experiment reveals, this response would be inadequate because the list on the board "contained" no "distinguishing features" and yet his students managed to find these poetic features anyway. In other words, what triggered their poetic interpretation was not the fact that the text *contained* poetic features, but rather the fact that the teacher had told them that they had a poem before their eyes. As Stanley Fish explains, "the meanings of the words and the interpretation in which those words were seen to be embedded emerged together, as a consequence of the operations my students began to perform once they were told that this was a poem" (FISH 1980: 326). Fish goes further and says that "interpretation is not the art of construing but the art of constructing. Interpreters do not decode poems; they make them" (FISH 1980: 327). So Fish largely disregards the context of text production to place greater emphasis on the context of *reception*, since it is reception that will turn the words written on a piece of paper or on a blackboard into a text as such – hence the witty name of his 1980 book, *Is There a Text in this Class?*, in the sense that without reception, there is indeed *no* text.

Britto's main criticism in his "Lycidas" stems precisely from this fact, namely that "[Fish] says poems are poems *solely* because they are read with the intention of reading poetry into them. He doesn't take into account the inten-

124 Fish's experiment resembles Rosemary Arrojo's experiment in her *Oficina de Tradução* (see II, 2.2) whereby half the class reads William Carlos Williams' poem "This is Just to Say" as a note, and the other half reads it as a poem.

tion to *write* poetry" (BRITTO 1995: 142 – his emphasis). On the one hand, Britto admits that he agrees with some of Fish's arguments. In fact, he even says that it is an "obvious fact that a poem's poeticalness (...) is not a property of the molecular structure of the ink it's written in or the paper it's written on", and also that "all these things make sense only in the cultural context of a system that has to be learned". However, Britto finds it absurd that this should lead to "the bombastic conclusion that the only thing that makes a poem a poem is the fact that you look at it with 'poetry-seeing eyes'" (BRITTO 1995: 146 – quoting FISH). Indeed, he stresses that "nothing justifies this quantum jump" (BRITTO 1995: 146).

Britto's paper consists of a fictitious, "more or less platonic dialogue" (BRITTO 1995: 142) between a young professor of English and literature who has just read Fish's text and is impressed by it, and a middle-aged, sceptical Brazilian literature professor who read the text as well and is not in the least enthusiastic about it. Thanks to this witty structure, Britto manages to anticipate most of the counterarguments one would bring forth in response to his own critique, presented of course through the voice of the sceptical professor. Therefore, the sceptical professor, as Britto's persona, conducts the dialogue in such a way as to unveil what he sees as the weaknesses in Fish's theory, often presenting alternatives to the dilemmas he encounters. The younger professor, in turn, presents the counterarguments.

Largely basing his arguments on Wittgenstein's concept of "Familienähnlichkeit", first presented in the philosopher's 1953 book *Philosophische Untersuchungen*, (cf. WITTGENSTEIN 1953: §67) Britto disputes the fact that things do not possess intrinsic characteristics, as Fish defends. According to Paulo Henriques Britto, just because certain elements of a "class" or "family" do not possess every single distinguishing feature normally attributed to that class of elements, it does not mean that it is not a member of that class or family altogether.

The main example that the two characters in Britto's paper discuss is William Carlos Williams' poem "This is just to say" (see II, 2.2). What the persona defends is that this modern poem, despite not containing all distinguishing features expected of a poem, still contains *some* poetic features, which means it can *still* be ascribed to the class we all know as "poetry". Britto, impersonated by the middle-aged professor, contends that this is the reason why Williams' poem is a poem, precisely because of its intrinsic characteristics, and not because one looks at it with "poetry-seeing eyes", as Fish suggests. Britto goes further and explains that in this light one can account for a fact Fish cannot, namely the number of intrinsic characteristics a poem possesses, for example, will determine how easily or quickly readers will recognise it as a poem or not. In order to illustrate his point, he mentions "Shall I Compare thee to a Summer's Day": "whether it is scribbled on the door of a subway train or written on a blackboard or printed on a piece of paper found in the garbage, [it] is recognized by any literate member of English-language culture as a poem" (BRITTO 1995: 145). "This is just to say", in contrast, is not perceived as a poem at first sight, and exemplifies how literature professors like Britto's two characters must often "engage in strenuous argument in order to convince people that a given text is a poem" (BRITTO 1995: 145).

As their fictitious dialogue advances, the young professor criticises the older professor for taking things (in this case, poems) in their essence, as if he had access to them without his reading, his interpretation, his construction of what they are. Britto's persona then replies that this (i.e. the fact that everything is a reading, a construction) is obvious, so obvious that it becomes "practically irrelevant" (BRITTO 1995: 146). To illustrate his point of view, the sceptic brings up the epistemological challenge of other minds, whereby one would not be able to tell whether other human beings actually have minds since one can account solely for one's own mind and behaviour. Yet, the middle-aged professor stresses that one *must* act as though one *thought* other human beings actually had minds. Therefore, "the other-minds problem is really a nonproblem, in the sense that it affects nothing, that it has no practical consequences whatsoever" (BRITTO 1995: 146). In other words, saying that one only has access to things (in this case, poems) through one's perception, one's reading, and that this reading or perception is not exactly the same as the thing *itself* (to which one *never* has access) is such an obvious assertion that it becomes irrelevant.

Another supposed weakness of Fish's theory revealed through the voice of the sceptical professor refers to the fact that if interpretive communities were indeed relevant and played an essential role in the shaping of minds and production of meanings, people like Fish, for example, would never have been able to come up with an idea that contradicts or is not part of the set of ideas circulating in this very same community. The following ironic and bitter remarks by the older professor clarify his point:

So are we supposed to conclude that Kant's mind was really determined by the institutional structure of the University of Königsberg? Or that Galileo's mind was determined by the institutional structure of seventeenth-century Italy, including the Aristotelian physics and the Ptolemaic astronomy he was brought up on? Come on, give me a break (BRITTO 1995: 147).

As he draws to his conclusion, Britto's middle-aged character tries to explain to his younger colleague that even though he agrees that meanings are not stable and interpretations vary greatly, he chooses to *pretend* he believes otherwise so as to withstand the pressure imposed by translating. He maintains that in the same way as one has to pretend other people are human (in spite of the problem of other minds), one has to pretend meanings can be stable and translations can be accurate, faithful to the author's intentions (BRITTO 1995: 148): "When I read or translate a text, for instance, I have to act – provisionally, anyway – as if I had full access to the text's objective meaning; otherwise infinite regression would stop me cold". This "acting" would of course be a fiction, but a necessary one according to him. Language and hence communication and translation would be a game of pretending and make-believe. Taking Fish's essay as an example, he explains how far the fiction goes (BRITTO 1995: 148-149 – my emphasis):

In the world we live in, Professor Fish writes his articles pretending (...) his readers will capture precisely the meaning he has in mind, (...) but wait a minute, we have to indulge in make-believe in even more ways than that: after all, we didn't read what Fish actually wrote, but only a translation of his article (...). We also have to pretend that the translator did a reasonably good job, so that we understand exactly what Professor Fish meant to say. That being the case, it seems more sensible (...) to assume the working hypothesis that the text does contain a stable, definite meaning, which may be identified with what the author meant to say and can be captured in a translation, and that the different readings of a text are functions of the individual differences between readers, who are not mere cake moulds that culture fills with exactly the same cake mix – even though I know it's impossible to prove that this is really the case, even though I know this is just a useful fiction. Because Fish's fiction (...) doesn't seem that ***useful*** *once you realize what its implications are.*

Faced with this scepticism and sarcasm, the younger professor then asks whether they are now back to rationalism, back to structuralism, pretending philosophers like Nietzsche, for example, never existed. His sceptical colleague retorts that this is by no means the case precisely because these ideas are only *fictions*. In other words, he knows meanings are not stable and translations cannot be said to be faithful or unfaithful in absolute terms; nevertheless, he chooses to stick with these fictions because they are *useful*, *more useful* than Fish's theory. The older professor clarifies that "it's one thing to assume an absolute conception of the subject or of meaning as was done in the nineteenth century, and something else again to hold that such notions cannot be taken to be absolutes, but that for certain *practical purposes* they remain *useful fictions*" (BRITTO 1995: 149 – my emphasis).

Therefore, when it comes to the work of the translator, Paulo Henriques Britto has his older character explain that "any attempt to translate a poem on the basis of the notion that it allows an infinite number of possible readings would be doomed from the outset" (BRITTO 1995: 149). He then makes an interesting analogy with painting (BRITTO 1995: 150): "It would be like trying to paint someone's portrait based on the notion that the person's face is constantly changing (...). If I am to translate a poem, the poem must sit still, at least for a while".[125]

* * *

In this third paper, similarly to the first two papers analysed in the present section, Paulo Henriques Britto seems to expect fierce radicalism and thus criticises Fish's theory on this basis. Britto discards the idea of "interpretive communities" because he takes them to an extreme and extends their scope to all realms of human behaviour, completely excluding creativity and individuality. Fish advocates that the community to which one belongs helps to shape the way in which meanings are produced. To Britto, however, this would mean that human beings are not capable of coming up with anything unless it has already been produced and acknowledged within a given community. One could also speak of a "quantum jump" here (cf. BRITTO 1995: 146). To the same extent that men organise themselves in communities and make use of language as a means of reference and communication (and therefore strive to under-

125 I will come back to this same analogy in IV, 2.

stand and be understood), they also tend towards change, innovation or "progress", as some might prefer to call it.

Kant was undoubtedly a product of the University of Königsberg in the sense that he belonged to a certain system, followed a number of rules to obtain his degree and to become a lecturer and later, a professor. He is even more obviously a member of his "interpretive community" when one takes into account the fact that he used *language* to convey his then innovative ideas. By teaching lectures and writing texts in certain formats, he (as any other member of his community) conformed to different sets of (mostly) tacit rules. Had it not been for these tacit rules, nobody would have found any relevance in what he had to say or write; worse, nobody would even have had access to what he wanted to say. So in summary, saying that interpretive communities limit one's consciousness does not necessarily mean human beings are only capable of repetition (which is a contradiction in itself), but rather that in order to be understood, one will unconsciously rely on certain communication norms, which in turn will gradually and constantly change, as *human* products.

When Arrojo (see II, 2.1) approaches the question of interpretive communities, she always stresses their power and the importance of being aware of their power. As a teacher, one must know and obey certain rules, but can also help to shape them as society evolves. As students, learning about their interpretive communities is the key factor that will allow them to change them later. As translators, knowing our interpretive communities is a paramount prerequisite to translating anything because this is the only way to know what is acceptable, what works in our community. But as Arrojo explains (ARROJO 1986/2002: 78), knowing these tacit rules in no way means blindly following them, but rather having privileged access to them in order to help to carve their contours.

Another important element that Britto appears to disregard is the impact of the "fictions" he defends on translation. Whereas he asserts that *without* these fictions of stable meanings and authors' intentions any translation task is doomed, the exact opposite could be claimed (as addressed at the end of Part I): it is precisely *because of* these fictions that translation has been perceived as doomed to failure for over two thousand years. It seems evident that translators who claim that their translations are "faithful" and "transparent" (cf. BOHUNOVSKY 2001; KATAN 2009), or who, confronted with two different translations of the same text, cannot tell which one is "the right one" (see I, 3), do not refer to *fictions*. Nor is it the case when so-called translation critics mechanically point out the alleged mistakes of the translator, claiming that a given translation is inaccurate and unfaithful – as one so often sees in literary sections of newspapers and magazines.

I have attended two conferences by Britto, first in the I Jornada de Estudos da Tradução at the Universidade Federal do Paraná in 2006, and then in the Congresso Ibero-Americano de Tradução e Interpretação, in São Paulo in 2007 (for more on them, see III, 3.3). His stimulating words on the unquestionable and obvious importance of the author's intentions, on the stability of meanings and on the need to seek perfect equivalents can be said to have had a twofold impact on the crowd. They certainly offered some "relief", since they freed the audience from the apparent "constraints" imposed by the ever growing discussion about the instability of meanings and the power of ideologies. "Relief" very much in line with Fish's "theory hope", in the sense that through Britto's words one could again believe in a fully scientific, almost mathematical (see III, 3.3) translation theory. On the other hand, this feeling can only end up feeding the vicious circle already described in Part I, i.e. the hope for a fully scientific translation theory leads to its own impossibility, which in turn projects itself onto translation practice, making it, too, seem impossible; from practical impossibility and dissatisfaction "theory hope" emerges again, and on it goes.

Finally, the *necessary* fictions Britto defends do not seem so necessary when one thinks about their nature. To take the example of Britto's persona in the paper in question, why do translators have to *pretend* they believe in equivalence, author's intentions and stable meanings? Why *pretend*? Should simply *different interpretations* not suffice? Why *must* one *necessarily* say that one such translation is good or bad because it respects the author's intentions or the original text? Can one not simply say "this translation more or less matches the way I have read this text and therefore it works well"; or "I see other elements as more important than the ones privileged in this translation and therefore I do not find it a good translation"?

The answer to the questions asked above is not an easy one. It seems that our logocentric tradition, our tradition of private property and copyright simply does not allow us to assign as much importance to interpretation and bias as to original intentions, texts and authors. As Britto very clearly reveals through his main character in the last paper analysed in this section (cf. BRITTO 1995), he himself feels his work as a translator is doomed to failure if he cannot believe that his interpretation perfectly matches what the original author meant to say, what the words in the text *actually* mean. Relying on his own interpretation as an experienced and renowned translator, acclaimed writer and successful professor would not be enough, and this is how far our essentialist tradition goes.

It appears that Britto's resistance to so-called postmodern ideas lies in the fact that he sees structuralism as necessary, *useful* and legitimate, although he agrees with a number of poststructuralist positions. His attitude reflects the structure of our own society, in which it is simply easier and more *useful* to pretend that certain fictions (such as truth, impartiality and reason) are actually possible. In Britto's case, it is quite clear that he takes them as fictions, as his work inarguably indicates. In the case of other translation scholars and students, however, Britto's "fictions" may actually represent their strong beliefs and convictions. Indeed, this is probably the reason why a number of scholars and practitioners all around the world are still looking for miraculous translation models and solutions; translation critics are still listing translators' mistakes; translation students are still wondering about perfect equivalence and authors' intentions. In this sense, this game of "make believe" has drastic consequences; consequences which we can no longer afford to have. I will develop this argument further in IV, 3.

But let us remember Britto's *standpoint* here, and let us keep in mind that he defends these so-called fictions because they are more *useful*. Useful in the sense that they can be applied to practice more easily and directly – to translating, marking papers, teaching translation and proposing translation exercises. Surely most translation teachers would not like to admit that their correction criteria are rather subjective; that they, too, do not know for a fact what the author meant to say. Likewise, for practical reasons, translators would not like to admit that their work is not definitive, finished, stable – would authors? Here is what Michael Cunningham has to say about it (CUNNINGHAM 2010):

A novel, any novel, if it's any good, is not only a slightly disappointing translation of the novelist's grandest intentions, it is also the most finished draft he could come up with before he collapsed from exhaustion. It's all I can do not to go from bookstore to bookstore with a pen, grabbing my books from the shelves, crossing out certain lines I've come to regret and inserting better ones. For many of us, there is not what you could call a "definitive text."

It is interesting to note that a writer as celebrated as Michael Cunningham can admit that his texts are not "definitive" – why is it that so many translators cannot?

Whether this kind of *usefulness* that Britto advocates is honest and ethical is now beside the point – we will come back to this question in IV, 3. In any case, discussions such as the one proposed by Fish certainly do not address translation practice – they do not contribute to practice nor do they have to. In this sense, judging them by their *usefulness* seems not only pointless, but also diminishing. But Britto appears to be mostly committed to practice-oriented issues after all (as addressed in III, 1 and 2), so it is only natural that he should tend to look for usefulness in Fish's and Arrojo's propositions. If we remember Britto's analogy presented in I, 4, Britto, Fish and Arrojo are lost in a forest, and all Britto wants is to get home. In this sense, his idea of a translation theory clearly presupposes the marriage of theory and practice, as we observed in Part I. Precisely because theory must necessarily lead to a more efficient practice, must necessarily be useful as far as practice is concerned, contributions such as the ones by Arrojo and Fish seem somewhat irrelevant. Particularly regarding Fish's considerations about what makes a poem a poem, Britto should feel strongly about some of the literary critic's suggestions as Britto is a poet himself, and thus is fully entitled to perceive his own practice differently.

But let us now move on to the next section, dedicated precisely to Britto's papers on the work of the poetry translator in comparison to that of the poet – an issue so close to his heart for obvious reasons. This section will help to shed light on his notion of translation practice, thus greatly contributing to our debate here. It will also clarify a few of his reservations as for Fish's remarks discussed in the present section.

3.2 The Craft and Art of the Literary Translator

Poetic influence, in its first phase, is not to be distinguished from love, though it will shade soon enough in revisionary strife.

Let me reduce my argument to the hopelessly simplistic; poems I am saying, are neither about "subjects" nor about "themselves." They are necessarily about other poems; a poem is a response to a poem, as a poet is a response to a poet, or a person to his parent.

Harold Bloom

Being a poet and literary translator, Britto is one of the only Brazilian translation thinkers to investigate the processes involved in the work of the literary or poetry translator, mostly in comparison to the work of the writer or poet. Indeed, in his paper entitled "Tradução e Criação", or Translation and Creation (1999), he compares the work of a poet to that of a poetry translator because "a much discussed issue in the field of translation studies is the nature of the difference between an original and a translation" (BRITTO 1999: 239). If we look back at Part I, we will remember Britto's long quotation in which he complains that

Some theorists maintain, for example, that the notion of original in translation is a myth, that literary translators should enjoy the same status as the author of the allegedly "original" work (and in their texts the word "original" always comes between inverted commas); the inferior role of a translation when compared to the original text would be a product of capitalism, of chauvinism, of logocentrism, of the oppression imposed by European cultures on peripheral cultures, etc.[126]

126 In what follows (see the entire quotation on pages 51-53), Britto also maintains that "As a response, translators then quite rightly claim that these very same theorists who defend the impossibility of distinguishing between translation and original text, the impossibility of accessing the author's intentions, the radical relativity of all and every value judgement of a translation, these theorists, when faced with a translation of their own texts, behave in the exact same way as the logocentric, Eurocentric, phallocentric authors they criticise: they expect their translators to reproduce in their translations exactly what they meant to say, demanding the very same notion of

In this light, Britto sets out to investigate the processes of poetry translation and creation so as to identify differences and similarities between them, and hence determine whether the so-called poststructuralist argument whereby "the notion of original in translation is a myth" has any truth to it (BRITTO 1999: 239). In a long footnote he explains that he derives this "poststructuralist" argument from Venuti and Levine, and then quotes the following passages by both authors: "Neither the foreign text nor the translation is an original semantic unity; both are derivative and heterogeneous" (VENUTI 1992: 7) and "there are no originals, only translations" (LEVINE 1992: 83).

In the Introduction he summarises the so-called structuralist and post-structuralist views on the matter of "originals" and "translations", the former being as follows:

> *the poetic text is a unique new production of a unique, conscious and creative subject, based essentially on his or her own living experience, whereas a translated poem, like any other translation, is a text produced on the basis of an original in a different language, with the intention of serving in its stead* (BRITTO 1999: 241-242).

As for the poststructuralist point of view, we can read the following descript-tion in Britto's text: "any text *t* refers back to a multitude of previously existing texts and is referred to by an indefinite number of subsequent texts, in the same language or in other languages" (BRITTO 1999: 242). He quite predictably admits that "I believe that the poststructuralist position (...) runs into serious trouble when applied to translation studies (and, indeed, most other fields)" (BRITTO 1999: 241). Despite that, he concedes that poststructuralist thought "has had the merit of forcing all who are involved in the discussion of theoretical issues related to translation (...) to reconsider a number of concepts that had been accepted unquestioningly for a long time" (BRITTO 1999: 241). At any rate, he adopts the poststructuralist perspective but announces that, in spite of that, both the more traditional and the more postmodern reader will probably be satisfied with the conclusions at which he will arrive.

'fidelity' to the 'original' of which they so fiercely disapprove". This argument against poststructuralist thought is indeed recurring and "by no means knock-down" (NORRIS 1982/2002: 185). In other words, poststructuralist thinkers are accused of at once denying that language and texts can possess any objective features and expecting their own texts to be read correctly. This is exactly Abrams' argument is his 1978 "How to do Things with Texts". I will come back to this question in IV, 2.

What he does next is thoroughly analyse all the different drafts he wrote both of a translation (of Wallace Steven's "Sunday Morning") and a poem of his own ("Pessoana" – cf. BRITTO 1997: 87), looking for similarities and differences in both processes. He then explains that source text authors rely on their own feelings, background and knowledge when writing a text, whereas translators take their own feelings, background and knowledge into account as well, but always in combination with one specific source text.[127]

Britto eventually comes to the conclusion that both processes – of poetic translation and creation – are characterised by a movement towards autonomy and a movement towards approximation. In other words, "autonomy" refers to the fact that poets strive to distance themselves from their own literary influences, whereas translators strive to distance themselves from the constraints imposed by source language, source text or even other texts that may influence them. "Approximation", on the other hand, concerns the fact that poets welcome these literary influences and use them in their texts through intertextuality, whereas translators strive to stick to the source text when writing the target text. Britto finds that the movement towards autonomy is undoubtedly more conspicuous in creation, while in translation "the two movements are more or less balanced" (BRITTO 1999: 250-251). Nevertheless, both movements are found in both processes. He then asserts that

> *the first source, or original, has a* ***controlling effect*** *on translation: every time the translation strays too far from the original, confrontation with the original pulls it back home. But when a poem is being written, the first source has no controlling effect on the new text. Quite the contrary* (BRITTO 1999: 251 – his emphasis).

Indeed, this is exactly what he observes by analysing his own processes of translation and creation: his new drafts of the translation clearly denote a pendular movement from autonomy to approximation, mostly ending in approximation. Unlike the translation, the new drafts of his own poem, which stretched over a period of six years, denote a movement towards autonomy. Incidentally, this was the only reason why he needed nine drafts to finish his own poem, i.e. in order to be able to "find [some]thing that hadn't been said before by Fernando Pessoa" (BRITTO 1999: 249). Therefore, Britto speaks of the

127 For a discussion on the relationship between original authors and translators through a poststructuralist perspective, see Part II (particularly Section 2.3 and the last paper of Section 2.4).

translation process as one that involves a centripetal force, very much in line with the idea of a controlling effect exerted by the source text, the "centre" of the process. In contrast, the creation process is characterised by a "self-consciously" centrifugal force, whereby the poet struggles to "overcome" these alien influences (BRITTO 1999: 249).

By resorting to a few mathematical formulae, Britto arrives at the final conclusion that "as long as we have access to the succession of versions resulting in the production of a text of arrival and to its possible texts of departure (...), we may say that the text of arrival in question either is or is not a translation" (BRITTO 1999: 252). The criteria he establishes for this definition are as follows: (i) source and target text must be in different languages; (ii) source text must have been written before target text; and (iii) there is a "controlling relation" (BRITTO 1999: 252) between source text (and source text *only*) and target text, which in turn occurs through a series of interventions in the target text "with the effect of approximating" it to the source text (BRITTO 1999: 252). His formulae are presented in the Appendix as they were translated by Britto (see footnote 121 on page 219).

Through his model and formulae one should be able to unquestionably determine whether a text is a translation or an original, since however many similarities there may be between the two processes, they *are* different. So going back to the issue of the supposedly poststructuralist argument of the original as a myth in translation, Britto explains that "even if much of the post-structuralist argument is accepted, it is still possible to point to the existence of clear-cut differences between the acts of 'translating' and 'creating' – that is, between writing translations and writing pure and simple" (BRITTO 1999: 252).

The greatest contribution of this paper is the questioning of originality. Through his argumentation, Paulo Henriques Britto unveils that the production of both source and target texts is always strongly influenced by various literary forces – i.e. the literary experience and background of both poet and translator. Therefore, one cannot speak of pure originality since every piece of writing is indeed derivative, at least to a certain extent. By acknowledging that, one no longer classifies translations as minor, derivative work. The difference, in Britto's view, is that in translation this strong compelling force exerted by the source text is, to a certain degree, healthily and consciously dealt with by the translator. In creation, on the other hand, these external literary influences are rather more unconscious, imprecise and, to a certain extent, unwelcome, since poets usually want to achieve some "originality". In conclusion then, Britto maintains that the processes behind works of poetry translation and poetry creation may even be similar, but there are clear-cut differences between them.

However enlightening Britto's paper may be regarding the processes of poetry translation and creation, presenting a rather innovative comparative approach of the two activities, a few issues should call the readers' attention. Firstly, the *mathematical* conclusion he draws. At the beginning of the paper, he justifies the choice of his own translation and poem by claiming that "since my analysis will rely on the *subjectivity* of the author/translator and refer to possible *unconscious* or *semiconscious* associations, use of my own work makes things much easier" (BRITTO 1999: 241 – my emphasis). For the purposes of this book it is beside the point whether he would be able to consciously describe the unconscious or semiconscious processes going on in his own mind; in any case, the nature of one such description would certainly not be mathematical, thus not leading to mathematical conclusions. But yet another question seems even more relevant than that: why should one need to resort to mathematical formulae when dealing with such a subjective, abstract issue? Or why would one even want to *mathematically* determine the difference between original and translation?

As this part of Britto's work indicates (as will the next one in the following section), he appears to share the "theory hope" diagnosed in many scholars, practitioners and students (as we saw particularly in Part I), hence his need for exact, absolute formulae. Despite his acceptance of a number of poststructuralist ideas – as he admits in the paper in question, but also in the papers examined in the previous section – he still relies, or pretends to rely, on structuralist principles because they are allegedly more *useful*, more legitimate.

Britto's mathematical conclusions lead to a vicious circle already familiar to those involved in translation studies today. Even if one took his formula seriously, disregarding the fact that the notion of equivalence is not explained, and neither are the notions of different languages (would different media count, such as book and film, picture and text?) and text (written, oral, texts taken as literary, non-literary?), one would only find exceptions and blind spots. What happens when one takes an old text and updates it into the "same" language – such as Shakespeare into modern English or Camões into modern Portuguese? That would not be included in his formula, but is it not a "translation"? How about books made into films, such as *The Hours* (2002 – book by Michael

Cunningham 1998) or *Atonement* (2007 – book by Ian McEwan 2001)?[128] And what about situations in which target texts must stray from their source texts in order to fulfil a different function in the target culture, as is the case in the translation of CVs, advertisements and so many other specific text types? His formula would then have to be enhanced, expanded to include more cases; nevertheless, it would never be able to comprise all minor exceptions, all the subjective and abstract aspects involved in work with language. The elaboration of a new, all-embracing formula would have to be postponed to the distant future, when science would finally bring all subjectivity to an end.[129]

Another interesting issue in Britto's paper is the assumption that so-called poststructuralist thinkers "relativiz[e] the differences between 'author' and 'translator' and between 'original' and 'translation' almost to the point of abolishing them" (Britto 1999: 241). Once again there appears to be an element of radicalism in Britto's expectations towards allegedly poststructuralist perspectives that is simply not pertinent. As mentioned here a few times, it *is* part of the poststructuralist repertoire to question and criticise dichotomies such as signifier versus signified, subject versus object, original text versus translated text; what is *not* part of poststructuralist thought is to abolish these dichotomies.

When Suzanne Levine claims that "there are no originals, only translations" (see above – LEVINE 1992: 83), she is drawing a parallel between the concept of Ursprache (as discussed by Steiner in *After Babel*), of the myth of an original language, and the idea of an original text. She mentions Borges' "Pierre Menard, Autor del Quijote" (see II, 2.2) as an example of this questioning of "sources" and "origins". But again her point seems to be to call attention to the illusory character of what we name "original" rather than to completely do away with the idea of an original.

As for Venuti, the derivativeness and heterogeneity he advocates as ineluctable marks of both source and target texts can by no means make the boundaries between original and translation disappear – but rather make evident that these boundaries are not as clear-cut as Britto's own paper appears to illustrate. It is as Rodrigues puts it: "A crítica pós-moderna à oposição tradicional entre

128 Britto translated this book into Brazilian Portuguese in 2008 for Companhia das Letras (*Reparação*).

129 Britto does seem to restrict his conclusions to poetry translation, so some of my arguments here may be misplaced. But if one proposes such a scientific, rigid approach, one must also detail the applicability of one's approach. Why only poetry and not other text types as well? What distinguishes poems from other texts? If his formula is so comprehensive and refers to such clear-cut differences between translation and original, should it not be applicable to any text type?

original e tradução não implica supô-los iguais; vale apontar que ambos são heterogêneos e plurais, que ambos são tecidos de diferenças" or the postmodern critique of the traditional opposition between original and translation does not imply that they are the same; one must bear in mind that both are heterogeneous and plural, both are made up of differences (RODRIGUES 1999: 222). I will come back to Rodrigues' argumentation later in this section, once we have analysed another paper by Britto on the same subject.

In any case, again the question of standpoints is paramount, outweighing all other questions. In "Tradução e Criação", Britto seeks to clearly and unequivocally determine the features that distinguish original from translation – so clearly and unequivocally (scientifically?) that he must resort to mathematical language to convey his conclusions. He is most certainly not operating on a level of epistemological reflections on the differences and similarities between translation and original – as Venuti or Levine appear to be. Instead, he draws examples from his practical experience as a writer and translator and deducts from them mathematical formulae that should allow one to determine with absolute precision what a translation is and what an original text is. Notwithstanding his references to so-called poststructuralist perspectives, his concerns and his conclusions are clearly practice-oriented. And there is absolutely nothing wrong with this – it all makes sense actually, considering that Britto has translated over 100 books, that his field of specialisation is linguistics and that his affinities lie closer to structuralism. It makes sense that he should want to produce mathematical formulae to be applied to concrete cases of poetry translation.

But let us look into another paper of his, still concerning poetry creation and translation. Based on the opening lecture he gave at the Universidade Federal de Juiz de Fora at the start of the first term of 2008, Britto wrote the paper – which strikes one more as a personal account than as a paper – "Poesia: Criação e Tradução" (2008b),[130] or Poetry: Creation and Translation. As he explains in the first paragraph of his paper, his main objective is to examine the notion of "subject" in poetic creation (BRITTO 2008b: 11), which he does by first looking into his own experience as a poet and translator. Again he is motivated by allegedly poststructuralist perspectives on translation, which threaten, in his view, to completely get rid of the traditional opposition subject versus object.

130 Unlike most papers in this section, this one has not yet been translated by Britto, so all translations of this particular paper were done by me.

His autobiographical journey may be roughly summarised as follows. Britto's interest in poetry began when he was still a teenager. He became an avid reader then and, a bit later, started writing his own poems. Back then he thought he wrote poems because of his "need to express himself" ("necessidade de expressão" – BRITTO 2008b: 11), but often found that he did not know the feelings he wanted to express until he actually wrote them down. In addition to that, he found that some of his poems expressed feelings far different from the ones that he assumed he had. This experience made him perceive his poetry writing as a need to *build* feelings, to *build* a persona rather than to merely *express* pre-existent feelings (BRITTO 2008b: 11). In other words, it was his writing that determined his poetic personality and not the other way round.

But yet another discovery would change his perception of poetry even more drastically: the Tropicália movement in the 1960s, particularly through the anthology of critical texts about Brazilian Popular Music organised by Augusto de Campos and entitled *Balanço da Bossa* (as already mentioned in III, 1). According to Britto, this book opened his eyes to literary theory and showed him how little he actually knew about poetry. More importantly than that, the book made him realise that his technique (using traditional metre and rhymes) was long out-dated, having been replaced by more visual, iconic texts without any syntax – indeed, concrete poetry was the style of the moment. Furthermore, the Tropicália movement opened his eyes to the project of the poet, who was no longer engaged in expressing personal feelings, but rather was a part of a larger aesthetic project whereby each poet worked as an engineer building texts that expressed this collective mission – and here the influence of Brazilian poet João Cabral de Melo Neto was paramount (as addressed in III, 1).

These discoveries led Britto to an impasse: for him poetry was a means to manifest or build a subjective identity for himself, and yet the great poets of the time argued that poetry had nothing to do with subjectivity. Not surprisingly, this predicament made him stop writing poems for five or six years (BRITTO 2008b: 12), and he only took it up again because of translation. Indeed, by translating he did not have to resort to his own persona. As a translator, he could claim that all the subjectivity and traditional structures had been chosen by the poets and not by himself. Of course his choices of poet and his translation choices still helped him to shape his own poetic self, but always through other personae. He summarises this point as follows: "Traduzir poesia foi para mim uma maneira de dar continuidade a meu projeto de construção de uma personalidade para uso próprio, só que utilizando sujeitos líricos alheios para esse fim" or by translating poetry I was able to carry on with my project of building a personality for myself, but I used other personae to reach my goal (BRITTO 2008b: 13).

Based on the experiences briefly summarised above, Britto asserts that one could easily come to the conclusion that there is no difference whatsoever between writing and translating poetry. After all, both processes involve the construction of a persona, a poetic self, and this construction takes place in more or less the same way. Moreover, as he quite rightly points out (again), neither poetry translation nor poetry creation are fully original activities. As his personal experience illustrates, he only started *writing* poetry once he had *read* enough poetry; he was only able to start building his persona thanks to all the other personae he had built from his reading of other poets. In this sense, "originality", "subject", "author" and "meaning" would be nothing but "ideological reifications" (BRITTO 2008b: 13). Similarly, the persona behind a poem by Fernando Pessoa would be as artificial as the persona behind a translation of the same poem.[131]

Although Britto agrees with the conclusions described above, he asserts that this sense of similarity between translation and creation is unacceptable. He presents the two following motives for his objection: firstly because poets/translators like himself always clearly distinguish between the texts they create and the ones they translate.[132] Secondly because even those who claim that creation and translation are "the same" – or rather have the same *status* – tend to attribute translations to original authors (as we saw in III, 3.1, when Britto criticises Rosemary Arrojo for quoting Nietzsche in Portuguese and not mentioning the translator). And here again Britto resorts to the concept of "fiction". For him, the differences between translating and creating may be fictions, but necessary fictions (BRITTO 2008b: 13). He then sets out to find theoretical arguments to justify why these fictions are absolutely necessary.

131 Indeed, as Britto very opportunely points out, this is what one can read in Fernando Pessoa's famous poem "Autopsicografia", thoroughly analysed in LEAL (2007b). The poem can be read as follows: a poet, who has his or her own feelings, writes about these feelings in a poem. These two sets of feelings, however, are different. The reader, who has his or her own feelings at a certain point in time *as well*, reads the poem and builds yet different feelings. At the end it is as though there were four different sets of feelings at the same time: the poet's, the ones put down on paper, the reader's and the ones the reader reads in the poem, all different.

132 See II, 2.3, where Lacan's translator, Forrester, claims in an indented quotation that one such distinction is by no means clear.

The first argument he presents concerns the fact that while so-called original texts are based on an indefinite number of previous texts, translations are based on one text only (BRITTO 2008b: 14):

um texto que resulta de um processo de criação poética remete a todos os textos poéticos lidos pelo autor e de algum modo citados ou glosados no seu poema; por outro lado, o texto resultante de um processo de tradução remete basicamente a um texto específico e definido, dito "original".

[a text derived from a process of poetic creation relates to all poetic texts read by the author and, in a way, quoted or glossed in the poem; a text derived from a process of translation, on the other hand, relates to one specific text, the so-called "original".]

The second argument regards the function of a persona and the relationship it bears to both poet and translator. Britto argues that when translators build a persona to translate a poem, the sole function of this persona is to get the translation done. In contrast, when poets build a persona in one of their own poems, this persona plays an important role in their lives as well; it becomes a part of the poet's personality. He concedes that the personae built by him in translation also influence his personality and help to build his own persona, the one in his poems. He nonetheless maintains that the difference between the two – i.e. the personae derived from translation and the one derived from creation – is quite clear, mainly because the former relates to his personal biography far less than the latter (BRITTO 2008b: 14). And here he quotes Foucault (part of the quotation below) to show precisely what it is that he disagrees with. In Foucault we read the following:

these aspects of an individual which we designate as making him an author are only a projection, in more or less psychologizing terms, of the operations we force texts to undergo, the connections we make, the traits we establish as pertinent, the continuities we recognize, or the exclusions we practice (FOUCAULT 1979: 150).

Unlike Foucault, Britto contends that "O poeta lírico elabora uma determinada persona poética para fins utilitários que vão além da literatura: ela é uma parte importante – em certos casos, vital – da sua personalidade", or poets build a given poetic persona for purposes that go beyond literature; indeed, this persona is an important, sometimes vital part of their personality (BRITTO 2008b: 15).

Even though Paulo Henriques Britto admits that personae and authors do not coincide perfectly, he stresses that the relationship between the two is tight-knit. He briefly examines Romanticism, and how back then poetic personae were perceived as the most authentic expression of their authors' personalities. Indeed, as Britto explains, it was not until the end of the 19th century that people became aware of how artificial, how construed the personae projected by literary works were. He claims that Modernism followed this trend; yet, he feels that it inherited from Romanticism the importance given to the interaction between the poetic persona and the author's self-image. As examples of this important interaction he mentions T. S. Eliot and Ezra Pound. For him, much of J. Alfred Prufrock corresponds to concrete facts of Eliot's life (BRITTO 2008b: 15). Similarly, he reads the same voice in Pound's *Cantos*, *The ABC of Reading* and his personal letters.

Before closing his paper, Britto goes back to Concrete Poetry, to João Cabral de Melo Neto. Today, decades later, he finds it easier to see João Cabral's subjectivity hidden behind the negation of subjectivity; the poet/engineer built his own personality by denying he was doing so. Britto then ends his paper by defending the concepts of "author", "persona" and "subject" as necessary, despite their construed or even fictitious nature. He concedes that in times when even the notion of "subject" is severely questioned, it is only natural to challenge the boundaries between "translator" and "author"; but still he feels that these boundaries are necessary for the reasons presented in his paper.

One has to agree with Britto's general reasoning here, but there are a few specific points that could be questioned. Going back to the first argument he brings forth to support why the "fiction" that distinguishes translation and creation is necessary (see previous page), it does not seem that the number of texts to which original and translation relate is a decisive factor. Translations must, indeed, bear a special relation to one specific text; otherwise they would not be called translations. However, this is by no means the same as saying that no texts other than the source text influence translators when they are translating. Britto himself presents a very interesting example in the article analysed earlier in this section (cf. BRITTO 1999: 246). As he was translating Stevens' "Sunday Morning", a verse by Fernando Pessoa "popped" into his mind. Later he noticed the similarities between the two poems – Stevens' and Pessoa's – and therefore justified this unconscious interference. Indeed, any reader of poetry who has tried to translate poetry certainly feels these apparently inexplicable influences, like hidden echoes in our minds. In this sense, of course translation usually takes one text strictly into account; yet, the influence exerted by other texts cannot be denied.

Correspondingly, poets sometimes identify very specific sources that influence the creation of a certain original work. Incidentally, Britto mentions an instance of that as well in the aforementioned paper (cf. Britto 1999). As he is attempting to write his own poem, "Pessoana" (cf. BRITTO 1997) – which in turn is a poem about Fernando Pessoa's traditional theme of the dissociation between knowledge and feeling – he realises that the two main sources of influence are indeed a poem by Pessoa, "Autopsicografia", and a poem by Sá de Miranda, "Cantiga VII" (BRITTO 1999: 249-250). Therefore, Britto's first argument in 2008b about the number and nature of the influences working on poetic translation and creation appears to be questionable. What seems more solid is his argument in the 1999 paper about the movements of "autonomy" and "approximation" that characterise both processes in terms of how each one handles these influences.

The second argument Britto presented in this last paper analysed above (BRITTO 2008b) in favour of a "fictitious" difference between poetry translation and creation concerns the importance of the translator's persona and the poet's persona for the artist, and here two aspects seem particularly relevant. Firstly, readers – or the public in general, if we think of artists in a broader sense – tend to have difficulty separating an artist's personal life from his or her work. So it is only natural for members of the public who are acquainted with aspects of an artist's personal life to find similarities and references in the artist's work. One cannot forget one's knowledge about a certain subject, say, for example, the biography of an author, when appreciating his or her work. Nonetheless, just how meaningful this interaction between the persona and the actual person can be is questionable, and therefore should probably not be taken as seriously as Britto suggests.

A curious example of this "dangerous" interaction is what happened with Chico Buarque, one of the most prominent Brazilian artists of all times, when his 1998 album *As Cidades* came out. Chico is well-known in Brazil for his brilliant lyrics during the military dictatorship that started in the 1960s. "Apesar de Você" is an example of a song that, despite its potentially offensive content, was miraculously not censored because the text was actually very ambiguous and subtle. Back then, Chico became a symbol of resistance, of political criticism and force, characteristics that forced him into exile in Italy in 1969. His career flourished in the 1980s and 1990s as well, with political issues playing a less important role in his songs as the dictatorship gradually became less harsh until it was replaced by a democratic regime. During these decades, Chico was an open supporter of the Partido dos Trabalhadores (PT – Workers' Party), with Luís Inácio Lula da Silva at its forefront.

When *As Cidades* came out in 1998, one of the songs, entitled "Injuriado", called the attention of the media. That year president Fenando Henrique Cardoso had been re-elected (the elections took place in 1997), beating Lula already in the first round. Not surprisingly (and actually quite rightly), many critics interpreted that particular song as another of Chico's typically political songs, with double meanings and hidden criticism, this time addressed to the newly elected president. However, Chico was not in the least bit pleased with what was being published about his song, and finally clarified that it was not addressed to anyone in particular, and that he had nothing to say to President Fernando Henrique.

Here one could ask a number of "what if" questions, such as "what if Chico, for whatever reason, had not been able to reply to the comments of the critics?", "what if Chico was lying and actually had meant to criticise Cardoso?", "what if he had not meant to criticise Cardoso, but after reading the critiques decided it was actually an interesting reading of his song, seeing as he was a supporter of PT?" The point of this example is to show that the interaction between persona and person is not that clear-cut. Perhaps what we take as actual truths of the person (as the critics did in this example with Chico Buarque, or as Britto does with Eliot and Pound) based on history, letters and facts, are nothing but nuances of the persona.[133] Of course this is an endless discussion in literary theory, one that cannot be addressed at length here. In any case, many of the considerations drawn here were inspired in Antoine Compagnon's *Literature, Theory and Common Sense* (1998/2004: 29-68 – translated by Carol Cosman).

133 In 2010, I attended the exhibition of Frida Kahlo's paintings at the Bank Austria Kunstforum in Vienna ("Frida Kahlo Retrospektive" – from September to December 2010). As a school teacher arrived with a number of pupils and started talking about the paintings, I could not help but pay attention to what they were saying. Indeed, the entire analysis of the paintings in question was built based directly on Frida's biography – her relationship with Diego Rivera, her miscarriage in the USA, her relationship with Trotsky, and so on. However much one such analysis may glamorise Frida's life, I wonder whether it does not end up impoverishing her works, her creativity and her art.

The second issue I would like to put forth is somewhat similar to the first one. Surely the long and constant process of construction of Britto's persona in his own poems is extremely important to him as a person. It probably does work as one of the key elements of his personality, his personal history. To the reader, however, this is of little importance. This is more of a personal experience to the poet than something to be taken into account by readers, translators or critics. It would be naive to assume that a careful reader, examining a poet's letters, biographies and facts surrounding his or her life, and then comparing all that to the artist's works, would come up with an accurate outline of the artist's personality and his or her poetic persona, or would come up with something with which the artist would actually agree. Of course studies like that may shed light on the artist's oeuvre, may assist translators and critics in their work, may enrich and enlarge the perception that a community has of a given artist.[134] But these contributions have to be taken with a pinch of salt, as construes, as products of perceptions delimited by time and context. Overvaluing the interaction between persona and author for the sake of his or her feelings seems exaggerated.

And here again we have the question of *standpoints*: Britto is certainly not very fond of what Fish (see III, 3.1) or Foucault, for example, say about the nature of literary texts or the (lack of) importance of the author because he, as a poet, perceives the entire process differently, from a different side. But then again Fish and Foucault are not speaking of the same as Britto – they are not concerned with the genesis of literary texts, as Britto appears to be, but rather are looking at literary texts from the point of view of their reception.

It also seems pertinent to point out yet again that it is *not* part of the poststructuralist reflection to suppose translation and original to be the same. Britto's paper (2008b) appears to be another example – as we saw in the previous section – of exaggerated criticism of the reception of deconstruction in Brazil. It is true that a number of poststructuralist theorists question the alleged superiority of original texts when compared to translated texts, and suggest their *status* should be the analogous. However, the theorists who oppose poststructuralism take this literally and accuse their academic opponents of defending a relationship of absolute equality between translation and original – hence their need to defend the difference between original and translation. As already mentioned, Britto is one of those theorists who oppose poststruc-

134 And, by the way, what I am doing with Arrojo's and Britto's life and work here may be said to go along these lines.

turalist perspectives and expect them to be extreme, to break with structuralism completely. Although in the paper in question (BRITTO 2008b) he does not make any reference to which theorists would claim translation and original – or author and translator – are the same, it might be another case of radical, unjustified criticism.

Regarding the relationship between original and translation within the reception of deconstruction in Brazil, it is as Cristina RODRIGUES (1999: 221-224) explains it. Postmodern thought forces one to question the hierarchical opposition traditionally attributed to translation and original for a number of reasons. As we saw in I, 5 and 6, this questioning starts with the very notion of sign, which can no longer be understood as a stable, constant relationship between signifier and signified. This new perspective makes the idea that meanings can simply be decoded from texts and that translating consists of transferring these decoded meanings no longer viable. Hence both translation and original are made up of conventional and arbitrary signs that are subject to interpretations and contexts. In this sense, the relationship between original and translation cannot by any means be described either as a relation of equivalence or as a relation of opposition. Instead, as Rodrigues defends, the nature of this complex relationship is of supplementation and mutual dependence (RODRIGUES 1999: 221-222).

Nevertheless, this does not mean that translators are free to interpret and translate texts however they want, neither does it allow for translation and original to be perceived as the same thing. This is another common misconception unfairly (and in a very ill-informed fashion) attributed to deconstruction, as Norris asserts (and here we could substitute "critics" with "translators"):

> *But if one thing is certain it is the fact that these readings [of Paul de Man and Jacques Derrida] bear no resemblance to the popular idea of deconstruction as a species of out-an-out hermeneutical license, a pretext for critics to indulge any kind of whimsical, free-willing of 'creative' commentary that happens to take their fancy* (NORRIS 1982/2002: 136).[135]

Precisely because signs are conventional and arbitrary, translators – and writers or speakers in general, for that matter – will struggle to learn and, to some

135 I will come back to this quotation and the issue of deconstruction being an "out-an-out hermeneutical license" in Part IV.

extent, conform to the rules valid in their community. This way, the author, rather than being the central element around which the translation process revolves, becomes one of the elements to be taken into consideration during the translation process, one of the elements within this complex activity of meaning production.

As for the difference between original and translation, Cristina Rodrigues quotes Culler and his ideas about enunciation and the system of distinction. As she explains, Culler stresses the importance of "difference" when it comes to signs, since it is precisely the differences between signs that allow us to define a sign:

> *The fact that these (...) signifieds are arbitrary divisions of a continuum means that they are not autonomous entities (...). They are members of a system and are defined by their relations to the other members of that system. If I am to explain to someone the meaning of* ***stream*** *I must discuss the difference between a stream and a river, a stream and a rivulet, and so on* (CULLER 1976/1986: 34 – his emphasis).

Culler then remarks that "signifieds (...) are nothing but the product or result of a system of distinctions" (CULLER 1976/1986: 35), and quotes Saussure's famous *Course*: "Their [the signifieds'] most precise characteristic is that they are what the others are not" (apud CULLER 1976/1986: 36). In this light, in order to create meaning – say, in a conversation or written text – one must employ a number of distinctions. As Rodrigues puts it, the first system of distinctions to be used in an utterance is that of the signifiers, employed to convey a certain form that can then be interpreted within the system of distinctions of the signifieds (RODRIGUES 1999: 222-223). This refers, of course, to a single linguistic system, a single language. But Rodrigues suggests we go a step further and look into Derrida's work ("la traduction pratique la différence entre signifié et signifiant" – DERRIDA 1972: 31) and apply Culler's ideas to translation. In other words, Cristina Rodrigues speaks of translating as using a system of distinctions – the source language – to produce a form that may be interpreted within another system of distinctions – the target language (RODRIGUES 1999: 223).

* * *

In the present section, Britto's opinion on the differences between the work of the poet and that of the poetry translator reinforce some of the views already discussed in the previous section. Even though he concedes that poststructuralist thought has contributed towards a more critical understanding of concepts that before used to be taken as absolute truths (such as the notions of equivalence, subject versus object, intention, amongst many others), he chooses to defend some of these former "absolute truths" for the sake of practicality and usefulness. And let us not forget that this behaviour is certainly very much in line with his work as a translator and translation professor whose interests tend towards translation practice. Britto appears to be reluctant to do away with these fictions because he cannot imagine his routine as a poet, translator and professor without them. But should he have to do away with them in the first place?

I will address this question in IV, 3. We shall now move to the third and last section of the present chapter, dedicated to what is perhaps the most controversial part of Britto's work: his model of translation criticism. For now, let us keep this question in mind as we advance into this last, longer section.

3.3 Translation Criticism

> *No centro [da relação transferencial] há um leitor ou tradutor que inevitavelmente chega tarde para ser o autor do texto que deseja e que, por isso mesmo, precisa tentar usurpar esse cobiçado lugar autoral num ato parricida pelo qual paga com uma culpa muitas vezes "invisível" e com uma declaração explícita de auto-apagamento e de respeito irrestrito ao original.*
>
> Rosemary Arrojo

In 2006, Paulo Henriques Britto took part in the I Jornada de Estudos da Tradução at the Universidade Federal do Paraná. Consisting mostly of translation and literature students and professors, the audience was dazzled by his dynamic model of poetry translation criticism. The reaction was not dissimilar to when he presented the same conference in the 2007 Congresso Ibero-Americano de Tradução e Interpretação, in São Paulo.

Claiming that the recent dissemination of poststructuralist ideas in the field of translation studies has turned translation criticism into a dubious, obscure field, Britto proposes a return to "positivist" principles. Guided by his fierce disapproval of postulates such as "one cannot say, in absolute terms, that a certain translation is better than another one", or "there is no such thing as author's intentions", just to mention two, he devises a model whereby source poem and target poem are analysed mathematically – in a seeming attempt to bring the subjectivity and polysemy commonly associated with the study of poetry to an end. Although he concedes that one can no longer speak of author's intentions or stable meanings as one did, say, in the mid-20th century, he maintains that, as a literary translator himself, he needs these allegedly logocentric postulates to be able to carry through his work. From his point of view, if one does not assume that the words in the text one is about to translate have stable meanings, intentionally and carefully chosen by its author, then one cannot read, never mind translate.

In these two conferences the object of Britto's analysis was two different Brazilian translations of John Donne's "To his Mistress Going to Bed" – one by Paulo Vizioli and the other one by Augusto de Campos.[136] His objective was to show that, in spite of poststructuralist reservations, one can mathematically determine what translation is better, more faithful to Donne's original. In the latter half of 2006, a paper with the contents of these conferences came out in the journal *Terceira Margem* ("Fidelidade em Tradução Poética: O Caso Donne" or Fidelity in Poetry Translation: The Donne Case). Before we analyse it, though, let us first look at another paper on the same issue that came out earlier, as a chapter of the book *As Margens da Tradução*, entitled "Para uma Avaliação Mais Objetiva das Traduções de Poesia" – or Towards a More Objective Evaluation of Poetic Translation.[137]

Therefore, in the present section Britto's works will be presented in chronological order since in many of these papers he makes references to his previous works. However, there shall be one small exception, namely a 2000 newspaper

---

136 The polemic surrounding these translations is also the object of Arrojo's paper "A que são Fiéis Tradutores e Críticos de Tradução? Paulo Vizioli e Nelson Ascher Discutem John Donne" (ARROJO 1993: 15-26), thoroughly analysed in II, 2.3. Britto's choice of this particular example might not have been random, as Arrojo's point in her paper is precisely that one *cannot* determine in absolute terms which translation is better – the very opposite of Britto's argument here.

137 This text was also originally a conference, delivered in Portuguese at the seminar As Margens da Tradução, at Universidade do Estado do Rio de Janeiro in 2001, and published a year later in a homonymous collection of papers. However, as most other papers in this section, this one was available in English (translated by the author himself) on Britto's webpage (see footnote 121 on page 219).

review that will be addressed after the presentation of this first 2002 paper. After the 2000 review, we will look into two papers published in 2006 and, finally, another paper published in 2008.

In his 2002 "Para uma Avaliação Mais Objetiva das Traduções de Poesia", Britto undertakes a similar project to the one in the 2006 conference and paper, but takes into consideration his own translation of a poem by Elizabeth Bishop, namely "Shampoo". Both Bishop's poem and Britto's translation can be found in the Appendix in the way they were presented in Britto's essay – including all analyses of formal aspects.

At the beginning of the paper, Britto announces that "[e]valuating the translation of poetry is a complex and delicate task. Poetic texts deal with language on all its levels – semantic, syntactic, phonetic, and rhythmic, among others" (BRITTO 2002: 54). Therefore his objective is to "sketch out a methodology for the evaluation of poetic translations, which requires a systematic examination of the different levels of language involved in the poem" (BRITTO 2002: 54). Given the paramount importance of the concept of "correspondence" in his methodology, Paulo Henriques Britto explains that he must first define it "with some degree of precision" (BRITTO 2002: 54).

In order to do so, he takes the poetic feature metre. In summary, what he does is propose "levels of correspondence" between the English iambic pentameter ("with a pyrrhic substitution in the fourth foot and a spondaic substitution in the fifth" – BRITTO 2002: 55) and a Portuguese metric structure. On the first and strongest "level of correspondence" we have "an identical sequence of stressed and unstressed syllables in Portuguese" (BRITTO 2002: 56). On the second and thus looser level of correspondence, the English iambic pentameter "corresponds" to the Portuguese iambic decasyllable. In other words, the English and Portuguese lines "have the same length and the same general stress pattern, but there is no exact one-to-one correspondence between the syllables of the two lines" (BRITTO 2002: 56). Next, on the third level of correspondence the only feature that remains the same is the number of syllables, with the English pentameter corresponding to the Portuguese decasyllable. Finally, the fourth level of correspondence is a generalisation, i.e. a long line in English corresponds to a long line in Portuguese.

Based upon these "levels of correspondence", Britto then defines the second most important principle in his methodology, namely the concept of "loss" (BRITTO 2002: 56): "the weaker the sense in which 'correspondence' is taken in a given translation – that is, the higher the level of generality on which it operates – the greater the loss". Unfortunately he does not explain how one such scheme would apply to what he calls the "semantic" or "phonetic level", for example, and simply states that "the same reasoning can be applied to other formal elements" (BRITTO 2002: 57).

Having defined the two pillars of his model, "levels of correspondence" and "loss", Britto proceeds to the analysis of Bishop's "Shampoo", identifying what elements – amongst rhyme pattern, metre, rhythmic structure and alliterations – seem more prominent. In his view, "when we translate a poem, we should attempt to reproduce those elements that are most regular in the original, since they are likely to be more conspicuous in the source language" (BRITTO 2002: 59). In the poem in question, the rhyme scheme, together with a "clearly (...) calculated" passage in the second stanza (BRITTO 2002: 59), strikes him as the most conspicuous features, and this is what he has in mind when carrying out his translation.

After presenting his translation, Britto finally begins the "evaluation of poetic translation", as the title of the article suggests. One by one, he looks into rhyme, metre, etc., and compares it to the original, striving to numerically establish the level of correspondence he achieved. Up until this point – more than half way through the paper – he does not account for possible language differences that would make such a strict notion of one-to-one correspondence not only impossible, but also frequently undesirable. However, when he gets to metre, he does mention that even though it was not a regular feature in the original poem, he tried to "reproduce" the distribution of short and long lines. And then he quickly remarks that "Portuguese words ten[d] to be longer than English words" (BRITTO 2002: 62), which results in the lines in his poem being longer than the ones in the original poem. Despite this "shift", he does not account for the level of correspondence achieved in this case: "strict correspondence" would mean to have the exact same number of syllables both in original and translation even when one of the languages contains longer words and a longer sentence-pattern? Or would that sort of adaptation to the nature of the languages in question count as "level one correspondence"?

Unfortunately, Britto does not elaborate much on these questions and moves on to the alliterations, when he admits that "this is one case where the sort of analysis I propose runs into difficulties: when the poetic resource in the source language does not exist, or exists in quite a different form, or is much less prominent, in the target language" (BRITTO 2002: 62). Here he explains that the use of alliteration is far more abundant in English than in Portuguese, and that in the poem in question, the distribution of alliteration is not at all regular. He is nonetheless quite surprised to find that "there is an abundance of sibilants in the first three lines of the first stanza of the translation, much as in the original" (BRITTO 2002: 62), and goes on to admit that "this result (...) is a product either of chance or of the translator's unconscious" (BRITTO 2002: 62). Concluding this issue Britto asserts that "this may perhaps be taken as a sort of compensation for the losses on other levels, an effect created in Portuguese to make up for those in the original which could not be reproduced" (BRITTO 2002: 62). I will address these various issues below, at the end of the presentation of this paper.

Relying on a traditional form-content structure, Britto proceeds to the analysis of the content but warns the reader that, due to space constraints, he will only examine the first three verses of the poem, as well as one particularly important verse in the second stanza. He begins his content analysis by asserting that "the first level of correspondence, a literal translation, is rarely possible in poetic translation". Once again it is not clear whether he means that this first level does not exist and must therefore be substituted by something else, or whether ideally literal translations should be used where possible. His analysis, though, indicates that literal translations should indeed be used where possible, as he proudly comments that in the first verses "[t]he rate of correspondence is relatively high: only the translation of 'spreading' and 'engordam' cannot be said to be literal" (BRITTO 2002: 63). As for the particularly prominent verse in the second stanza, where two long Latinate terms contrast with everyday, Anglo-Saxon vocabulary, Britto concedes that for obvious reasons one-to-one correspondence is simply impossible in this case, and then moves on to present other possibilities, lower in the correspondence hierarchy.

In his conclusion, he claims that before examining the "degree of loss" between original and translation, one must first ask three questions, as follows: "how relevant is the feature in the original?", "is the maximum degree of correspondence feasible?", and "how desirable is an exact match?" (BRITTO 2002: 66). Indeed, this attitude resembles that of his predecessors, particularly Mounin and Newmark (see I, 5). The answer to these questions will depend on a number of factors, three of which being the languages in question, the purpose of the translation and the way the translator interprets the text, i.e. what elements strike him/her as prominent and what do not. These factors, in turn, are inevitably intertwined and greatly depend on the interpretive community to which the translator belongs. In any case, not only are these three points highly specific, varying from text to text, context to context, translation brief to translation brief, and translator to translator, but they are also subjective, meaning that no mathematical formulae or absolute levels of correspondence apply. And yet, despite his resemblance to Mounin or Newmark, for instance, he closes his paper on a cheerful note:

I have presented here no more than a preliminary sketch of a method; many details remain to be spelled out. Still, I believe that my proposal amounts to a promising way to arrive at less subjectivistic forms of evaluating poetic translation, relying on more objective data and making it possible to quantify value-judgements expressed through such concepts as "correspondence" and "loss" (BRITTO 2002: 66).

I will comment on Britto's conclusions at the end of the present section.

At any rate, in the same paper, Britto makes reference to a very short newspaper article that is in fact a translation critique. The article is called "Uma Forma Humilde" (A Humble Form) and was originally published in Portuguese in *Jornal de Resenhas* no. 60, in *Folha de São Paulo* (11th March 2000) as a review of Paulo Vizioli's[138] translation of Oscar Wilde's *The Ballad of the Reading Gaol*. Most newspaper reviews of translations consist of lists of "mistakes" (that is, if they mention the translation at all), and Britto's certainly is an exception as he actually compliments aspects of Vizioli's translation, as well as other translations done by him. According to Britto, "Vizioli reproduces quite faithfully (...) formal features of the original: he even reproduces the internal rhymes that are sometimes to be heard between the second and the fourth beat of the four-foot lines" (BRITTO 2000: 7). The "mistake", however, seems to lie in Vizioli's choice of metre:

But the stanza he uses – alexandrines alternating with hexasyllables – is simply not a standard form in Portuguese poetry, and certainly not in our popular poetry. And nothing could be more alien to the folksiness of the English ballad form than the general effect of Vizioli's ponderous lines, more strictly metered than the original (BRITTO 2000: 7).

138 Paulo Vizioli is the same translator whose translations of John Donne are the object of a paper by ARROJO (1993) and another one by BRITTO (2006b) – see page 257, as well as II, 2.3 and footnote 136 on page 251.

Britto believes that even though there is no "close Portuguese equivalent" to the English form chosen by Wilde (in this case Vizioli's choice being the closest possible), there are "better solutions" in Brazilian Portuguese. Nevertheless, although these "better solutions" are "formally not equivalent to the English ballad, [they] may be said to correspond to it functionally" (BRITTO 2000: 7). Moreover, adopting one of Britto's "better solutions" would "force the translation to take daring liberties (...). But such a translation, though less faithful to the letter of the original, would certainly be closer to its spirit" (BRITTO 2000: 7). And in order to justify why Vizioli's choice is "less than successful" and "disappointing", Britto relies on the "tone sought by Wilde", on the "spirit" of the poem, as well as on the "special meaning" the ballad *obviously* had to Wilde, given his circumstances. In Britto's opinion, choosing the ballad, the "humblest of all English poetic forms", was Wilde's way of dealing with his imprisonment and subsequent loss of reputation (BRITTO 2000: 7).

As Britto himself remarks, Vizioli's choice of metre was the closest "correspondent" in Brazilian Portuguese to the English ballad, thus achieving Britto's first "level of correspondence" (cf. BRITTO 2002). In this case, however, a third or even fourth level of correspondence would be more desirable than the first level, according to Britto. In other words, by looking at his model, one can contradictorily remark that a "less faithful" translation is a "better solution" than a "more faithful" translation. Indeed, is this not the case of most translations in general?

The considerations made by Britto are undoubtedly relevant and pertinent. Even without having studied Vizioli's translation carefully, one might agree that a different choice of metre might have been an interesting decision. On the other hand, it is obvious that to Vizioli, creating a similar metric and rhyme scheme was far more important than adapting the metre to a more Brazilian form. Furthermore, as Britto himself points out, Vizioli is well-known for his sophisticated, sometimes nearly "literal" translations (see footnote 138 on page 255). This is Vizioli's style and, like it or not, it has made him a successful, renowned translator. From this point of view, it seems that a more ethical translation review should mention Vizioli's style and comment on how this translation is another instance of it. Suggesting a new metre should be absolutely no problem, but instead of justifying it for being closer to the "spirit" of the poem (did only Britto have exclusive access to this "spirit"?) or to "the tone sought by Wilde", Britto could simply explain that he, personally, would have translated it differently, that he feels Wilde's life at the time influenced his choice of metre, that he interprets this choice as the most meaningful characteristic of the poem, etc. These justifications – particularly coming from such an experienced translator and brilliant poet – are certainly far more plausible than saying that this is what Wilde "meant" to say. Not to mention the fact that using Wilde's intentions and biography as justifications for translation decisions make Vizioli sound unfairly ill-informed, as though Britto had taken one look at the poem and had figured out what Wilde had meant to say, whereas Vizioli had simply been incapable of doing the same.

But let us now go back to that 2006 paper on the two Brazilian translations of Donne's "To His Mistress Going to Bed", "Fidelidade em Tradução Poética: O Caso Donne"[139] (BRITTO 2006b – see footnotes 136 and 138 on pages 251 and 255, respectively). This paper is actually at once a declared critique of another essay by Arrojo ("A que são Fiéis Tradutores e Críticos de Tradução? Paulo Vizioli e Nelson Ascher Discutem John Donne" – see pages 155-159) and a step further on his model of criticism of poetry translation. At the beginning of the paper he announces that poststructuralist thought currently exerts a certain influence on the field of translation studies in Brazil, and then associates this perspective with what he calls contemporary scepticism in its most radical version ("versão contemporânea do ceticismo em sua versão mais radical" – BRITTO 2006b: 239). Largely based on Arrojo's abovementioned paper, Britto argues that this poststructuralist view holds that

> *os textos não possuem significados estáveis que correspondam a intenções que seus autores tivessem (...); só temos acesso a nossas próprias leituras dos textos. Assim, quando dizemos que uma dada tradução é fiel ao original, estamos dizendo apenas que nossa leitura dessa tradução é fiel à nossa leitura do original; nada podemos afirmar sobre os textos em si. Entende-se, pois, que não haja consenso absoluto a respeito dos méritos relativos de duas traduções de um dado texto (...)* (BRITTO 2006b: 239).
>
> [texts do not possess stable meanings that correspond to their authors' intentions (...); all we have access to is our own reading. In this sense, when we say that a given translation is faithful to the original, all we are saying is that our reading of the translation is faithful to our reading of the original; we can say nothing about the texts themselves. One may

139 As this paper has not been translated by Britto yet, all translations presented here were done by me.

conclude, then, that there can be no consensus in terms of the relative merits of two translations of the same text (...).]

Therefore in his view, poststructuralist thinkers would argue that translation criticism is pointless because it is always marked by the critic's personal perspectives, preferences, bias and subjectivity. Not satisfied with this conclusion, however, he draws a parallel between the Humanities and the Natural Sciences, claiming that if Arrojo's propositions are valid for translation studies, they should be equally valid for all other fields as well. He then mentions string theory[140] as an example of an issue around which there is no consensus in the academic community, but despite that is very much discussed and vehemently defended by a few members of these circles. Thus, his conclusion is that there is no reason why one should *not* apply rational arguments and a method of test and refutation of hypotheses to the comparative analysis of poems. In fact, poststructuralist objections aside, he contends that these so-called scientific methods can and should be employed in all fields of knowledge (BRITTO 2006b: 141).

Paulo Henriques Britto then turns to the translations of Donne's poem and announces right away that the vast majority of the translators and translation scholars he knows tend to favour Campos' translation over Vizioli's. In view of this fact, his main goal is to objectively and mathematically determine why one such preference exists by firstly carefully comparing each translation to Donne's original, and then the translations between themselves.

Before he sets out to analyse all three poems, nonetheless, he anticipates again the so-called poststructuralist answer to the question of why there tends to be a preference for Campos' translation. He sarcastically speaks of Fish's "interpretive communities" (see III, 3.1) and says that Campos probably reflected the values and tacit rules of his community more successfully than Vizioli did, hence more members of the community identify with Campos' style than with Vizioli's. He speaks of cheap Foucauldianism ["foucaultianismo vulgar" – BRITTO 2006b: 141] to justify why Campos might have more disciples, more institutional power than Vizioli. Finally he asks, rather ironically, "Não seria possível arriscar a hipótese de que o respeito que ele [Campos] goza entre os tradutores de poesia se deva a características de seu trabalho de tradu-

140 For more on this, Britto recommends the reading of Jim Holt's 2006 paper "Unstrung", published in *The New Yorker* (2nd October 2006) and available at http://www.newyorker.com/printables/critics/061002crat_atlarge (last accessed in April 2006).

tor, características essas que podemos depreender através da análise de suas traduções?", or could we not perhaps consider the hypothesis that Campos may enjoy a certain status amongst poetry translators because of characteristics of his work as a translator, which we can derive from the analysis of his translations? (BRITTO 2006b: 141)

In the following pages, he presents both translations side by side with the original, using different symbols (boldface, italics, underline) to mark the following features: (i) words in the original that do not seem to have been transposed in the translation (omissions), (ii) passages whose meaning has been altered significantly, and (iii) words and expressions that have absolutely no correspondent in the original (additions) (BRITTO 2006b: 144). Verse by verse, Britto carefully and meticulously examines both poems in terms of form and content, often choosing what translator handled what situation better, numerically keeping track of gains and losses.

Unsurprisingly, at the end of his analysis he comes to the conclusion that Campos' translation can be mathematically proven to be superior to Vizioli's. A key factor appears to have greatly influenced Britto's judgment, namely the translators' choice of metre. Donne used heroic couplets in his poem, whose closest equivalent in Portuguese would be the so-called decasyllables – since both contain 10 syllables. Campos did opt for the decasyllables; Vizioli, in contrast, chose to use dodecasyllables, which is perceived by Britto as a legitimate strategy because Portuguese word and sentence patterns tend be longer in comparison to English. In this light, choosing a lengthier verse pattern should allow Vizioli to avoid omissions in his translation, whereas Campos might have trouble conveying all ideas present in the original with a shorter choice of metre. In reality, however, Britto found that both translations contain a similar number of omissions (13 in Vizioli's and 15 in Campos'), and that the number of additions is more than seven times as high in Vizioli's (15) as it is in Campos' translation (two).

In his conclusion, Britto admits that there may be readers who disagree with specific steps of his analysis, whose opinions may vary as far as particular aspects of his method are concerned – such as "I would not consider this to be an omission" or "this does not strike me as a *significant* meaning alteration", etc. Despite this, however, he argues that just because there can be no absolute consensus on something, it does not mean that absolute dissension should automatically ensue (BRITTO 2006b: 52). As far as criticism of poetry translation is concerned, Britto argues that (BRITTO 2006b: 52) *é possível utilizar o discurso racional para fazer avaliações e tecer considerações em torno de traduções, fazendo referência a certas propriedades dos textos traduzidos com relação às quais há um certo grau de acordo entre um bom número de pessoas envolvidas na atividade de traduzir. Dadas duas traduções de um mesmo texto, A e B, cotejem-se A e B com o original e uma com a outra, linha a linha, sílaba a sílaba, examinando e pesando as diferenças, para se chegar a uma conclusão baseada em fatos (...).*

[it is possible to use rational discourse to assess the quality of translations, making reference to particular textual properties about which there is a high level of agreement amongst translation specialists. Faced with two translations, A and B, of the same text, one can compare them to the original and then to each other – line by line, syllable by syllable, pondering the differences – so as to reach a conclusion based on facts (…).]

Britto then once again praises the scientific method – with its falsifiable hypotheses, careful data analysis and formulation of rational arguments – as the most suited method both for the Natural Sciences and the Humanities; he goes even further and suggests that this is the *only* way to build knowledge – "É assim – e *só* assim – que se constrói o conhecimento" (BRITTO 2006b: 52 – my emphasis).

As in the previous papers analysed in this chapter so far, Britto's objections to poststructuralist thought on translation appear to arise mostly from the two aspects already mentioned here, i.e. his keenness on theories that are *useful* as far as practice is concerned, and his expectation of so-called poststructuralist perspectives to be radical. These two elements mark his attempts to write translation theory as well, as it is quite evident that his utmost concern is to systematise translation practice – very much in line with his standpoint as a literary translator. From this point of view, poststructuralist thought has very little to contribute, as it is very rightly not deemed *useful* by Britto. These questions have come up a few times so far and will still be very much present in the two final papers that compose this chapter. I will return to them at the end of the chapter, as well as in Part IV.

In another 2006 paper, called "Correspondências Estruturais em Tradução Poética"[141] (or Structural Correspondence in Poetry Translation), Britto revisits the topic of poetry translation criticism. Again he opens his paper by asking how one can determine whether the formal elements of an original poem and those of its translation correspond to one another (BRITTO 2006a: 53). He takes it for granted that the reader is acquainted with his 2002 paper on the same topic, and adds that he intends to particularly investigate how this "poetic transposition" takes place when the elements in question are not a part of the "conventional repertoire" of formal resources of a given language, but rather are specific to a certain poet or poem (BRITTO 2006a: 53).

He takes another poem by Elizabeth Bishop ("Cirque d'Hiver"[142]) as his object of study, together with his own translation of it, and carries out a typically linguistic analysis of original poem and translation, dividing the texts into form and content. As in his 2002 paper, Britto establishes what elements appear to stand out in the original poem, and then sets out to skilfully recreate them in his translation. Indeed, his conclusions go back to this initial objective (BRITTO 2006a: 64):

ao trabalhar com um poema que utiliza uma forma regular ad hoc, é necessário antes de mais nada fazer uma avaliação dos diferentes recursos formais em jogo e de sua inter-relação com os aspectos semânticos. Como é impossível recriar em outro idioma de modo exaustivo todas as características de forma e significado de um poema, essa análise detalhada servirá para que se possa determinar quais elementos devem ser reproduzidos da maneira mais fiel possível, por serem os mais importantes.

[when working on a poem with an ad hoc regular form, one must first and foremost examine the different poetic resources in question, as well as the relationship they bear to the semantic aspects. Since exhaustively recreating all and every aspect of the form and content of a poem is impossible, this detailed analysis shall help to determine which elements are the most important ones and hence must be translated as faithfully as possible.]

141 The translated version of this article is not available on Britto's webpage, so I will translate the Portuguese quotations into English myself.

142 As in Britto (2002), the analysis of this poem and its translation can be found in the Appendix.

As for the feasibility of this task, he concedes that "naturalmente, nem mesmo essa meta mais modesta será atingida por completo (...). Em tradução, temos sempre de nos contentar em atingir apenas em parte a meta almejada" or of course not even this very humble objective can be completely fulfilled (...). In translation, only partially fulfilling the desired objective is already enough (BRITTO 2006a: 64).

Despite his careful, detailed analysis and interesting translation, Britto does not take the discussion much further than he did in his 2002 paper. He seems to run into the same difficulties by saying that the "most important" features of a poem must be translated "faithfully" (see indented quotation above). As already pointed out previously, what is perceived as "most important" by some may be perceived as less important by others. Moreover, as Britto himself admits, strict fidelity is mostly not desirable in poetry translation (nor is it in translation in general, for that matter), so what does "as faithful as possible" mean exactly? As in the previous papers, his analysis appears to indicate that "faithful" means "literal"; nevertheless, there are cases in which a lower "level of correspondence" is more effective than "level one correspondence" (cf. BRITTO 2002). So what is "faithful" after all?

It is precisely the notion of "faithfulness" or "fidelity" that Britto discusses in another paper, entitled "A Tradução para o Português do Metro de Balada Inglês" (2008a)[143] or The Brazilian Portuguese Translation of the English Ballad. In this text he completely disregards his model based on "levels of correspondence" and "gains and losses" (cf. BRITTO 2002) and simply shows, in a very direct and absorbing way, how strict levels of correspondence are often *not* desirable when it comes to poetry translation.

The object of this paper is again the English ballad (cf. BRITTO 2000), this time in a poem by Emily Dickinson. He starts the paper by addressing the concept of "fidelity" (BRITTO 2008a: 25 – his emphasis):

Em tradução de poesia, o desejo de ser fiel à forma do original pode nos levar a adotar soluções insatisfatórias. Muitas vezes a noção de fidelidade não deve ser entendida como reprodução de uma forma poética de um idioma num outro, não só porque nem sempre tal coisa é possível, mas também porque o ***significado*** *de uma determinada forma no idioma de origem pode não ser o mesmo no idioma para o qual se está traduzindo.*

143 The translation of this particular paper is also not available on Paulo Britto's personal webpage, so all translations here were done by me.

[When it comes to translating poetry, trying to remain faithful to the form of the original text may lead to unsatisfactory solutions. The notion of fidelity must not be understood as the mere reproduction of the poetic form of one language into another language, not only because this is not always possible, but also because the *meaning* of a particular structure in the original language may differ from that of the target language.]

Britto goes on to thoroughly analyse first the structure of the English ballad and then the way Dickinson used it in her poem. He later turns to the translation by asking the following key question: "o que devemos fazer ao traduzir um poema em metro de balada se quisermos ser fiéis à forma do original?", or when translating a ballad, what should one do in order to remain faithful to the form of the original text? (BRITTO 2008a: 27). The first answer that springs to mind, Britto says, is to find the "closest formal equivalent" possible in the target language (BRITTO 2008a: 27), in this case six and eight-syllable verses. But as he very aptly points out, a poem with six and eight-syllable verses would not strike a Portuguese-speaking reader as a typical poetic form, unlike the English ballad, which in turn would immediately be recognised as a typical poetic form. Britto's way out of the dilemma is the following: one should look for a typical poetic form in Portuguese that, though not strictly equivalent to the English ballad, evokes similar meanings. For him, ballads are a symbol of simplicity and spontaneity, a poetic form that contrasts with the complexity normally associated with poetry. In this case specifically, Britto sets out to translate a ballad by Dickinson (J870)[144] whose main characteristic, from his point of view, is precisely the contrast between form and content. According to him, this is the main feature that he has in mind when writing his translation.

Britto concludes his paper by saying that even though he identified a number of "losses" comparing original and translation, he is satisfied with the form he has chosen (BRITTO 2008a: 32). He also quite rightly points out that other poetic forms – including the option with six and eight-syllable verses – might have worked well, too. What he means to emphasise, however, is the fact that there may be functional correspondence between poetic forms that are not strictly analogous in terms of structure ("pode haver correspondência funcion-

144 Britto uses Thomas H. Johnson's numbering system. In his paper, Britto presents original and translation side-by-side in a table, where a number of poetic features are analysed. This table is available in the Appendix.

al entre formas poéticas que não sejam estritamente análogas do ponto de vista estrutural" – BRITTO 2008: 32).

* * *

This last section about Britto's works on literary criticism unveils a rather contradictory model of poetry translation and criticism of poetry translation. While on the one hand his considerations on the work (mostly *his* work) of the literary translator are highly interesting and enlightening, bringing numerous contributions that could only be made by someone as experienced as he is, on the other hand his attempts to formalise such knowledge and experience into formulae and rigid precepts is not without its difficulties. The voice that appears to stand out, particularly in the papers of this last section, is the voice of the poet and translator, and not that of the academic.

In these papers one can clearly see the combination of the two factors surrounding the concepts of theory and practice analysed in Part I, namely a predominantly essentialist tendency and a practice-oriented standpoint. And both these factors are revealed mostly through his choice of method, very much in line with the "traditional" sciences. The use of mathematical language and inflexible rules lends an air of "science" to his reflections, and so does the exclusively practice-oriented character, the "utilitarian purpose" (see I, 2) of these same reflections. His works analysed here seem to distinctly convey his aspirations as an academic-practitioner: Britto strives to scrutinise and systematise his own practice as a translator so as to devise general rules with a view to helping other translators and translation teachers to do their jobs more easily, more efficiently. In this light, and remembering the allegory Britto so often mentions in his talks (see I, 4), all he wants is to get home, and he certainly does his utmost to help others to get home, too.

Surely if Britto were to compile all these papers into one work, one model of criticism of poetry translation, he would very competently flesh out many of the issues that here, in these papers, appear to only have had their surfaces scratched. In this sense, perhaps what comes across as inadvertent incoherences – such as the issue of whether one should seek level one correspondence or not – may be nothing but a matter of formulation, of clearer writing. In addition to that, he may have come to regret a few of these theoretical propositions made in the late 1990s and early 2000s, which is perfectly natural. But still it is important to point out that the expectations of the reader faced with one such model, one such theory that promises to systematise poetry translation are indeed quite high. As numerous voices to which we listened in Part I exemplified, many of Britto's readers would be eager for watertight, clear-cut categories and absolute rules, and hence they would not tolerate the smallest incoherence or illogicality. As addressed in Part I a few times, they would envision the perfect marriage of theory and practice – as Britto seems to do as well. Any flaw in this direct application of theory to practice and their divorce would immediately follow, feeding back the vicious circle of "theory hope", as well as the "suspicious feeling" (see I, 3) many practitioners and students are said to have as far as translation theory is concerned (cf. BENEDETTI & SOBRAL 2003/2007: 92-93).

Going back to the expectations sketched at the end of III, 2 in terms of Britto's standpoint and his interests and motivations in relation to translation theory, one can say that these expectations were indeed pertinent, at least to a certain extent. Britto's academic papers do convey a sense of "utilitarian purpose" (see I, 2), of direct applicability to translation practice, very much in line with his own work as a practitioner and translation professor. His keenness on linguistics is also quite noticeable in his papers, not only because of his predominantly scientific, rigorous methods and aspirations, but also in view of the countless typical linguistic categories of which he makes use with no further questioning or criticism – such as "semantic level", "phonetic level", "form" and "content", amongst others. And it is also through his scientific method and his interest in linguistics that his essentialist tendencies are revealed, his ambitions to bring subjectivity and abstractedness to an end as far as poetry translation is concerned. His attempt, it seems, is indeed to make jobs like his – translating, criticising translations, teaching translation, correcting translations, devising translation exercises, etc. – easier, more systematised, objective and automatic.

On the other hand, his eagerness to dialogue with thinkers like Arrojo, Fish and Foucault, for example, appears to denote a different kind of interest, at least in a very incipient fashion. Unlike him, these thinkers do not tend to have practice-oriented issues as their top preoccupations, and their notion of theory seems more akin to that notion, discussed in Part I, of a stimulus for reflection with no direct practical purpose in mind. It is true that he then criticises or discards some these thinkers' ideas precisely for their lack of *usefulness* as far as practice is concerned, but it is in these moments of philosophical debate and confrontation that Britto's voice as an academic speaks the loudest and contributes the most.

In the next and final part of this book, I will address the validity and the place of theories such as Britto's within the field of translation studies. Additionally we will look into some of the questions raised in the present chapter, as follows: (i) to what extent the *usefulness* Britto repeatedly advocates when criticising theories is sound, (ii) to what extent his determined defence of certain concepts as *fictions* (such as equivalence, author's intentions, loss, etc.,) is ethical in translation studies, (iii) why scholars like Britto (and others, as addressed in I, 3 and II, 2) expect fierce radicalism of poststructuralist theories, immediately discarding them on these grounds, and (iv) to what extent the methods of the Natural Sciences are suited to the Humanities, to translation studies (a question raised in Part I as well).

Part IV: Closing Remarks to an Open Debate

You see, pure unity or pure multiplicity (...) is a synonym of death.

Jacques Derrida

So far innumerable crucial issues have arisen in the present book, issues that do not solely concern translation studies in Brazil, but rather affect our basic notions of theory, practice, higher education, professionalism and research. Not to mention many other, more specific issues surrounding the nature and impact of poststructuralist thought in the Humanities, along with the contours of translation studies as a discipline, its institutionalisation and its methods. As it has been made clear thus far, I have no pretensions to providing universally applicable answers to the questions asked in Parts I, II and III, nor do I intend to promote the reconciliation of the antagonistic sides involved in the various debates proposed here. In fact, by bringing these issues up again and commenting on them I will somewhat contribute to the already patent heterogeneity that appears to prevail in this book.

Looking back at the previous pages, it becomes clear that the issues that need to be addressed here are intertwined to such an extent that one can hardly imagine them separately. For this reason, I ask for the readers' understanding as regards the structure of the present part – a structure that, like translation, seems to obstinately resist systematisation. Four chapters will make up Part IV, as follows. Chapter 1 will be devoted to a few analogies and anecdotes – some already mentioned, others new – that illustrate particularly the following questions: what makes translators professional and what is the role of higher education. The reflections carried out in this first chapter will also brush on the subject of the aim and contours of translation theory, an issue that will nevertheless be addressed at length in Chapter 3.

Chapter 2 will be dedicated to poststructuralist thought and deconstruction. The objective is to clarify three main common misconceptions surrounding these poststructuralist tendencies, namely that they are (i) overly radical and (ii) pessimistic, and that (iii) they make way for people to indulge in random, inexplicable behaviour under the pretext of their subjectivity, unconscious or whatever they might fancy mentioning. As it will be made clear, beneath these misconceptions is a strong unwillingness to give in to, but mostly a profound ignorance of, poststructuralist thought. In this sense, the chapter will also address some of the criticism – often unjustified – of which these perspectives have been the target.

Next, Chapter 3 will approach translation studies more specifically, discussing, amongst other topics, the issue of its institutionalisation both as it happened in Europe and as it might happen in Brazil; the question of interdisciplinarity in translation studies; whether there should be clear "utilitarian purposes" in translation studies, i.e. whether research should be exclusively applied and theory should be useful; as well as the issue of the methods of the Natural Sciences and their applicability to translation studies. This chapter will also address some of Britto's claims in terms of his attack to poststructuralist perspectives for their lack of usefulness, and in terms of his defence of certain concepts (such as equivalence and author's intentions) as *necessary* fictions.

Finally, in Chapter 4 we will investigate the impact of poststructuralist thought in translation studies and the future of translation studies. Here, the notion of the "savage horde" proposed by Freud (see II, 2.3) will be crucial. Are we, in translation studies, a savage horde, too? And if so, should we ideally eventually reach homogeneity and unanimity? This will be the leitmotif of the chapter, exploring the issue of global, universal translation theories and the wish for reconciliation. Accordingly, in this chapter we will address the question of whether the negligible impact of poststructuralist tendencies in translation studies should remain negligible.

The main goal of Part IV is to show how all these issues are interrelated and, to a certain extent, inseparable. Additionally, I intend to stimulate reflection and contribute to the various debates that have been the subject of so many academic events and publications in translation studies (but not exclusively) in the past few years. As already stated, the intention is not to arrive at definite answers; at best, I hope to inspire the reader to reformulate certain questions, or at least to see the same questions in a different light.

1. Translation, Higher Education, Professionalism: Analogies and Anecdotes

Our language, even if we are pleased to speak it, has already substituted too many articulations for too many accents, it has lost life and warmth, it is already eaten by writing. Its accentuated features have been gnawed through by consonants.

Jacques Derrida (translated by Gayatri C. Spivak)

I, 4 opened with that forest analogy proposed by Britto (see Part III). This analogy illustrated the question of standpoints, i.e. whether one wants go get home as quickly as possible, or whether one is speaking of one's understanding and perception of the world. It would be impossible to argue in favour of the former or the latter; the point is simply that they are different, circumscribing utterly disparate, irreconcilable standpoints – at least momentarily. The question is what higher education should be about – or, formulated in narrower terms, what should be the point of translator and interpreter training, what makes translators and interpreters professional. And here another analogy, this time proposed by Wagner in the aforementioned *Can Theory Help Translators?* (CHESTERMAN & WAGNER 2002: 4-5), comes in handy.

Wagner uses an interesting analogy about the training of medical students to address the question of translator training – which, in her view, should be practice-oriented, primarily fulfilling the needs of the market. She explains that a doctor, faced with a malaria epidemic, must be able to rapidly and efficiently diagnose the disease and treat it successfully. For her, relying solely on "probabilistic laws" and the "observation and imitation of what other professionals do" would not suffice (CHESTERMAN & WAGNER 2002: 4). Instead, this doctor should have proper prescriptive training, whereby s/he is taught to apply the techniques and strategies devised by his/her predecessors. Wagner goes further and says that

How can we translators lay claim to professional status, and assert ourselves as professional and not charlatans, if our research scientists fight shy of real-world problems and the advice that would help us to solve them? There seem to be no clear guidelines on how to select people for

translator training, how to assess a translation, how to specify the purpose of a translation, how to measure and thus ensure reader satisfaction (CHESTERMAN & WAGNER 2002: 4).

But what Wagner seems to forget is that *someone* must first *devise* diagnostic strategies and a treatment to diseases like malaria. *Someone* must be able to do *research* and think *beyond* existing methods for medicine to break new grounds. *Someone* needs to be able to then teach incoming doctors and stimulate them to go further than their precursors have managed to. If doctors were trained merely to mechanically apply what their predecessors achieved in the past, who would be able to develop new treatments or improve current techniques?

But more importantly than that, let us consider in the last quotation above the underlying notion of language and translation implied by Wagner's medicine allegory, as well as the feasibility of her demands. Wagner appears to believe that going about translating a text or treating a patient is somewhat analogous. Like doctors that should immediately be able to tell what illness is afflicting a patient and what treatment will be most effective, translators should be able to look at texts and determine the most adequate translation. One such skill and awareness should come initially from training: doctors learn about the typical symptoms of a disease together with the most effective treatments available. If they find an infection, for instance, caused by particular bacteria, they will know that a certain type of antibiotics will probably inhibit the growth and hopefully even destroy those bacteria. Similarly, translators should be able to deal with languages in the same objective and scientific fashion: faced with a particular text, a particular group of words, they should know what group of words in the target language best corresponds to it. Indeed, they should have been given "clear guidelines" to "assess a translation", to "specify the purpose of a translation" and to "measure and thus ensure reader satisfaction" (CHESTERMAN & WAGNER 2002: 4) just like doctors have "clear guidelines" to "assess" clinical states and "ensure" healing.

Well, but if the objectivity of medicine is already highly questionable, are Wagner's expectations as regards language and translation not too farfetched? After all, curing ailments is far from purely objective. We go to numerous doctors until we feel that we agree with how they propose to go about treating us. Faced with the *same* symptoms, the *same* test results and the *same* complaints, doctors manage to come up with *different* opinions and suggest *differ-*

ent treatments. So much for the strict objectivity of their training. Also, medicine will work *differently* in different patients or even in the same patients in different moments of their lives. Some might experience side effects, others might feel no improvement at all, others might even develop additional problems as a result of the treatment. It all depends on an endless series of factors – some of which are quite palpable, like age, weight, eating habits; others, in contrast, remain in the mysterious realms of abstractedness and subjectivity, such as moods, feelings, one's outlook on the treatment, along with several others that we probably lack words to name. And this is without going into placebo research or alternative medicine at all and keeping to allopathic medicine.

Are there patients who have had such a matter-of-fact experience treating health problems like Wagner suggests in her analogy? There always seems to be personal history in the way, always individual empathies and antipathies, always unforeseeable effects and strange developments. However much we may want to believe that medicine is flawless and purely objective, it is highly questionable that Wagner herself does not have her own anecdotes to tell about how illusory these beliefs are. But let us go back to translation studies. Why should we expect it to be able to function in a strictly objective manner? Why should we hope translator training to fulfil all and every market need? If, like doctors, translators will face highly changeable settings constantly subject to external influences and personal preferences, how can we expect their training to prepare them to mechanically deal with *every single possible* text or situation that might come up?

These objectivity aspirations on the part of Wagner (and numerous others involved with translation) entail a strong "theory hope", and this hope is unrealistic. As mentioned here repeatedly, "theory hope" drives people to believe in a kind of ascetic pre-babelic state (as Arrojo would argue), where languages and meanings are exact and constant, and human beings are capable of full detachment from the world around them. As these ambitions do not in the least bit "match" reality, these "hopers" are stimulated to hope further, regarding their "current" state of chaos and "probabilistic laws" with scepticism and envisaging a brighter, more disciplined future. And "theory hope" is certainly present in different areas, too. Let us not forget the much sought-after correlation between wave and particle in physics, for example – the dream of a unified theory of matter and energy that would transcend Heisenberg's uncertainty principle. It is just that translation, because of its peculiar nature, tends to lay bare in a more obvious way the discrepancies between its "reality" and these scientistic hopes.

Of course translator/interpreter training and translation theory should not completely disregard the market; of course translation strategies and other so-called practice-oriented methods should be a part of the curricula of translation courses (see IV, 3). However, we must first acknowledge the limited and narrow character of these strategies and methods, so that we do not expect them to be able to comprehend every single aspect of the profession, both in the present and in the future, everywhere in the world. In other words, beyond the possibilities they open, all strategies and methods have limitations. Of course they can be applied in many situations, but cannot be taken as universal and infallible. Faced with a highly changeable scenario, translators had far better work on their awareness of the environment surrounding them than on minute (but not "probabilistic") laws and strategies to be applied to *a* particular text in *a* particular situation.

And here the question of the purpose of higher education enters the scene again (see I, 2). It seems that any professional had better invest in a broader understanding of his/her area, its history, the values that inform it and its relationship with cognate areas. This kind of knowledge appears to be more fitting to equip pupils with the tools that they need to perform their jobs with autonomy and to help to carve the contours of their jobs in the future. So let us to go back to Aristotle and rethink the whole idea of an education. Should we be taught to do or to think? Well, certainly both, though it seems easier to come up with actions and strategies based on awareness and reflection, than to depart from mechanical actions and achieve greater awareness. Surely those taught how to think will probably be more versatile, autonomous and enterprising when it comes to performing a job than those taught to objectively apply strategies to mock situations without much reflection. It is like in Britto's forest analogy: those who would advocate the illusory character of cardinal points know how to use them to their benefit; on the other hand, I am not sure whether those who strictly want to get home have ever stopped to think about the construed nature of cardinal points.

Let us remember the example of self-taught translator Vera Pereira quoted in I, 3. She claims not to be able to tell which of two translations of the same text is *the* correct one, and she reluctantly feels suspicious of the authority of the original author who, in her view, sometimes suggests translations that are inadequate. Having received no training whatsoever, she does not claim to miss objective strategies – like Wagner – but rather the *understanding* of her role in society, of how language and translations work, and of the role of the original author. In this case, translator training should be able to provide her with *awareness* of these issues, with a source of *reflection*. This broader outlook on her job and area is what would make her a professional translator in the full sense of the term. And let me emphasise here that this awareness is not restricted to knowledge of translation strategies; this awareness would most likely not even affect the way she actually translates as it is not supposed to be useful in practical terms.

Vera Pereira can be perceived – to use yet another allegory – as a competent musician who has never learnt music theory. She can read music and perform beautifully, but she does not understand the relationship between the tones and chords she plays. She probably memorised them individually and learnt how to *use* them, but she does not know how they are formed, where they come from, how they work. Studying theory now would probably not have any impact whatsoever on her playing, for she plays well as it is, she is experienced. What it would do is enhance her *understanding* of her practice, her *awareness* of what she does.

Let us remember another medical analogy, this time proposed by Mona Baker (see I, 5). She very rightly claims that without theory and theorisation translators will "never be seen as anything but witch doctors and faith healers" (BAKER 1992: 4). In other words, she defends institutionalisation and professionalisation as ways to legitimise and raise the status of our profession. But if these measures – theorisation, institutionalisation and professionalisation – lead to "theory hope", the status of the profession will *not* be raised. As already pointed out in different parts of this book, theorisation rooted in essentialism invariably leads to "theory hope", which inevitably sets a vicious circle in motion whereby the present "chaos" is discarded in the name of a scientific future. And this scientific future entails transparent renderings, translator's invisibility, full objectivity and mathematical faithfulness – and universalism, of course. Well, if we are investing all our hopes in the possibility of full invisibility and objectivity, in other words, if we expect our "science" to advance to a level in which it will deal with language and subjectivity mathematically, how can we expect the status of our profession to be raised? If it were actually possible for our subjectivities and interventions to be fully erased, if we could reduce our conduct to mechanical and fully objective actions, than we might as well be replaced by computers.

Mona Baker is certainly right when she advocates that theorisation will lead to the elevation of the translator's status in society. Nevertheless, we appear to be barking up the wrong tree here since this theorisation should not lead to the *denial* of our presence as translators, but rather the very *opposite*. I very much agree with Arrojo when she asserts that it is through the questions raised by poststructuralist thought that we will achieve the much sought-after goal of raising the status of our profession (see Part II). After all, it is poststructuralist reflection that enables us to legitimise our *interference*, our *presence* and our *visibility*. In any case, I will come back to the question of poststructuralist thought in translation in the next chapters. For now let us simply bear the idea in mind that this theorisation defended by Baker in her analogy – and by Wagner in hers, too – should be more awareness-raising than practice-oriented. And even when it is practice-oriented – as some of it is bound to be – it must be awareness-raising as well. Let us not forget Derrida's analogy of the faithful guardian stood at the university gates: yes, we shall embrace the market needs and accommodate them as well as possible in our courses, but let us be critical of them, let us dare influence them, let us constantly deconstruct them in our classes (cf. DERRIDA 1983: 20 –see I, 2).

If businesses would rather have incoming employees fresh out of university with highly specific training, so that they do not need to provide these newcomers with additional instruction, we might as well openly and directly affiliate universities to businesses, as is already the case in many areas (see I, 2).[145]. How willing are we to bow to these demands? Should we not perhaps slowly "show" the market that they are better off with students who were taught how to *think critically*, than with students who were trained to *perform* one specific task? Deconstructing these demands in class and understanding the dynamics and motivations underneath them might prove far more fruitful than obediently following them and striving to fulfil them through universal (and not "probabilistic") laws.

In one of the chapters of Jean-François Lyotard's 1979 *The Postmodern Condition*, entitled "Education and its Legitimation through Performativity" (LYOTARD 1979/2006: 92-99 – translated by Geoffrey Bennington and Brian Massumi), he defends a similar idea. Lyotard describes how the "performativity principle" is now crucial as far as higher education is concerned. In other

145 A sad sign of this trend is the recent (December 2013) abolition of the Austrian Federal Ministry of Science and Research, whose remit has been taken over by the Ministry of Economics and Family Affairs.

words, performing a task is key, i.e. applying techniques and technologies with a view to their use in the market. So much so that the chief question now asked by "the professionalist student, the State, or institutions of higher education" is indeed "what *use* is it?", which often translates as "it is saleable?" and "is it efficient?",[146], he claims (LYOTARD 1979/2006: 95 – my emphasis). This leads to a notion of competence that is strictly "operational",[147], as well as to a notion of education as the mere "reproduction of skills" (LYOTARD 1979/2006: 96). But then, Lyotard maintains,

If education must not only provide for the reproduction of skills, but also for their progress, then it follows that the transmission of knowledge should not be limited to the transmission of information, but should include training in all of the procedures that can increase one's ability to connect the fields jealously guarded from one another by the traditional organization of knowledge (LYOTARD 1979/2006: 96).[148]

In other words, if we look back at Wagner's analogy, we should try and go beyond the level of the "reproduction of skills", beyond the scopes of questions like "what *use* is it?" or "is it *efficient*?"

Moreover, to remain within the medical analogy, successful, acclaimed doctors and translators do not achieve their success strictly because of their objectivity, neutrality and *performance*, but rather because of their *subjectivity* and *interference*. We tend to like or dislike doctors mostly for subjective factors – whether they confirm our expectations as for our treatment, whether they inspire our trust, whether they understand our claims in a way that satisfies us, whether we feel safe in their hands and find comfort in their judgement, whether they share our values. If everything were so black-and-white, so ob-

146 If we look back at Part I, particularly at Chapter 1, we will see that my entire experience with higher education has been marked by questions such as these.

147 In I, 2, there was a quotation by Derrida in which he made a strikingly similar remark. Firstly he asked: "does the university have as its essential mission that of producing professional competencies, which may sometimes be external to the university?", and then "is the task of the university to ensure within itself – and under what conditions – the reproduction of professional competence by preparing professors for pedagogy and for research who have respect for a certain code?" (DERRIDA 1983: 17). As a reply, he asserted that one such acceptance of "the reproduction of professional competencies" should "be accompanied at least by a movement of suspicion, even of rejection with respect to the professionalization of the university in these two senses, and especially in the first, which regulates university life according to the supply and demand of the marketplace and according to a purely technical ideal of competence" (DERRIDA 1983: 17).

148 I will come back to the question of interdisciplinarity implied in this quotation in IV, 3.

jective and universal; in other words, if everything came down to performing efficiently, why would we seek different doctors and second opinions?

But let us now advance into the second half of this chapter. Two similar anecdotes, one involving me and another one involving Brazilian scholar Maria Paula Frota, will help to develop this notion of translation theory and translator training further. As these anecdotes concern the teaching of literary translation, I will address some of the questions asked at the end of Part III, apropos of Britto's model of poetry translation. This issue will then be raised again in IV, 3.

As I was doing a course on Irish Poetry during the last year of my BA in English and Portuguese at the Universidade Federal do Paraná, I had the chance to work with a very experienced literary translator and writer, undoubtedly one of the best teachers I have ever had. Towards the end of the course we were given the assignment to translate Sebastian Barry's poem "The Smell":[149]

Buttery, vanilla...
Tobias, moments old,
my nose against his head,
sniffing, trying to make out

the soft word of his smell
couched in memories
of clover maybe,
crushed, inhaled

but elusive, also
the cow's milk years ago
skited by Auntie Anne
for a joke from the udder,

that odour of inside skin,
interior, private,
lightless, no language,
something lovely of that order.

149 Barry's book from which this poem was taken appears to be unavailable. As he has published two volumes of poetry, *The Water Colourist* (1983) and *The Rhetorical Town* (1985) – along with numerous volumes of fiction and drama – and as I am sure that this poem is not in *The Water Colourist*, I suppose it is in *The Rhetorical Town*.

I remember discussing the poem in class on the day we were supposed to hand our versions in. The key question that the teacher asked us on that occasion was who "Tobias" was. The class was divided: some believed Tobias was a new-born calf because of the cow in the third stanza and all the other animal-like descriptions. Others, in contrast, felt it was a new-born baby, maybe the persona's child. Personally, I did not find the question extremely relevant precisely because it seemed to me that this mixture between animals and humans was what made the poem so special. Nevertheless, deep down I liked to think Tobias was a calf for my own personal reasons.

When I was growing up I used to go to my aunt's farm in Minas Gerais, a state in central Brazil. Since it was a proper farm, going there was always a wonderful experience. We got to play with animals, drink freshly squeezed milk and eat fresh vegetables. One particular day, however, really marked me forever. I must have been about four or five. A cow was about to give birth, and the children were allowed to watch as my uncle assisted it. All I can remember is the strong smell of milk I always felt around the stable (indeed, "buttery, vanilla"), as well as the indescribable display of love between cow and calf. I hardly recall any blood – which is quite strange – but rather the way the cow put its "nose against [the calf's] head / sniffing, trying to make out / the soft word of his smell", and how the calf patiently looked for the udder and sucked the milk eagerly and gently at the same time.

I did not share this story with my classmates and teacher back then, but we talked about whether the idea of the smell of a stable sounded positive or negative to us, and I remember finding it curious that some students actually felt disgusted by the idea of a stable, whereas to me it was one of my fondest childhood memories. When discussing the best translation for the title of the poem, these personal preferences proved to be paramount, as some preferred words in Portuguese with slightly negative connotations, whereas others chose positive words. But the question in which the teacher was most interested was indeed who Tobias was. At the time she told us that she was rather disappointed that none of us had "discovered" the actual "truth" behind the poem; that a simple, careful Google search would have "revealed" that Tobias is actually the name of Barry's son, therefore "proving" that the poem is "obviously" autobiographical. She took the opportunity to say that the work of a translator is usually detective-like; that one must put the pieces of the puzzle together if one wants to produce an "accurate" translation.

Incidentally, I was writing my BA thesis at the time – "The German Functional Approach in Translation: Christiane Nord's Model Applied to the Translation of Three British Contemporary Short-Stories" (originally written in Portuguese). Working with Nord's model of text analysis (cf. NORD 1988/2005) I had already come across the extratextual factor entitled "author's intention", and had already had my doubts about it. As I thoroughly explain in Leal (2006), I had just made quite an extraordinary discovery about one of the short-stories I was translating. In a given passage of the story – whose style was extremely prosaic and simple – the author made use of perfect metre, making that particular excerpt stand out. As Britto puts it in one of his papers, it was a "clearly (...) calculated effect" (cf. BRITTO 2002 – see III, 3.3). When asked about it, however, the author (Anne Cassidy) was rather surprised herself, claiming that she never intended to insert poetic feet into her story.

After exchanging a couple of e-mails,[150] Cassidy and I arrived at the conclusion that two factors must have contributed to the "production" of those poetic feet. Firstly because originally she had written the story to be read out loud in a story-reading club, and this must have unconsciously influenced her. Secondly because I had been trained to read and analyse poetry, and hence would probably find poetic features even in the manual of a washing machine.[151] The question that remained unanswered was whether I, as the translator, should strive to make that excerpt stand out as well, thus being faithful to my interpretation, or whether I should do as Nord (1988/2005: 53-57) recommends in her model and strictly follow the "sender's intentions". At the time I decided that I should translate the poem as I had read it since one's interpretation is all on which one can normally rely. After all, in most cases translators have no access to the original author, and their experience and knowledge should suffice. In any case, my Irish poetry professor at the time would probably not have accepted my position. And I wonder what Britto would do if the authors he translates came up to him and said that what he takes as "clearly calculated effects" – based on which he mathematically justifies his choices – were actually not intentional at all. Would he then rely on his interpretation as a legitimate justification or simply change his translations to match the authors' intentions? And do unconscious intentions count as well?

150 The full e-mail exchange is included in the BA thesis (cf. LEAL 2005).

151 This incident resembles Fish's experiment in "How to Recognise a Poem When You See One" – see II, 2.2, as well as III, 3.1 (cf. FISH 1980: 322-337).

Speaking of Britto's model (see III, 3.3), along with all the issues raised here about essentialist models as a whole, we still have to add a number of subjective factors for which these theories simply do not account. As this experience with Barry's poem illustrates, one relates to everything around one (including texts) in one's own way, so what I may find prominent in a given text may be considered less prominent by a different reader, or even by the same reader at a different point in time. It is beyond doubt that one cannot mathematically calculate the joy or anguish of reading a poem, for example.

Similarly, if one speaks of strict "correspondences" and "losses" in translation (as Britto does), how does one account for the pleasure of reading a poem, for example, in one's mother tongue? In Britto's scheme of gains and losses, should that not count as a gain? It is as Madame de Stäel puts it: "lors même qu'on entendrait bien les langues étrangères, on pourrait goûter encore, par une traduction bien faite dans sa propre langue, un plaisir plus familier et plus intime" (STÄEL 1821/2004: 142). Models such as Britto's do not take into account that reading a text in one's own mother tongue will set different processes into motion, will create new impressions and feelings.

Another subjective element largely neglected in Britto's model is what he himself calls "products of chance" or of "the translator's [or author's] unconscious". In his translation of Bishop's "Shampoo", a number of alliterations "found their way" into the text either by chance or as a product of his unconscious. What happens if some of the poetic devices Britto takes as "clearly calculated" in Bishop's original were products of chance or of her unconscious? Would they be less relevant then? And how can one mathematically account for the translator's and reader's unconscious?

Indeed, these issues surrounding Paulo Henriques Britto's model – as well as any translation theory influenced by essentialist views – unveil his "theory hope" in the same way as Emma Wagner's medicine allegory unveils hers. Before we return to these questions and their implications for translation theory and translator training, let us analyse the last anecdote that will make up the present chapter – which by the way is remarkably similar to my experience reading Barry's "The Smell".

In 2000 Maria Paula Frota[152] published her PhD thesis, conducted under Rosemary Arrojo (see II, 1). The book, entitled *A Singularidade na Escrita*

152 Frota is a colleague of Britto's at the Pontifícia Universidade Católica do Rio de Janeiro. It may be that the second teacher involved in her anecdote here is indeed Britto, as he usually teaches the poetry translation courses. But this is of course speculation and not really relevant.

Tradutora: Linguagem e Subjetividade nos Estudos da Tradução, na Linguística e na Psicanálise (Singularity in Translational Writing: Language and Subjectivity in Translation Studies, Linguistics and Psychoanalysis), is certainly one of the most interesting volumes on translation ever published in Brazil. Right at the beginning of the book she makes a brief digression that, according to her, illustrates her entire PhD project (FROTA 2000: 23). The anecdote goes as follows: one of her undergraduate students of the course Translation Theory questioned her one day about the discrepancies between the rather open theoretical discussions they had been carrying out in her class, and the ever so closed correction criteria of Frota's colleagues. What particularly bothered the student was the translation of a single verb in a poem by Sylvia Plath, "Kindness". The verb was "picking up", which the teacher in question claimed was closer to "catando" in Portuguese, in the sense of lifting or collecting. Yet the student had translated it as "colando", in the sense of "bringing together" or "sticking together" (FROTA 2000: 24-25). The subject of the verb was "sugar", and the object was "pieces".

Maria Paula Frota then encouraged her student to explain why she was so determined that "picking up" must be translated as "colando", contrary to her teacher's advice. The student explained that, when she was a child, her mother (who baked cakes for a living) had always taught her that sugar, or rather frosting, was the best remedy to "stick" the "pieces" of cakes "together"; that no matter how fluffy or crumbly the cake was, frosting worked as an effective glue (FROTA 2000: 24). This brief digression explains why student and teacher looked at the same English verb and had different pictures in mind; indeed, each had their own personal history with that word and subject, and both "fit" to the poem somehow. Frota summarises the question as follows:

> *proponho apenas que encaremos o "colando" como um acontecimento singular, expressão de um momento em que tradutora e texto constituíram-se simultânea e mutuamente. Uma expressão singular porque efeito de um trânsito particular de* ***um certo sujeito*** *entre os elementos estruturais da linguagem; efeito de uma certa vinculação subjetiva com esses elementos – um trânsito e uma diferença que implicam uma diferença para além da polissemia já codificada de* ***picking up*** (FROTA 2000: 25 – her emphasis).
>
> [I suggest we perceive "colando" as a singular happening, as the expression of a moment in which both translator and text constituted them-

selves simultaneously and mutually. It is a singular expression because it is the result of a particular transit of *a particular subject* amongst the structural elements of language; it is the result of a certain subjective link to these elements – a transit and a difference that imply a difference beyond the known polysemy of "picking up".]

In other words, the student in question could not leave her personal history outside the translation process – and who can anyway? And this personal history cannot by any means be perceived as a mere whim. The student in question probably grew up watching her mother bake cakes and listening to her advice, so this connotation of a verb like "picking up" came to form her very understanding of the verb.

Abstracting now from questions such as "what translation is better" or "what translation is more faithful" – because these questions are completely beside the point – let us consider the implications of these two anecdotes to translator training. Both poetry translation teachers (mine and Frota's colleague) appear to have been entirely committed to the questions above – i.e. faithful translations, equivalents and author's intentions. Yet this does not seem to have contributed much to the students' understanding of language and translation, to their awareness of their role, the teacher's role and the author's role both in translation and in society – as proposed in the first part of this chapter.

The first teacher – my teacher – categorically declared that a Google search had revealed the truth of the poem. The second teacher – Frota's colleague – crossed the student's translation out without further inquiries to favour a more "literal" translation. But when it comes to translation, should pupils not work precisely on their *awareness* of these language and cultural phenomena? If anything, these teachers' attitudes led students to forever seek the truth of poems, to forever seek perfect, literal translations – whatever "literal" may mean. It would certainly have been significantly more fruitful if these teachers had taken these opportunities – these discrepancies, these *differences* – to analyse the values and motivations lying beneath these dissimilar interpretations. Instead of merely learning that Sebastian Barry's son is named Tobias and that the word "picking up" translates as "catando" in Plath's "Sweetness", they would have increased their awareness as for the different values that inform the members of their communities. Furthermore, they would have learnt from where different interpretations sometimes come, and that power relations – in this case teacher-student and author-translator – are not the be-all and end-all of these situations. This sort of reflection could equip students with the knowledge and awareness to expand these ideas and to adapt them to new situations, new texts.

In other words – and going back to Britto's forest analogy – translator/interpreter training or even higher education in general can and should be about teaching students how to get home quickly and safely, but not without acknowledging the illusory and changeable character of cardinal points; not without relativising that my North, here in Vienna, may be someone else's South, say, in Oslo; not without emphasising that cardinal points are *human* construes and not absolute truths above and prior to mankind; not without the willingness to deconstruct these construes and to acknowledge our presence in them. In other words, higher education should not be restricted to teaching pupils how to get home.

If our laws in translation studies are only "probabilistic" and not absolute, as Wagner claims (CHESTERMAN & WAGNER 2002: 4), perhaps we should have less to do with "laws" after all. Having been a translation student myself, I would far rather have been told that translating is subjective, abstract and hence no model or law can embrace *all* its specificities and peculiarities. That would certainly have been less frustrating than spending years studying translation models only to come to the conclusion that they do not comprehend the complexity imposed by translation.

I will come back to some of the issues raised in this first chapter in what follows. Let us now move on to Chapter 2, dedicated more specifically to post-structuralist thought and deconstruction.

2. Poststructuralist Thought: Three Common Misconceptions

The answer is simple enough: this definition of the deconstructionist is false (that's right, false, not true) and feeble; it supposes a bad (that's right bad, not good) and feeble reading of numerous texts, first of all mine, which therefore must finally be read or reread.

Jacques Derrida (translated by Samuel Weber)

In the previous chapter, I briefly mentioned that it is probably through the questions raised by poststructuralist thought that we will be able to eventually raise the status of translation – both as a profession and as a discipline. Throughout this book I have made clear that poststructuralist reflection offers a thought-provoking niche for translation studies, one that allows us to reformulate the age-old questions that have haunted us for so long – such as the questions of fidelity and equivalence, to mention only two. But there seems to be a lot of resistance as regards poststructuralist thought in translation – and not only in translation. This resistance appears to stem from three main misconceptions, briefly summarised as follows: poststructuralist thought is (i) too radical and (ii) pessimistic, and works as (iii) an excuse for whatever sort of behaviour. Christopher Norris, for example, in the 1991 afterword to his 1982 *Deconstruction: Theory and Practice*, claims that

> *Often it [the word deconstruction] is used with a strongly* ***negative*** *connotation: thus 'deconstruct' = 'take things apart' (literary texts, philosophical arguments, historical narratives, truth-claims or value systems of whatever kind) in a spirit of* ***game-playing nihilist abandon*** *and without the least concern for constructing some better alternative* (NORRIS 1982/2002: 135 – my emphasis).

So in the present chapter let us address these misconceptions and to try and understand where they might come from.

Amongst translation studies scholars I have only very rarely come upon instances of thorough criticism of poststructuralist thought. As I have argued elsewhere (LEAL 2010b), the attitude of those who do not openly defend these postmodern trends tends to reveal either listlessness, in most cases, or antagonism, in a few cases. This is perhaps one of the reasons why Britto's criticism is highly interesting; unlike many of his peers, he invests time and effort in this undertaking, which contributes greatly to the debate. It is curious to observe – not to say sad – that Derrida, for example, is criticised in various cognate areas for his "extravagant style" and "counter-intuitive conclusions" (NORRIS 1982/2002: 144), as opposed to logical and "structured" discourse, which in turn appears to prevent him from being taken seriously. In translation studies, nevertheless, criticism to poststructuralist thought tends to remain confined within the modest limits of hasty and impressionistic conclusions drawn from second hand readings.

Let us take, for example, Chesterman's view on deconstruction expressed in the abovementioned *Can Theory Help Translators?*, already quoted in I, 3: "I will bypass here the deconstructionist argument that there is no objective (...) in the first place" (CHESTERMAN & WAGNER 2002: 9-10); "Some postmodern/deconstructivist thinkers (...) stress that (...) the author is dead (...). Outside texts: nothing. Further: meanings are not fixed but endlessly shifting and deferred, all is indeterminate, everyone interprets a text in their own way (...)" (CHESTERMAN & WAGNER 2002: 24). For anyone acquainted with poststructuralist tendencies, these remarks strike one as hasty and ill-informed – though Chesterman's competence is of course not in question, but rather his understanding of these views he so obviously despises.

Derrida himself speaks with utter contempt of similar remarks made by equally acclaimed scholars. The epigraph that opens the present chapter is an example, written as a counterreply to Searle (DERRIDA 1988/1992: 146). Another instance is his reaction to Walter Jackson Bate's and Willis J. Bennett's comments on deconstruction published in the early 1980s, both in celebrated American media. In a long footnote, Derrida (translated by Catherine Porter and Edward Norris) exemplifies his assertion that people do not read his texts at all before criticising deconstruction (DERRIDA 1983: 15 – quoting Bennett and Paul de Man):

Among many possible examples, I shall mention only two recent articles. They have at least one trait in common: their authors are highly placed representatives of two institutions whose power and influence hardly need to be recalled. I refer to "The Crisis in English Studies" by Walter Jackson

Bate, Kingsley Porter University Professor at Harvard [Harvard Magazine, Sept./Oct. 1982], and to "The Shattered Humanities" by Willis J. Bennett, Chairman of the National Endowment for the Humanities [Wall Street Journal, Dec. 31, 1982]. The latter of these articles carries ignorance and irrationality so far as to write the following: "A popular movement in literary criticism called 'Deconstruction' denies that there are any texts at all. If there are no texts, there are no great texts, and no argument for reading." The former makes remarks about deconstruction – and this is not by chance – that are, we might say, just as unnerved. As Paul de Man notes in an admirable short essay ["The Return to Philology," Times Literary Supplement, December 10, 1982], Professor Bate "has this time confined his sources of information to Newsweek magazine... What is left is a matter of law-enforcement rather than a critical debate. One must be feeling very threatened indeed to become so aggressively defensive."

But before we look into how Chesterman (and Bate and Bennett) could possibly have come to such a conclusion, let us focus on the radicalism that appears to be the source of their protest.

As pointed out repeatedly in Part II, there does seem to be a tendency for those committed to poststructuralist perspectives to express themselves in a way that may come across as radical and even aggressive. I shall not belabour this issue much further as I have already done so at length in Part II. But although I can grant this much, a careful reading of texts by poststructuralist thinkers should suffice to unveil far more than sheer radicalism. Even Arrojo's work which may, at first sight, strike one as overly militant, offers plenty of resources for the reader to overcome this initial impression.

NORRIS (1982/2002) offers great insight into the matter of deconstruction and its seeming radicalism, finding various opportune responses in Derrida's work to these claims of extremism. Apropos of Derrida's debate with Austin, for instance, Norris maintains that

Derrida's point is certainly not to argue that we should (…) suspend all the usual (properly philosophical) criteria of truth and falsehood, logical rigour, consequential argument, (…) and henceforth treat his writings as belonging to a realm of generalised 'undecidability' where those standards can no longer be taken to possess the least relevance or critical force (NORRIS 1982/2002: 145).

Likewise, Norris remembers that Derrida by no means denies (and even insistently *affirms*) "that we have to think in accordance with classical logic if we are not only to make adequate sense of (...) texts but also to locate their symptomatic stress-points – the moments of *aporia* or logical tension – where such thinking meets its limit" (NORRIS 1982/2002: 163 – his emphasis). In this sense relativising and deconstructing are a far cry from abolishing and destroying. And this point bears repeating as it reflects the very idea of deconstruction – and to a certain extent of poststructuralist thought as a whole. Exercising deconstruction is not about systematically getting rid of and replacing existing entities; rather, it is in Norris's words "that to which we aspire – individually and collectively – while acknowledging its unattainability in practice and also the fact that it cannot be gainsaid or rendered historically obsolete by the melancholy records of failures and setbacks to date" (NORRIS 1982/2002: 167).

It is this notion that I have tried to defend in the previous chapter. Instead of a full reform of higher education, for example, the idea is simply to "acknowledge" the "unattainability" of what we "aspire to" "in practice", and still strive to achieve it within the given limits available to us. This is by no means radical, as the works of scholars like Rosemary Arrojo might appear to suggest. And the reader was warned several times in Part II that this seemingly radical discourse of hers, in the hands of readers less acquainted with poststructuralist thought, appears to fall to pieces once she gets to more "practical" matters. Her suggestions regarding translator training, specific translation exercises and even translation strategies and translation criticism are far from radical.[153] Nevertheless, like Derrida's critics, those who oppose her ideas seem to overlook these passages and still accuse her of radicalism. Norris exhorts that we take notice of these clearly *not radical* passages in Derrida's work (NORRIS 1982/2002: 153) and quotes a part of the following excerpt by Derrida:

(...) the value of truth (and all those values associated with it) is never contested or destroyed in my writings, but only reinscribed in more powerful, larger, more stratified contexts. And that within interpretive contexts (that is, within relations of force that are always differential – for

153 Let us not forget that this is because deconstruction is not an approach, not a methodology; it is a practice in itself and therefore does not make recommendations as for other practices. In the abovementioned work, Norris argues that "Deconstruction is (...) an activity of reading which remains closely tied to the texts it interrogates and which can never set up independently as a method or system of operative concepts" (NORRIS 1982/2002: 31).

example, socio-political-institutional – but even beyond these determinations) that are relatively stable, sometimes apparently almost unshakable, it should be possible to invoke rules of competence, criteria of discussion and of consensus, good faith, lucidity, rigour, criticism, and pedagogy (DERRIDA 1989: 146).

So to come back to Chesterman's view on deconstruction – which is by no means singular in translation studies – his remarks strike one as a hasty, superficial sort of blanket criticism. It is precisely as Derrida puts it in the epigraph that opens this chapter as regards Searle's view of deconstruction – it is "false" and "bad".

Of course in Arrojo's case one may claim that she plays her part in the production of these hasty conclusions. She does resort to far too many binary oppositions – often without much relativisation – and relentlessly defends one over the other. Her systematic use of quotation marks also seems to point in this direction – i.e. of an exaggerated will to subversion. Britto (cf. BENEDETTI & SOBRAL 2003/2007: 92-93 – see I, 3) is the one to complain about her insistent use of quotation marks around words like original, author and fidelity. Well, it is true that these words are key when it comes to translation, but if we were to be so pedantic about the possible connotations of every word, we would have to write everything between quotation marks. Incidentally, Christopher Norris said that when his 1982 book *Deconstruction: Theory and Practice* came out, it caused a lot of huffing and puffing amongst "Derrida acolytes" because Norris reportedly missed some of the most "basic principles" of deconstruction (NORRIS 1982/2002: 162). A pivotal problem, they held, was Norris' use of binary oppositions – beginning with the subtitle of the book, "theory and practice", which "any self-respecting deconstructionist ought to have spotted a mile off" (NORRIS 1982/2002: 162). But if this use is careful and well aware of its implications, there is no need for constant concern.[154] And surely Arrojo's use of binary oppositions is also careful, but since she resorts to them

154 And let us not forget here that the reception of deconstruction is widely varied and, embedded in the larger context of poststructuralist thought, is the source of a lot of disagreement. Roustang's use of the term "savage horde" (see II, 2.3) might well be used to describe those committed to the postmodern condition as well. As the focus of this book lies on translation, I did not enter these quarrels surrounding poststructuralism and merely treated it as a relatively unified line of thought. But as I hope to have made clear already in the Interchapter in Part I, this is an illusion – and Norris' 1982 book, particularly the 1991 Afterword and the 2002 Postscript, go into these disagreements in great detail (cf. NORRIS 1982/2002: 134-178).

far too often, those readers less familiar with poststructuralist thought might get the wrong impression.

As already pointed out at the end of Part II about poststructuralist thought, radicalism is at once out of the question and the only possible way to convey ideas. The question is *what radicalism* and radicalism *to whom*. When we look at flatly common sense assertions about these poststructuralist tendencies, radicalism usually takes on a "practical" form. As numerous examples presented here have illustrated, in these assertions deconstruction would claim that texts do not exist, that reading is impossible and useless, that all claims to objectivity and truth must be abolished, etc. This kind of practical radicalism that some perceive in poststructuralist thought is simply misplaced because those claims are never made – and it would defeat the very objective of poststructuralist thought if they were made.

But there is another radicalism in question here – the one of discourse. Poststructuralist discourse may be *perceived* as radical because it strives to reveal the moments of aporia particularly of canonical texts. It is as Derrida explains in the indented quotation of pages 286-287: deconstruction entails, at once, on the one hand the respect to a text and to classical logic and, on the other hand, critical reading in the sense of a will to subversion. This radicalism in its *discourse* is inevitable, but let us not forget that respect for texts and respect for classical logic are not abandoned in the name of this radicalism. So all in all, if we ask whether poststructuralist thinkers are unjustly accused of radicalism, we have to answer "yes and no". Yes, because they never suggest radical measures to be put into practice as far as practices other than *discourse* are concerned. Yes, because even when poststructuralist discourse is radical, it does not discard classical logic and disregard the text under deconstruction. No, because in comparison to other types of discourse – discourse that is not labelled as poststructuralist – poststructuralist discourse *does sound* radical.

As I make these efforts to relativise the radicalism of which poststructuralist thought is often accused, I hope to be also calling the readers' attention to the fact that claims to pessimism and nihilism are likewise usually unjustified and exaggerated. There are many examples of thinkers who dismiss Derrida's ideas on these grounds too – Norris mentions chiefly Habermas and Searle (NORRIS 1982/2002: 155). Critics like Gross and Levitt, for example, appear to perceive nothing in postmodern thought beyond radicalism and pessimism. In an entire chapter devoted to this issue – appropriately named "The Realm of Idle Phrases: Postmodernism, Literary Theory, and Cultural Criticism" (GROSS

& LEVITT 1994: 71-106) – words like "nihilism", "negation", "scepticism", "radicalism", and "moral blankness" echo throughout the text. Gross and Levitt maintain, for example, that

> *Derrida's deep epistemological pessimism has infected his disciples as much as have his stylistic eccentricities. Deconstructionism holds that truly meaningful utterance is impossible, that language is ultimately impotent, as are the mental operations conditioned by linguistic habit. The verbal means by which we seek to represent the world are incapable, it is said, of doing any such thing. (...) There is no reality outside the text, but texts themselves are vertiginously unstable, inherently self-contradictory and self-cancelling* (GROSS & LEVITT 1994: 76).

They go further and claim in astonishment that Derrida's position offers "little *cheer* to the would-be revolutionary or *radical* reformer", and yet it became popular amongst them – though "in contrast to literary critics, few serious philosophers have had much *use* for [Derrida's] ideas" (GROSS & LEVITT 1994: 76 – my emphasis). Gross and Levitt too, as numerous other thinkers influenced by essentialism quoted here, associate the idea of optimism with usefulness, very much in line with scientism.

The question of pessimism has already been addressed at the very the end of Part I, when the title of this book was explained for the first time. One can only attribute pessimism and nihilism to poststructuralist thought if one has a limited view of these tendencies, if one hastily takes relativisation as abolishment, and questioning as negation. Beyond this level of the "unnerved remarks" (cf. DERRIDA 1983: 15) and flatly common sense assertions, the questions proposed by poststructuralist thought are a very positive source of reflection.

Though the chapter dedicated to analogies was the previous one, let us quickly recall Britto's analogy of the painter trying to paint a portrait that changes incessantly (see III, 3.1). His point was that, as a translator, he needed to assume that words had fixed meanings and that meanings were stable, otherwise his work would be doomed from the onset. But if we know that this is by no means real, if we know that countless elements will alter the image we are painting – light, angle, canvas, paint type, brush, use of colour, age and conservation of the painting, amongst many others – is it not somewhat relieving to think that *difference* is an inevitable part of the process? Is it not realistic to be sure that no two things can be *the same* – in Britto's analogy person and portrait, original and translation?[155] In this light, if poststructuralist thought helps one to perceive one's obsessions with stability, objectivity and sameness more critically, why should it be deemed pessimistic? And we have to concede that the process of deconstruction of our convictions can be rather painful; let us ask ourselves nonetheless whether going through painful processes is not a necessary part of maturing. Should learning be exclusively "cheerful" and "useful" in order for it to be worth one's while?

Of course it is quite clear from where these claims to pessimism might come, even though they are not particularly thought-out. Britto's fictions (see III, 3) are a corollary of these claims. They may certainly represent cold comfort to some, as is Britto's case. Nevertheless, let us take the challenge of perceiving the glass half empty *and* half full; in other words, let us strive to see at once the pessimistic and optimistic entailed by these different views. In this sense, the very same issues that may be perceived as sources of optimism and "cheerfulness" in essentialist thought – its usefulness, its objectivity, its universalism, its constancy – may unveil their eminent drawbacks and unfeasibility and quickly turn to dust. I will come back to this issue and Britto's fiction in IV, 3.

In addition to allegedly being too radical and pessimistic, poststructuralist thought is also often (contradictorily) understood as an "out-and-out hermeneutic license", as Norris puts it, whereby "critics" (or in our case, translators/interpreters) can "indulge any kind of whimsical, free willing or 'creative' commentary that happens to take their fancy" (1982/2002: 136)[156]. How can pessimism and full liberation possibly go hand in hand? Well, the same critics that appear to discard these tendencies for their lack of "cheer" and "usefulness" and for their excessive "negativity", also seem outraged by the "hermeneutic license" it bestows on its followers. In other words, it is often said that poststructuralist thought allows one to find excuses for whatever sort of behaviour, legitimising even the most unpardonable gesture. Indeed, many would claim that if meanings are unstable and every act of translation involves the subject who translates – together with his/her subjectivity, personal history

155 This analogy is also quite meaningful in terms of the purpose of higher education. In plastic arts, learning fairly objective techniques on how to paint a portrait, for example, is undoubtedly paramount. Nevertheless, being knowledgeable and aware of these countless elements that influence one's painting plays just as pivotal a role. And here we should think not only in terms of "palpable" elements like the ones mentioned above, but also as far as the values and history of the values that inform the production and reception of one's work are concerned.

156 This excerpt has already been quoted in III, 3.2.

and ideology – then this translator may translate a given passage *however* s/he pleases, for virtually *anything* is possible.

This view, particularly of deconstruction as an excuse for unjustifiable behaviour, goes as far as to become an argument against poststructuralist thought as a whole because of Paul de Man's renowned papers, written during the Nazi occupation of Belgium.[157] As Norris very aptly explains, several critics, such as David Hirsch, Roger Kimball and David Lehman – but also GROSS & LEVITT (1994: 76-77) – used de Man's writings as a "platform for advancing their blanket view of deconstruction as a[n] (...) enterprise bent upon covering the tracks of wartime collaborators" (NORRIS 1982/2002: 159). Further, they claimed that a deconstructive view "was one that could always find excuses for condoning any amount of bad behaviour" (NORRIS 1982/2002: 173-174). In other words, these critics took deconstruction as de Man's excuse for writing offensive texts, claiming that it may work as a justification for any kind of extravagant reading or interpretation on the part of its acolytes.

The issue of the truth or falsehood of interpretations is indeed a sensitive one within poststructuralist thought because different currents perceive it in different ways. As already pointed out in footnote 154 on page 287, since this book does not focus specifically on the various tendencies within poststructuralist thought, but rather on its reception particularly within translation studies in Brazil, I shall not belabour this issue much further. In any case, to my knowledge there is no poststructuralist tendency that advocates absolutely "free" and idiosyncratic interpretations. As we saw in Part II, Arrojo defends the idea of the conventionality of signs, ascribing the limits of interpretation to these constantly renewed tacit pacts that regulate every human undertaking. In other words, different groups at different moments will consider a particular interpretation adequate or inadequate in a given context. This emphasis on the social and conventional aspect of language is very much in tune with Stanley Fish's notion of interpretive communities, a thinker to whom Arrojo often turns in her works (cf. FISH 1980: 338-355). But Fish, though labelled as a poststructuralist, is by no means a deconstructionist.[158]

157 For more on this, see de Man's posthumously published *Wartime Journalism, 1939-1943*, as well as the volume edited by Werner Hamacher, Neil Hertz and Thomas Keenan in 1989 entitled *Responses: On Paul de Man's Wartime Journalism*.

158 And here a critic such as Norris would have his objections, as he argues that particularly Fish and Rorty take poststructuralist thought in a significantly different direction in comparison with Derrida and de Man, for example. His disagreements, particularly with Rorty and his readings of Derrida, are explored at length in Norris' *Deconstruction and the Unfinished Project of Modernity* (NORRIS 2000). The similarity between this debate and Freud's "reception" by Lacan and

In deconstruction the issue of the limits of interpretation is perceived differently – or rather its focus appears to lie in a different element. As Norris insists, "deconstruction involves no slackening or suspension of the standards (logical consistency, conceptual rigour, modes of truth-conditional entailment, etc.) that properly determine what shall count as a genuine or valid philosophical argument" (NORRIS 1982/2002: 149). In his view, a common misconception often associated with the idea of radicalism discussed above is that Derrida's readings of canonical texts resort to a "wholly unfamiliar (...) species of textualist 'logic'", and that this logic in turn would "license all manner of wild interpretive games" (NORRIS 1982/2002: 149). Norris stresses that Derrida's (and de Man's) readings are but "meticulously faithful" and rigorously follow the text's logic, taking it to its extreme and thus revealing meanings not traditionally attributed to those texts. He very rightly concedes that Derrida and de Man do produce various heterodox or counter-intuitive arguments, but always based upon a detailed reading of the text in question and always referring back to specific excerpts of the text (NORRIS 1982/2002: 149-150).

Norris opportunely quotes a passage by Derrida in which he addresses his well-known readings of Rousseau. Whereas on the one hand Derrida speaks of "critical readings", of readings as a "critical production" of "signifying structures", on the other hand he defends a certain respect for the text's "classical exigencies", and herein lie the limits of interpretation (DERRIDA 1967/1997: 158 – translated by Gayatri C. Spivak – his emphasis):

> *To produce this signifying structure obviously cannot consist of reproducing, by the effaced and respectful doubling of commentary, the conscious, voluntary, intentional relationship that the writer institutes in his exchanges with the history to which he belongs thanks to the element of language. This moment of doubling commentary should no doubt have its place in a critical reading. To recognize and respect all its classical exigencies is not easy and requires all the instruments of traditional criticism. Without this recognition and this respect, critical production would risk developing in any direction at all and authorize itself to say almost anything. But this indispensable guardrail has always only* ***protected****, it has never* ***opened****, a reading* (cf. NORRIS 1982/2002: 150).

Laplanche is probably no mere coincidence – see II, 2.3.

Similarly, Norris quotes Paul de Man in his preface to Carol Jacob's 1978 *The Dissimulating Harmony: Images of Interpretation in Nietzsche, Rilke and Benjamin*: "reading is an epistemological event prior to being an ethical or aesthetic value. This does not mean that there can be a true reading, but that *no reading is conceivable in which the question of its truth or falsehood is not primarily involved*" (apud NORRIS 1982/2002: 151 – my emphasis).

So when it comes to the so-called limits of interpretation, though emphasis may be added to the suasive-rhetorical aspect of language (as is Fish's case), or to the double gesture of subversive "critical production" and respect in the sense of traditional criticism (as in Derrida's case), no "out-and-out hermeneutic license" is granted to anyone. In this sense, this third common misconception – along with the supposed radicalism and pessimism of poststructuralist thought – may also be perceived as the product of hasty conclusions drawn from not very careful and thorough readings of the texts produced by poststructuralist thinkers. Without proper care, one can promptly see how a remark such as "this does not mean that there can be *a* true reading" may easily turn into "any reading is acceptable". Between the two assertions, nevertheless, there is an entire philosophical tendency that cannot be taken for granted.

All in all, these three misconceptions surrounding poststructuralist thought discussed here appear to stem from a generally ill-informed perception of these tendencies. Particularly in translation studies, it seems that few thinkers have gone to great lengths to carefully criticise works labelled as poststructuralist, quoting canonical works and presenting counterarguments. As argued at the beginning of this chapter, a general sense of listlessness and even unwillingness to hear anything remotely related to poststructuralist thought seems to prevail. As different quotations by Derrida, de Man and Norris have illustrated, even those thinkers who appear to have put time and effort into criticising deconstruction in particular, have done so only half-heartedly, revealing not the most thorough and careful reading of the works in question.

3. Translation Studies: Independence and Utilitarian Purposes

Thus Derrida plays the role of an arch-debunker, a latter-day sophist or wily rhetorician whose special gift is to dance rings around those earnest seekers-after-truth (...) who still make believe that such problems exist, or that theirs is the discipline best equipped to solve them.

Christopher Norris

In Part I, the main questions that arose were the following: (i) should the institutionalisation of translation studies take place in Brazil as it did in Europe, i.e. with little interdisciplinarity and a lot of independence from cognate areas?; (ii) in order to be taken seriously, must the institutionalisation and professionalisation of translation studies entail large amounts of scientism?; (iii) should research in translation studies be exclusively applied and, likewise, are the methods of the Natural Sciences applicable to translation studies?; and (iv) should translation theory be useful to translation practice and hence market-oriented?

Parts II and III revealed very different perspectives as far as these questions are concerned. Issues of standpoint were particularly crucial, since scholars tend to be more willing to engage in theoretical discussions without obvious bonds to practice than practitioners, who in turn tend to expect theory to resolve their practical difficulties. However, these standpoints are not to be taken isolatedly and in absolute terms because only in comparison with other standpoints and embedded in a context can they be said to have any meaning. If we take the example of Andrew Chesterman, for instance, in comparison with Emma Wagner (cf. CHESTERMAN & WAGNER 2002), he seems far more disposed to take theorisation as an awareness-raising activity than as a strictly practice-oriented matter. Nevertheless, in comparison with Rosemary Arrojo (cf. ARROJO & CHESTERMAN 2000), Chesterman appears to emphasise that theory should have its roots in practice, or at least strive to systematise and improve practice, as opposed to Arrojo, who stresses that theory should first and foremost raise one's awareness about one's area.

In this debate, one's position as regards essentialist perspectives also played a pivotal role. If we take this last example of Chesterman and Arrojo, it is clear that precisely because of their overt affinities with essentialist and anti-essentialist views, respectively, they cannot reconcile their ideas on the contours of translation studies and on the aim of translation theory. A similar dynamics appears to have taken place between Rosemary Arrojo and Paulo Henriques Britto, both as far as standpoints are concerned and in terms of their relationship with essentialist perspectives. Most of their disagreements concerning the questions proposed above appear to stem from their utterly disparate standpoints – one as a renowned, accomplished scholar and the other as an acclaimed writer and profuse translator. Moreover, Britto's declared affinities with essentialist views seem to prevent him from seeing any purpose in translation theory and translator training other than to prepare pupils for the market. Arrojo, in contrast, appears to be critical of these "utilitarian purposes" and defends theorisation as a means to overcome the age-old essentialist prejudices that have tormented translation for so long.

Between different standpoints and theoretical affiliations – along with personal dispositions, political and institutional interests, different cultures and contexts, and so many other factors – there is a gargantuan constellation of possible views on these questions, some utterly conflicting and others rather complementary. As already argued, determining *the* correct and most appropriate view and then expecting it to become unanimous is *not* the way to approach this debate, since these questions of institutionalisation, professionalisation, research and theory invariably reflect the heterogeneous values and motivations of those behind them.

Well aware of my own standpoint and theoretical affinities, I feel more inclined towards Derrida's ideas, expressed in his 1983 paper discussed in I, 2. A double gesture of protection and openness is necessary, and here this image should be twofold: it concerns not only the protection from and openness to market demands, but also the protection from and openness to cognate areas – especially those traditionally associated with translation, such as linguistic studies and literary studies.[159] This way of thinking does not entail major institutional changes since both the universities that declared independence for translation studies decades ago (like the Universität Wien, where translation studies has no link whatsoever with the language, literature and linguistic

159 Cf. LYOTARD (1979/2006: 96) for this idea of interdisciplinarity.

departments) and the universities in which translation is the sub-area of a sub-department in the Humanities (like most universities in Brazil) can accommodate this protection and openness in their existing curricula. It is of course a matter of awareness on the part of the faculty of these universities, awareness of this need for a constant double gesture – of openness and protection, of rejection and acceptance, of assentation and criticism; in other words, of relentless deconstruction.

In this light, Arrojo's seemingly categorical views on the urgent independence of translation studies, particularly from linguistics, may seem inadequate. Aware though I am of her motivations in light of her standpoint as an academic (see II, 1), her gesture seems as potentially extreme as the ones behind the full segregation of translation studies as it took place in Europe.[160] It should not be *either* openness *or* protection; it should be *both at once*. Arrojo's outcry for *a place* for translation studies within higher education at all is understandable. However, this outcry would probably not end with the establishment of sub-areas in sub-departments but would most likely advance to maximum protection, until translation studies makes up a fully independent branch at university. In any case, it is *not* the complete institutional separation that is *necessarily* bad: as long as those involved in it are aware of the need for interdisciplinarity and make room for it, as long as both openness and protection are exercised at the same time, then full separation is not an issue. The challenge for rapidly developing countries like Brazil is now to find a balance between extreme scenarios – whereby translation *has* a safe niche at university that is *not* completely segregated from cognate areas.

In the specific case of Brazil, the fact that there still is a course called Letras – in which literary studies and linguistic studies live side by side in relative harmony – is a consolation, since within this course students can still profit from a little interdisciplinarity. As translation studies has been emerging precisely from these departments of Letras, the question is whether or not this urge for protection and independence will propel it beyond the boundaries of these departments. And though I hope not, I have to admit that this is probably what is going to happen.[161]

160 And let me emphasise here that I do not disagree with this segregation, nor do I find it harmful. Indeed, it was certainly the only possible way for translation studies to go back then. Nevertheless, now, decades later, we can look back and rethink some of the premises of this independence and be more critical of it.

161 Witness the establishment of PGET (Postgraduation in Translation Studies) at the Universidade Federal de Santa Catarina, which per se is extremely positive, but whose development might eventually lead to full segregation.

So to the same extent that translation should become institutionalised by a double gesture of simultaneous openness to, and protection from, cognate areas, it should become professionalised by a double gesture of simultaneous openness to, and protection from, market demands. As already argued in IV, 1, even when translator training is openly practice and market-oriented, it should be critical of these "utilitarian purposes", it should deconstruct them, it should seek to carve their contours differently. And here is where the relationship between structuralist and poststructuralist tendencies is laid bare: because they must at once oppose and complement each other, they mutually and relentlessly feed each other. This is why some of Arrojo's remarks in Part II may sound radical, particularly when she seems to fully discard the contributions by thinkers she deems essentialist. And I say *seems* because she does occasionally concede that her intention is by no means to do so, though perhaps not often enough. Teachers committed to the postmodern condition *need* essentialist models to deconstruct with their pupils, to read deconstructively and spot their moments of aporia. Only through essentialist precepts can we "acknowledge" the "unattainability" of what we "aspire to" "in practice" (cf. NORRIS 1982/2002: 167 – see IV, 2) and thus practice deconstruction.

In this light, let us address the question of whether translation theory should be useful, as Britto, for example, so ardently defends. As already pointed out at the end of Part III, Britto's intentions to derive insight from his vast experience with translation practice and to systematise it in theoretical precepts are valid and legitimate. As such an experienced translator, he certainly could not do otherwise – and should not. What may strike one as not as legitimate are his claims for absoluteness and universality, for truth and objectivity. For his productions are but the undertaking of a sole individual, embedded in a context and marked by bias and ideology. Of course his models may be *useful* to some, and being useful is in no way a problem. But these models should acknowledge their own local and finite character rather than lay claim to full objectivity.

And here Britto's "fictions" acquire utmost importance. Being well-informed of poststructuralist contributions and even embracing several of them, Britto – unlike many of his peers – is not naïve enough to actually argue that perfect equivalence and author's intentions are palpable and attainable. So he appears to hide behind his "fictions", i.e. he concedes that these essentialist elements are illusory, but claims that he cannot do without them – hence they are "*necessary* fictions". But as argued in Part III, defending these fictions as necessary is not very ethical of him, particularly considering that we live in a world built upon such fictions. Defending them is the same as encouraging pupils to pursue them endlessly and hopelessly, very much in line with our age-old essentialist tradition of unattainable goals and unfeasible tasks. Believing in these fictions is what allows Britto to criticise translations in absolute terms under the false pretences of objectivity and neutrality. And, as a matter of fact, when Britto arrives at his conclusions of what translation is mathematically better, he does not refer to fictions; instead, he categorically states, in a very factual fashion, who is better and who is worse – regardless of contexts and points of view.

Of course Britto can defend these concepts as fictions, and here we can even draw a parallel between his attitude and Derrida's. In a way, Derrida takes certain notions – such as logic and reason – as fictions because he uses them to his benefit and is, at the same time, critical and aware of their illusory character. But as Britto's work illustrates, rather than call attention to the *illusory* character of his "fictions", he emphasises their *necessary* character. This is when he runs into ethical problems – as already thoroughly explained at the end of Part III.

But for now let us focus on the idea of the usefulness of theorisation. As Britto's example illustrates, theories may aspire to be useful so long as they acknowledge the essentialist gesture behind this aspiration. All in all, theories influenced by essentialism seem to lack honesty – honesty about their local and limited context, honesty about the theorist's motivations and impartialities, honesty about the feasibility of their ambitions. Only when theories stop promising "the world" – i.e. universality, neutrality, scientific rigour and objectivity – will those involved with translation bring their "theory hope" to an end. If they carry on coming across models that allow them to harbour expectations for the taming of subjectivity and the systematisation of language, they will carry on hoping and ultimately feel frustrated that these objectives were not achieved after all. The vicious circle will continue for as long as there are theories to feed it.

But theories – and this is key – need not be exclusively practice-oriented. Defending the solely useful character of theory in translation studies would be similar to claiming that after reading, say, Kant's *Critique of Pure Reason*, one should be able to do things differently, to apply Kant's ideas to practical situations and thus lead a different life. It would be similar to claiming that employing scientists to study the rings of Saturn is a waste of time because they do not have any practical impact on the Earth today. And legitimate though practice-oriented theorisation may be, theorisation rooted in reflection should be the priority of translator training. Let us not forget the double gesture: even when theory is practice-oriented, it can foster reflection, criticism, questioning – rather than mere obedient assentation.

When Britto and Arrojo speak of translation theory, or when Chesterman and Arrojo speak of translation theory, they do not seem to be speaking of the same thing – and indeed they *are not*. They cannot possibly reconcile their views when one defends theorisation as a source of reflection, whereas the other claims that theorisation should lead to more efficient practice. Of course the former will look at the latter with disdain claiming that latter's view leads to no reflection and greater awareness; similarly, the latter will look at the former with disdain because the former's view leads to no improved and more efficient practice. But as already stated in Part I, one view need not exclude the other. What they need is to be acknowledged as *different* and *irreconcilable*, and thus no longer be unjustly criticised for not being what the other one is.

This ambivalence of the term theory is by no means exclusive to translation studies. Engineers, for example, will have numerous theoretical courses with no direct application whatsoever to their future jobs. The aim of these "useless" courses is to teach them to think in a certain way, to make their reasoning more acute, to raise their awareness of their discipline. Of course they will also have various so-called practice-oriented courses, in which specific guidelines are taught to be applied to concrete situations of their future jobs. Both "theories" need not exclude each other, though the harmony in which they live may be questionable. In physics, for example, the so-called theoreticians tend to belittle the so-called practitioners because they (the practitioners) are the ones in charge of the hands-on matters. Practitioners, on the other hand, also tend to look down on theoreticians precisely for their lack of regard for hands-on matters. Surely these "quarrels" are more of a joke than serious conflicts, but the question of *difference* is and will always be present (cf. RÚJULA 2008).

This brings us to the final questions we should approach in this chapter, namely the ones surrounding research in translation studies. Even in various areas of the Natural Sciences theoretical works – i.e. not applied works – are very much welcome, so I do not see why there should be a near obsession with applied and/or empirical research in translation studies. Influential though Popper's model may be, it is by no means the be-all and end-all of research, certainly not in translation studies. The methods of the Natural Sciences may even be fit for research in the Humanities. But here the same reservations as regards practice-oriented theories apply. If one *must* have falsifiability as a chief criterion, then let it not be presented in absolute terms, without taking the researcher and his/her circumstances into account.

More important than that, however, is the acknowledgment that research which does not promote the marriage of theory and practice is just as legitimate. If we embrace the idea that there is theorisation in translation studies that does not necessarily take practical issues of translation into account – but rather *a discourse on translation* – then it should be perfectly acceptable for research to be devoted to these discourses, without any regard to translation practice.

Taking a second look at the dozens of theses defended in the Translation Studies Programme at the Universidade Federal de Santa Catarina (see I, 1), the only one that hints at a theoretical concern that exceeds the limits of the marriage theory-practice is entitled "Os Clássicos Árabes da Teoria da Tradução" (Classics of Arabic Translation Theory), by Manhal Kasouha (KASOUHA 2009).[162] A closer look at the text, nevertheless, reveals that the student translated these classic texts into Portuguese *based on Antoine Berman's theories*. In other words, the emphasis did not lie on these thinkers' discourses on translation, but rather on the application of Antoine Berman's ideas to the translation of the texts in question. Because the Universidade Federal de Santa Catarina has become a reference for translation studies in Brazil, being the only one in the 2000s to offer a programme on translation *strictu sensu*, it seems that this model of applied, empirical research predominant there has become a reference in translation studies in Brazil too. And this prevalence of applied research is not a trend exclusive to this university, nor is it particularly typical of translation studies only in Brazil – as I have argued elsewhere (LEAL 2010b).

So all in all I would like to urge the reader to address the questions suggested in the present chapter with Derrida's double gesture. Let us be in favour of the independence of translation studies from the departments of language, but let us ensure that it communicates with cognate areas as well. Let us produce translation theories with a view to their practical applications, but let us acknowledge the limits and circumstances of these theories. Let us accommo-

162 As already remarked in I, 1, all theses defended in this programme ("Pós-Graduação em Estudos da Tradução") at the Universidade Federal de Santa Catarina are available on their website: www.pget.ufsc.br (last accessed in January 2011).

date market needs as well as possible in translator and interpreter training, but let us be sharply critical of them. Let us use the methods of the Natural Sciences in our research, but let us unveil the limitations of these methods. Let us embrace applied, empirical research in translation studies, but let us not forget that they are not the sole modes of research of our area. And this call to action should in no way imply a will to unanimity, since circumstances and motivations are and will always be *different*.

But the question of unanimity and consensus will be addressed at length below, in the next and last chapter of this book. So let us now analyse how some of the issues discussed so far relate to the question – so popular these days – of global theories.

4. Consensus, Homogeneity and the Savage Horde

Only when one sees itself as a gang of killers, as an assemble of madmen or, according to Freud, as a savage horde, does a psychoanalytical society take on the only form suited to it, the only image that it can uphold without misrepresenting psychoanalysis. If one looks for the effects of psychoanalysis, one can see them in that any group of psychoanalysts carries within it the principle of its own disintegration. If the group is stable and functions well, however, it is a proof of the contrary: it has definitely abandoned the Freudian discovery.

François Roustang
(translated by Ned Lukacher)

Why all this cultural busyness, colloquia, interviews, seminars? Just so that we can be sure we're all saying the same thing. About what, then? About alterity. Unanimity on the principle that unanimity is suspect.

Jean- François Lyotard
(translated by Georges van den Abbeelen)

Let us begin the present chapter by looking into Roustang's words and Freud's image of the "savage horde" in psychoanalysis. The key question that will permeate this last chapter is indeed whether we in translation studies cannot help but be a "savage horde" too, and whether those thinkers affiliated to poststructuralist thought are nothing but a "savage horde". And the implications of the idea of savagery go straight to the heart of the question – in such vogue these days – of global theories, of standardised methods, of a unified terminology for translation studies. Let us take the adjective "savage" here not necessarily as wild but rather as a state of disagreement, conflict and lack of unanimity.

As already briefly remarked in Part I, there have been growing concerns in translation studies because of its stubborn heterogeneity. This heterogeneity, many claim, would afflict not only the overall terminology employed in the area, but also its methods and even its name. In this light, this heterogeneity is perceived as an impediment to the establishment both of translation studies as an academic discipline and of the translator and interpreter as proper professionals. In Part I I mentioned José Lambert's 2007 paper as an example of this concern over heterogeneity. In fact, as I thoroughly explain in Leal (2010b), the wish for standardisation and uniformisation in translation studies seemed rather conspicuous at the 2009 CETRA Summer School at the K.U.Leuven, coming across particularly incisively in Lambert's, Delabastita's and Gambier's seminars.[163]

Lying beneath these issues of homogeneity and consensus is the question of reconciliation. Words such as "global", "unified" and "standard" entail unanimity and thus reconciliation of conflicting views. In a way, it is as Arrojo puts it in her attempt to find shared ground with Chesterman (cf. ARROJO & CHESTERMAN 2000 – see II, 2.4): the establishment of translation universals presupposes the imposition of a privileged, predominant group over countless others, less powerful. Less powerful probably because of the languages they speak and the countries they come from. Let us first explore the idea of heterogeneity, and then I will come back to the question of power.

Within poststructuralist thought, the issue of heterogeneity is described in a more "positive" light, or as an unavoidable trait of humanity. Lyotard, for example, includes the rejection of a consensus in his very definition of the postmodern (LYOTARD 1988/1997: 15 – translated by Julian Pefanis, Morgan Thomas and Don Barry – my emphasis):

The postmodern would be that which in the modern invokes the unpresentable in presentation itself, that which refuses the consolation of correct terms, ***refuses the consensus of taste permitting a common experience of nostalgia for the impossible****, and inquires into new presentations – not to take pleasure in them, but to better produce the feeling that there is something unpresentable.*

In other words, a consensus would be understood as a shared perspective through which people can entertain their nostalgia for the impossible, the unattainable; it would be a sort of all-embracing "theory hope". Similarly, Der-

163 José Lambert did write a response to my 2010 paper, which in fact was a critical account of the event and mentioned this question of the global versus the local. Unfortunately he did not elaborate much on this issue in his reply (cf. LAMBERT 2010).

rida places the idea of heterogeneity close to the heart of deconstruction. In response to Thomas Busch's question, "Is there a place for unity after deconstruction?", Derrida asserts that

I think we do have to choose between unity and multiplicity. Of course deconstruction (...) insisted not on multiplicity for itself, but on the heterogeneity, the difference, the dissociation, which is absolutely necessary for the relation to the other. What disrupts the totality is the condition for the relation to the other (CAPUTO & DERRIDA 1997/2004: 12-13).[164]

As already mentioned a few times in the present book, Derrida's very reading of Saussure is emblematic of this defence of heterogeneity. Derrida is interested precisely in what Saussure leaves out of linguistics, in the differences Saussure either excludes or conveniently flattens. In this sense, Derrida's deconstructionist reading is always marked by a move towards heterogeneity and away from consensus and unity – away from the "structure" as a homogenising whole. To avert misunderstandings, though, in his response to that same question Derrida insists that he is not calling for the destruction of unity (CAPUTO & DERRIDA 1997/2004: 13 – his emphasis):

Now this does not mean that we have to destroy all forms of unity wherever it occurs. I have never said anything like that. Of course, we need unity, some gathering, some configuration. You see, pure unity or pure multiplicity (...) is a synonym of death.[165] *What interests me is the limit to every attempt to totalize, to gather,* ***versammeln*** *(...).*

This is indeed the question at stake in translation studies: what is the limit of this attempt to totalise, to standardise? How much consensus is necessary in order for the discipline to be sufficiently acknowledged? Once again this question is similar to so many others addressed in the present part. How much must the university open itself to the outside so that it does not close in on itself? How much must translation theory attend to practical needs so that it does not completely disregard the market? How much must translation studies

164 In the first part of this book, Derrida is interviewed by Busch on several issues. The interview takes place directly in English, so this is why I have not mentioned a translator here. In the second and longest part of the book, Caputo comments on various questions raised in Derrida's interview.

165 See the epigraph that opens Part IV.

open itself to other areas so that it is not fully segregated from them? The answers to these questions are all very similar and call for a double gesture, though they are not at all precise or palpable.

Even within Freud's "savage horde" there is a certain amount of consensus. Yes, "we need unity, some gathering", says Derrida. Of course some homogeneity must be achieved for there to be heterogeneity at all. There must be some sort of *community* for one to be able to perceive its heterogeneity, its impossibility. It is as John Caputo explains in his commentary on the interview with Derrida mentioned above: the idea of a "universal community" is a contradiction because communities must have "an inside and an outside" (CAPUTO & DERRIDA 1997/2004: 108). Moreover, within deconstruction a closed community would be inconceivable since deconstruction is the very "preparation for the incoming of the other" (CAPUTO & DERRIDA 1997/2004: 108). In this sense, in Caputo's words Derrida exhorts against "the guard that communities station around themselves to watch out for the other" (CAPUTO & DERRIDA 1997/2004: 108).

Let us explore the idea of a community a bit further. This is what Derrida has to say (DERRIDA 1992/1995: 355 – translated by Peggy Kamuf):

I don't much like the word community. (…) If by community one implies, as it is often the case, a harmonious group, consensus and fundamental agreement beneath the phenomena of discord and war, then I don't believe in it very much and I sense in it as much threat as promise. There is doubtless this irrepressible desire for a "community" to form, but also for it to know its limit – and for its limit to be its opening.

So, in other words, any attempt in translation studies to form one such "universal community" must be well aware of the danger of flattening differences and repressing heterogeneity. Furthermore, this community should be well aware of its limits and, as Derrida suggests, this awareness of its limits should not cause it to close in on itself but rather to open itself. So even though a certain degree of consensus might be necessary in order for there to be "a gathering" at all, full consensus should not be a question – because "unanimity is suspect" (see Lyotard's epigraph in this chapter). It is in this sense that translation studies can be seen through a poststructuralist light as a savage horde (see Roustang's epigraph in this chapter).

And even when such small doses of consensus are admitted, issues of power and imposition remain at the centre of the debate, for imposition is inevitable and has its impact on every aspect of society. One of the crucial issues at stake at the moment in translation studies is the widespread use of English as lingua franca. As I have addressed this issue at length elsewhere (LEAL 2010b and 2013), I will only briefly mention it here. Inspired by one of Mary Snell-Hornby's latest works (cf. SNELL-HORNBY 2010), it seems that a double gesture is required here as well. Of course we cannot decidedly affect the dominance of the English language, which manifests itself everywhere and exceeds the domains of language. What we *can* do is mitigate its impact on our area, an area that should par excellence embrace linguistic diversity rather than repress it. We can embrace this dominance to a certain extent (we have no choice after all), but not without being critical of it, without fostering its deconstruction in our circles. Yet there still are numerous publications and events in translation studies that accept contributions in English *only* – the CETRA Summer School being an example of it.

And again those who oppose this critical attitude towards the dominance of the English language do so on grounds of *usefulness*. I remember at the 2009 CETRA Conference "The Known Unknowns of Translation Studies", after Snell-Hornby's seminar precisely on this issue, Daniel Gile jokingly started his seminar in French instead of in English, and then quickly remarked in English that however interesting Snell-Hornby's words may be, they cannot be put *to practice*. But rather than suddenly allow the use of any language at conferences and in journals, Snell-Hornby appears to be calling our attention to an issue that would otherwise go unnoticed; she is asking for our *awareness* and *critical attitude*. Dismissing an opportunity to reflect upon something simply because it is not *useful* as far as practice is concerned seems unthoughtful. So in translation studies we are not willing to turn to interpreting at our academic events; in translation studies we are not willing to deal with linguistic heterogeneity. What does that say of translation studies? What does that say of its awareness of its limits?

The question of language policies should go far beyond the "useful" and the "easily applicable to practice". In a book entirely dedicated to the then latest (early 1990s) developments in Europe, Derrida asks the following question:

> *What philosophy of translation will dominate in Europe? In a Europe that from now on should avoid both the nationalistic tensions of linguistic*

difference and the violent homogenization of languages through the neutrality of a translating medium that would claim to be transparent, metalinguistic, and universal? (DERRIDA 1991/1992: 58 – translated by Pascale-Anne Brault and Michael Naas).

Needless to say, Europe has opted for the latter. In fact, the word "translation" is not even used at all, as texts are taken to have been produced "originally" in each of the EU languages. In this sense, the attitude of the European Union is not so different from the attitude at CETRA, for example. In both instances translation is not a question: language is merely taken as a neutral vehicle of ideas (in the Lockean sense – cf. LEAL 2013, 64-66), regardless of the languages in question. Nevertheless, in both translation is *constantly* an issue – in the case of the EU into one of its official languages and, in the case of CETRA, into English. What they have in common is that both choose not to speak about translation, not to question it, not to address its implications.

And here I ask the reader not to fail to notice the similarity between this debate and that of closed communities. Derrida very aptly claims that he dislikes the term "community" because it presupposes consensus and harmony on the surface while hiding disagreements and wars underneath the surface. But because the term "community" is used, consensus is often taken for granted and disagreements are often not addressed. If we take this question of language policies, both the EU and an institution like CETRA resort to a similar dynamics. Neither will speak of translation because it is fairly consensual that both *avoid* translations – CETRA by having everything in English and the EU by having everything "written" in each of their languages. This is what ensures their unity, what closes their communities. But underneath the surface translation is there, whether they like it or not.

So to come back to the questions proposed at the beginning of the chapter above, the desire for consensus, for the formation of communities will always be strong and legitimate – and healthy, at least to a certain extent. This desire may be behind the establishment of translation theories and schools, of societies and organisations of various kinds. It may be behind the choice of language or languages to be used. And in postmodern times this desire need not in any way be repressed, but rather its implications must be acknowledged – as Derrida defends (see penultimate indented quotation above). In a way, if we wish to be honest as for our disagreements, as for our inevitable heterogeneity, then we have to embrace the idea of the "savage horde" (see Roustang's epigraph above). We can but strive to reach a consensus, but that should not prevent us from acknowledging the impossibility of this consensus. We can but hope for reconciliation, though we should be aware of the fact that full reconciliation is neither feasible nor desirable.

At the heart of the reconciliation debate is the question of the impact of poststructuralist thought in translation studies – the last question in this chapter. Though I have no well-regarded numeric source to quote, the influence that the so-called poststructuralist views has exerted in the area appears to be negligible. Having been in intense contact with three different universities, two in Brazil and one in Austria, I have always been one of the only ones – if not *the* only one – interested in poststructuralist thought. My impression of the academic events in which I took part is no different. Having attended numerous conferences in a few different countries, I fear poststructuralist thought has featured in very few of them. The vast majority of my peers from countless different countries and in the most varied contexts have almost never expressed the least interest in these perspectives. To be honest, I can only think of *one* fellow student in Brazil who dedicates her research to poststructuralist thought in translation (cf. PETRY 2009).

But should this impact remain negligible? Would popularity and consensus not "defeat the objective" of poststructuralist thought? As already mentioned, this consensus and popularity are remotely likely to be achieved. In order for poststructuralist thought to prevail, say, in translation studies, a vast number of thinkers would have to be willing to reconsider their *entire* tradition, to deconstruct their own convictions, and this is just not going to happen. Of course poststructuralist thinkers also strive for consensus in their own manner; but they *are* aware that "unanimity is suspect" (see Lyotard's epigraph in this chapter). So however much scholars committed to poststructuralist thought might wish to be able to convince other people to think the way they do, this attempt is immediately followed by the questioning of the attempt, by a feeling of suspicion. And here we are reminded of Derrida's statement about unity (see epigraph to Part IV): "pure unity or pure multiplicity (…) is a synonym of death". Striving for pure consensus and being intolerant of difference and heterogeneity is not an option. Likewise, not attempting to reach a consensus at all would be the same as condemning oneself to silence, which is equally not an option.

So can the impact of poststructuralist thought be felt today in thinkers whose research is not necessarily committed to it? Though on the surface this impact seems negligible – at least in a more overt fashion – I wonder whether there has been an impact at all. Is the fact that people now take notions such as "author's intentions" and "fidelity" with a microscopic pinch of salt all post-structuralist thought is going to achieve in translation studies? Back in the early 1980s Norris wrote that

> *In the hands of less subtle or resourceful readers deconstruction can become – it is all too clear – a theoretical vogue as uniform and cramping as the worst New Critical Dogma. At best it has provided the impetus for a total revaluation of interpretive theory and practice; the effects of which have yet to be fully absorbed* (NORRIS 1982/2002: 17).

Nearly 10 years later, Norris seemed more able to assess the effects of this way of thinking (NORRIS 1982/2002: 134):

> *It [deconstruction] is a term that now comes readily to novelists, politicians, media pundits, pop journalists, TV presenters, newspaper columnists (…) and others with an eye to intellectual fashion or a taste for debunking such pretentious jargon. What they mostly have in mind (…) is a vague idea of 'deconstruction' as the kind of thing that academics typically get up to when they question commonsense truths and values that everyone else takes pretty much for granted. And when the word is applied with non-derisory intent (…) it tends to mean simply 'criticism of received ideas' (…).*

As far as translation studies is concerned, a similar trend to the one described by Norris may apply. With the exception of those who openly defend deconstruction and those who openly attack it (but sometimes they too), most people do show signs of "criticism of received ideas" amidst their listlessness to poststructuralist thought. Today it is not uncommon for thinkers with clear essentialist affinities to say that we cannot speak of fidelity and equivalence in the same way that we used to. Nor is it rare – though less common – for thinkers like that to say that they want to render their context and bias clear to their audience in papers or at conferences, so that the audience takes their views embedded in this context. In any case, the words "deconstruction" or "post-modern thought" never accompany these gestures, but rather they appear to stem from some new kind of common sense – more or less as Norris puts it.

At the Universität Wien, for instance, it is rather curious that the students are usually more than ready to concede that there is not just one correct interpretation of a text, that reading is always conditioned by context, and that our personal history plays a fundamental role in reading too. When it comes to translation, nonetheless, their views are out-and-out essentialist: one has to respect the author's original intentions, one's job is to find the most suited equivalents in the target language, etc. So this tells me that their "enlightened" views about reading are probably derived from some instance of common sense, and for this reason are not particularly well thought-out. Hence they seem initially unable to expand these views to translation, for example.

In this sense, I agree with Norris that deconstruction has affected common sense slightly. And let us bear in mind that Norris drew those conclusions nearly 20 years ago, so this impact should now be more clearly visible and crystallised than in the early 1990s. What he described then appears to have developed into a nameless trend, some sort of common sense view of which everyone seems to be aware and which has no greater implications in one's way of thinking. Amongst translation scholars it usually takes the shape of ideology as it seems quite consensual today that ideology is everywhere and cannot be stopped. Amongst students it appears to be some kind of urge to freedom from older, stricter views on what is right or wrong, acceptable and unacceptable. But in both cases its implications are not far-reaching but rather confined to these limited areas. In other words, a scholar who would defend the importance of ideology might still uncritically defend original texts as *the* source of insight for translation decisions (see I, 5). Similarly, a student who defends the plurality of interpretation might claim that translations have to obey the author's intentions.

I am not sure whether the impact of poststructuralist thought will remain restricted to the terms described above. But one thing is certain: its defenders will continue trying to call attention to it while knowing that reaching a consensus would not be desirable.[166]

166 In two recent papers (cf. LEAL 2012a and 2013) I develop the issues discussed in this chapter further.

Final Remarks

I would like to close this book by returning to its title, “Is the Glass Half Empty or Half Full?” Though in general this work stresses the heterogeneity of translation studies, its title hints at a similarity, an approximation of seemingly antagonistic sides. For although we would like to claim that Rosemary Arrojo's glass is half empty while Paulo Henriques Britto's glass is half full, let us not forget that the glass is the same and that the difference lies in perspective.

When we examined the different concepts of theory and practice back in Part I, it became clear that disparate dimensions seemed to be at stake. Some spoke of theory as a set of precepts that systematises and controls practice. Others, conversely, defended theory as a sort of awareness-raising reflection. From this second perspective, practical issues were not at all in question. So much so that these awareness-raising reflections, particularly when labelled as poststructuralist, should not have any impact whatsoever in one's practice, but rather on one's *perception* of this practice.

Let us not forget that deconstruction is not a method, that it is not radical as far as practices are concerned and that it does not propose concrete changes to existing practices, nor does it suggest that new practices be adopted. If we take Arrojo and Britto's example again, it is easy to see these different perspectives. Their attitudes as far as translation criticism, for instance, is concerned, are not strikingly different in practical terms. Both choose what they like and dislike, what they believe works and does not work, what they would do differently and what they would not. This is why Arrojo's *Oficina de Tradução* is so *useful* regardless of the reader's philosophical affinities.

But what *is* utterly different is their *perspectives* of their practice, their *discourses* about their practice. And this refers to another dimension altogether – a dimension of reflection, of discourse, of *theoretical* thinking. Poststructuralist thought has a lot more do to with this dimension than with practical issues. And it is precisely this dimension of “thought” that sometimes seems lacking in our theories, in our higher education.

But let us return to the title of this book. The challenge lies not only in perceiving the glass half empty *and* half full, but also in understanding the simultaneous difference and similarity, distance and approximation, entailed by the image of the glass. This paradoxical gesture, this double gesture, is the idea that I would like the reader to keep in mind.

It would be somewhat disappointing if the criticism that this book were to receive were limited to the modest boundaries discussed in Part IV – that it is not a strictly scientific work, it is not useful in terms of translation practice, it does not provide absolute and universally-applicable answers, it contains large amounts of controversy, subjectivity and heterogeneity, and it leaves a lot of room for discussion.

My suggestions for future research include the investigation of the different poststructuralist tendencies and their interface with translation studies – an issue very briefly mentioned here. In addition to that, it would be interesting to see more analyses of works by different scholars that take the scholars' standpoints into account. In other words, it might be fruitful to try and understand the interests and motivations lying beneath a particular work before studying the work itself.

References

ALTIERI, Charles (1979): "Presence and Reference in a Literary Text: The Example of Williams' 'This is Just to Say'", *Critical Inquiry* 5 (3), 489-510.

ANDERMAN, Gunilla & ROGERS, Margaret (2000): "Translator Training between Academia and Profession: A European Perspective". In: SCHÄFFNER, Christina & ADAB, Beverly (eds.): *Developing Translation Competence.* – Amsterdam: John Benjamins, 63-73.

ARONOWITZ, Stanley (1996): "The Politics of the Science Wars", *Social Text* 46/47, 177-197.

ARROJO, Rosemary (1986/2002): *Oficina de Tradução: A Teoria na Prática.* – São Paulo: Ática.

ARROJO, Rosemary (ed.) (1992/2003): *O Signo Desconstruído: Implicações para a Tradução, a Leitura e o Ensino.* – Campinas: Pontes.

ARROJO, Rosemary (1993): *Tradução, Desconstrução e Psicanálise.* – Rio de Janeiro: Imago.

ARROJO, Rosemary (1996a): "Os Estudos da Tradução na Pós-Modernidade, o Reconhecimento da Diferença e a Perda da Inocência", *Cadernos de Tradução I*, 53-69.

ARROJO, Rosemary (1996b): "Postmodernism and the Teaching of Translation". In: DOLLERUP, Cay & APPEL, Vibeke (eds.): *Teaching Translation and Interpreting 3: New Horizons.* – Amsterdam: John Benjamins, 97-104.

ARROJO, Rosemary (1998a): "Os 'Estudos da Tradução' como Área de Pesquisa Independente: Dilemas e Ilusões de uma Disciplina em (Des)Construção", *D.E.L.T.A.: Documentação de Estudos em Linguística Teórica e Aplicada* 14 (2), 423-454.

ARROJO, Rosemary (1998b): "The Revision of the Traditional Gap between Theory and Practice and the Empowerment of Translation in Postmodern Times", *The Translator: Studies in Intercultural Communication,* 4 (1), 25-48.

ARROJO, Rosemary (2000a): "Modernidade e o Desprezo pela Tradução como Objeto de Pesquisa", *Alfa: Revista de Linguística* 44, 71-87.

ARROJO, Rosemary (2000b): "O Tradutor Invisível por ele Mesmo: Paulo Henriques Britto entre a Humildade e a Onipotência", *Trabalhos em Lingüística Aplicada*, 36, 159-165.

ARROJO, Rosemary (2004): "Translation, Transference and the Attraction to Otherness: Borges, Menard, Whitman", *Diacritics: A Review of Contemporary Criticism* 34 (3/4), 31-53.

ARROJO, Rosemary (2005): "Deconstruction, Psychoanalysis and the Teaching of Translation", *TranScribe: The Teaching Translation Journal* 1, 25-35.

ARROJO, Rosemary & CHESTERMAN, Andrew (2000): "Shared Ground in Translation Studies", *Target: International Journal of Translation Studies* 12 (1), 151-160.

AUBERT, Francis (2003/2007): "Introdução: Diálogos da Prática com a Teoria". In: BENEDETTI, Ivone C. & SOBRAL, Adail (eds.): *Conversas com Tradutores: Balanços e Perspectivas da Tradução.* – São Paulo: Parábola, 7-16.
BAKER, Mona (1992): *In Other Words: A Coursebook on Translation.* – London and New York: Routledge.
BARTHES, Roland (1979): "From Work to Text". In: HARARI, Josué V. (ed. and transl.): *Textual Strategies: Perspectives in Post-Structuralist Criticism.* – Ithaca: Cornell University Press, 73-81.
BENEDETTI, Ivone C. (2003/2007): Prefácio. In: BENEDETTI, Ivone C. & SOBRAL, Adail (eds.): *Conversas com Tradutores: Balanços e Perspectivas da Tradução.* – São Paulo: Parábola, 17-32.
BENEDETTI, Ivone C. & SOBRAL, Adail (eds.) (2003/2007): *Conversas com Tradutores: Balanços e Perspectivas da Tradução.* – São Paulo: Parábola.
BERLIN, Isaiah (1953): *The Hedgehog and the Fox: An Essay on Tolstoy's View of History.* – London: Weidenfeld & Nicolson.
BEST, Steven & KELLNER, Douglas (1991): *Postmodern Theory: Critical Interrogations.* – New York: The Guilford Press.
BICHSEL, Peter (1995): *Ein Tisch ist ein Tisch. Eine Geschichte.* – Frankfurt/Main: Suhrkamp.
BLOOM, Harold (2003): *A Map of Misreading.* – London: Oxford University Press.
BOHUNOVSKY, Ruth (2001): "A (Im)possibilidade da 'Invisibilidade' do Tradutor e da sua 'Fidelidade': Por um Diálogo entre a Teoria e a Prática de Tradução", *Cadernos de Tradução VIII*, 51-62.
BORGES, Jorge Luis (1964): "Pierre Menard, Author of the *Quixote*". In: BORGES, Jorge Luis: *Labyrinths: Selected Stories and Other Writings.* Transl. James E. IRBY. – New York: New Directions Publishing Corporation, 36-44.
BORGES, Jorge Luis (1972): *El Hacedor.* – Madrid: Alianza Editorial.
BORGES, Jorge Luis (1985): *Dreamtigers.* Transl. Harold MORLAND and Mildred BOYER. – Austin: University of Texas Press.
BORGES, Jorge Luis (1986): *Ficciones: El Aleph, El Informe de Brodie.* – Caracas: Biblioteca Ayacucho.
BOSI, Alfredo (1992/2006): *Dialética da Colonização.* – São Paulo: Companhia das Letras.
BRITTO, Paulo H. (1995): "Lícidas: Diálogo Mais ou Menos Platônico em Torno de 'Como Reconhecer um Poema ao Vê-lo', de Stanley Fish", *PaLavra* 3, 142-150.
BRITTO, Paulo H. (1997): *Trovar Claro.* – São Paulo: Companhia das Letras.
BRITTO, Paulo H. (1999): "Tradução e Criação", *Cadernos de Tradução IV,* 239-262.
BRITTO, Paulo H. (2000): "Uma Forma Humilde", *Jornal de Resenhas - Folha de S. Paulo*, São Paulo, 11th March 2000.
BRITTO, Paulo H. (2001): "Desconstruir para quê", *Cadernos de Tradução VIII,* 41-50.
BRITTO, Paulo H. (2002): "Para uma Avaliação Mais Objetiva das Traduções de Poesia". In: KRAUSE, Gustavo Bernardo (ed.): *As Margens da Tradução.* – Rio de Janeiro: FAPERJ/Caetés/UERJ, 54-67.

BRITTO, Paulo H. (2006a): "Correspondências Estruturais em Tradução Poética", *Cadernos de Literatura em Tradução 7*, 53-69.
BRITTO, Paulo H. (2006b): "Fidelidade em Tradução Poética: O Caso Donne", *Terceira Margem 15*, 239-254.
BRITTO, Paulo H. (2007a): "É Possível Avaliar Traduções?", *Tradução em Revista* 4, 1-12.
BRITTO, Paulo H. (2007b): *Tarde*. – São Paulo: Companhia das Letras.
BRITTO, Paulo H. (2008a): "A Tradução para o Português do Metro de Balada Inglês", *Fragmentos* 34, 26-33.
BRITTO, Paulo H. (2008b): "Poesia: Criação e Tradução", *Ipotesi* 12, 11-17.
CAMPOS, Jorge Lúcio de: "Trovar Claro, de Paulo Henriques Britto", *Jornal de Poesia*. Available at http://www.revista.agulha.nom.br/jlucio13.html (last accessed in December 2010).
CAPUTO, John & DERRIDA, Jacques (1997/2004): *Deconstruction in a Nutshell: A Conversation with Jacques Derrida*. – New York: Fordham University Press.
CARVALHO, Silvânia (2008): *Biografia e tradução:* Madame de Stäel - A Study of her Life and Times. MA thesis. – Florianópolis: Universidade Federal de Santa Catarina.
CATFORD, John C. (1965): *A Linguistic Theory of Translation*. – London: Oxford University Press.
CHAMBERLAIN, Lori (1988): "Gender and the Metaphorics of Translation", *Signs: Journal of Women in Culture and Society* 13, 457-472.
CHESTERMAN, Andrew (2005): "Consilience and Translation Studies", *Revista Canaria de Estudios Ingleses* 51, 19-32. Available at http://www.helsinki.fi/~chesterm/2005l.Consilience.html (last accessed in January 2011).
CHESTERMAN, Andrew & WAGNER, Emma (2002): *Can Theory Help Translators?* – Manchester: St. Jerome.
COELHO, Fabrício (2008): *Ensaio sobre a tradução da conferência "Über die Verschiedenartigkeit des Naturgenusses und eine Wissenschaftliche Ergründung der Weltgesetze", de Alexander von Humboldt*. MA thesis. – Florianópolis: Universidade Federal de Santa Catarina.
COMPAGNON, Antoine (1998/2004): *Literature, Theory and Common Sense*. Transl. Carol COSMAN. – Princeton: Princeton University Press.
COUTINHO, Nana (2008): *A Tradução Teatral* Widowers' Houses *de George Bernard Shaw – uma tradução comentada*. MA thesis. – Florianópolis: Universidade Federal de Santa Catarina.
CULLER, Jonathan D. (1976/1986): *Ferdinand de Saussure*. – New York: Cornell University Press.
CUNNINGHAM, Michael (2010): "Found in Translation", *New York Times*, 2nd October 2010.
DERRIDA, Jacques (1967/1997): *Of Grammatology*. Transl. Gayatri C. SPIVAK. – Baltimore: Johns Hopkins University Press.
DERRIDA, Jacques (1972): *Positions*. – Paris: Minuit.

DERRIDA, Jacques (1972/1982): *Margins of Philosophy.* Transl. Alan BASS. – Chicago: The University of Chicago Press.
DERRIDA, Jacques (1983): "The Principle of Reason: The University in the Eyes of Its Pupils" (transl. Catherine PORTER and Edward P. MORRIS), *Diacritics: A Review of Contemporary Criticism* 13 (3), 2-20.
DERRIDA, Jacques (1985): "Des Tours de Babel". Available at http://www.jacquesderrida.com.ar/frances/tours_babel.htm (last accessed in January 2011).
DERRIDA, Jacques (1985/1988): "Letter to a Japanese Friend". Transl. David WOOD and Andrew BENJAMIN. In: WOOD, David & BERNASCONI, Robert (eds.): *Derrida and Différance.* – Coventry: Parousia Press, 1-5.
DERRIDA, Jacques (1988/1992): *Limited Inc.* Transl. Samuel WEBER, Alan BASS and Jeffrey MEHLMAN. – Illinois: Northwestern University Press.
DERRIDA, Jacques (1991/1992): *The Other Heading: Reflections on Today's Europe.* Transl. Pascale-Anne BRAULT and Michael NAAS. – Bloomington: Indiana University Press.
DERRIDA, Jacques (1992/1998): *Points... Interviews, 1974-1994* (ed. Elizabeth WEBER, transl. Peggy KAMUF). – Stanford: Stanford University Press.
DERRIDA, Jacques (1996/1998): *Resistances of Psychoanalysis.* Transl. Peggy KAMUF, Pascale-Anne BRAULT and Michael NAAS. – Stanford: Stanford University Press.
DERRIDA, Jacques (2001a): "The Future of the Profession or the University without Condition (Thanks to the "Humanities", what could Take Place Tomorrow)". In: COHEN, Tom (ed.): *Jacques Derrida and the Humanities: A Critical Reader.* – Cambridge: Cambridge University Press, 24-57.
DERRIDA, Jacques (2001b): *L'Université sans Condition.* – Paris: Galilée.
DIZDAR, Dilek (2006): *Translation: Um- und Irrwege.* – Berlin: Frank & Timme.
DONNELLAN, Keith S. (1966): "Reference and Definite Descriptions", *The Philosophical Review* 75 (3), 281-304.
EASTERBROOK, Neil (1994): "'Somehow Disturbed at the Core': Words and Things in William Carlos Williams", *South Central Review* 11 (3), 25-44.
ELLIS, Brian D. (2001): *Scientific Essentialism.* – Cambridge: Cambridge University Press.
FAWCETT, Peter (1997): *Translation and Language: Linguistic Theories Explained.* – Manchester: St. Jerome.
FERREIRA, Izacyl Guimaraes (2010): "*Macau*, de Paulo Henriques Britto: Niilismo, Humor e Metapoesia", *Jornal de Poesia.* Available at http://www.revista.agulha.nom.br /izacyl12.html (last accessed in December 2010).
FISCHER, Michael (1979): "Charles Altieri's 'Presence and Reference in a Literary Text'", *Critical Inquiry* 6 (2), 343-345.
FISH, Stanley (1980): *Is There a Text in this Class? The Authority of Interpretive Communities.* – Cambridge: Harvard University Press.
FISH, Stanley (1989/1999): *Doing What Comes Naturally: Change, Rhetoric and the Practice of Theory in Literary and Legal Studies.* – Durham: Duke University Press.

FORRESTER, John (1990/1994): *The Seductions of Psychoanalysis: Freud, Lacan and Derrida*. – Cambridge: Cambridge University Press.
FOUCAULT, Michel (1966): *Les Mots et les Choses*. – Paris: Gallimard.
FOUCAULT, Michel (1979): "What is an Author?" In: HARARI, Josué V. (ed. and transl.): *Textual Strategies: Perspectives in Post-Structuralist Criticism*. – Ithaca: Cornell University Press, 141-160.
FRANCO, Bernardo Mello (2010): Entrevista com Paulo Henriques Britto, *Jornal de Poesia*. Available at http://www.revista.agulha.nom.br/bmfranco1.html (last accessed in December 2010).
FRANKLIN, Sarah (1996): "Making Transparencies: Seeing through the Science Wars", *Social Text* 46/47, 141-155.
FREUD, Sigmund (1905/2001): *Drei Abhandlungen zur Sexualtheorie*. – Frankfurt am Main: S. Fischer.
FREUD, Sigmund (1959): *Collected Papers* vol. 5. Transl. Alix STRACHEY and Joan RIVIERE). – New York: Basic Books.
FRIJHOFF, Willem (1996/2003): "The Purposes of Universities". In: RÜEGG, Walter (ed.): *A History of the University in Europe* vol. II. – Cambridge: Cambridge University Press, 43-110.
FROTA, Maria Paula (2000): *A Singularidade na Escrita Tradutora: Linguagem e Subjetividade nos Estudos da Tradução, na Linguística e na Psicanálise*. – Campinas: Pontes.
FUCHS, Stephan (2001): *Against Essentialism: A Theory of Culture and Society*. – Cambridge: Harvard University Press.
GALLOP, Jane (1984): *The Daughter's Seduction: Feminism and Psychoanalysis*. – Ithaca: Cornell University Press.
GHIRARDI, José Garcez (2000): *John Donne e a Crítica Brasileira: Três Momentos, Três Olhares*. – São Paulo: AGE Editora.
GROSS, Paul R. & LEVITT, Norman (1994/1998): *Higher Superstition: The Academic Left and its Quarrels with Science*. – Baltimore: Johns Hopkins University Press.
HASSAN, Ihab (2000): "From Postmodernism to Postmodernity: The Local/Global Context", *Critical Issues Series* 3, Sydney, 2000. Available at http://www.ihabhassan.com /postmodernism_to_postmodernity.htm (last accessed in November 2010).
HOUAISS, Antonio et al. (2001): *Dicionário Eletrônico Houaiss da Língua Portuguesa* versão 1.0. Editora Objetiva.
HULME, Peter (2004): "Professor Arthur Terry: Doyen of Catalan Literary Studies", *The Independent*, 4th February 2004.
JAKOBSON, Roman & HALLE, Morris (1956/2002): *Fundamentals of Language*. – Berlin: Walter de Gruyter.
KASOUHA, Manhal (2009): *Os Clássicos Árabes da Teoria da Tradução*. MA thesis. – Florianópolis: Universidade Federal de Santa Catarina.
KATAN, David (2009): "Translation Theory and Professional Practice: A Global Survey of the Great Divide", *Hermes: Journal of Language and Communication Studies* 42, 111-153.

KROTOSKI, Aleks & HUNSINGER, Jeremy (eds.) (2012): *Learning and Research in Virtual Worlds*. – London and New York: Routledge Chapman & Hall.
KROTOSKI, Aleks (2013): *Untangling the Web: What the Internet is doing to you*. – London: Faber & Faber.
KUHN, Thomas S. (1970): *The Structure of Scientific Revolutions*. – Chicago: The University of Chicago Press.
LAGES, Susana K. (2002/2007): *Walter Benjamin: Tradução e Melancolia*. – São Paulo: Edusp.
LAMBERT, José (2007): "Translation and the Globalization of the Modern World". In: KITTEL, Harald et al. (eds.): *Übersetzung, Translation, Traduction: Ein internationales Handbuch zur Übersetzungsforschung* vol. 2. – Berlin: Walter de Gruyter GmbH & Co., 1680-1700.
LAMBERT, José (2010): "Lambert"s Notice on Leal's CETRA Discussion", *Scientia Traductionis* vol. 7, Florianópolis, UFSC, 235-242.
LEAL, Alice (2005): *Funcionalismo e Tradução Literária: O Modelo de Christiane Nord em Três Contos Ingleses Contemporâneos*. BA thesis. – Curitiba: Universidade Federal do Paraná.
LEAL, Alice (2006): "Funcionalismo e Tradução Literária: O Modelo de Christiane Nord em Três Contos Ingleses Contemporâneos", *Scientia Traductionis* 2, Florianópolis, UFSC, 1-9.
LEAL, Alice (2007a): *Funcionalismo e Tradução Literária: Quatro Projetos para a Tradução de* The Years, *de Virginia Woolf*. MA thesis. – Florianópolis: Universidade Federal de Santa Catarina.
LEAL, Alice (2007b): "Poeta Fingidor e Tradutor Traidor", *Caderno Seminal Digital* 7, 22-37.
LEAL, Alice (2010a): "Anthropophagy and Translation". In: AZADIBOUGAR, Omid (ed.): *Translation Effects: Selected Papers of the CETRA Research Seminar in Translation Studies 2009*. – Leuven. Available at http://www.kuleuven.be/cetra/papers/papers.html (last accessed in March 2011).
LEAL, Alice (2010b): "Being a CETRA Student: A Critical Account of the 2009 Summer School", *Scientia Traductionis* vol. 7, Florianópolis, UFSC, 190-206.
LEAL, Alice (2012a): "De jure Mehrsprachigkeit und de facto Einsprachigkeit in der EU". In: SNELL-HORNBY, Mary & KADRIĆ, Mira (eds.): *Die Multiminoritätengesellschaft. Beiträge zum Symposium "Sprache, Identität, Translationswissenschaft"*. – Berlin: SAXA Verlag, 45-54.
LEAL, Alice (2012b): "Equivalence". In: GAMBIER, Yves & VAN DOORSLAER Luc (eds.): *Handbook of Translation Studies* vol. 3. – Amsterdam: John Benjamins, 39-46.
LEAL, Alice (2013): "The European Union and Translation Studies: Unity, Multiplicity and English as a Lingua Franca (ELF)", *Translation Spaces* 2, 63-80.
LEFEVERE, André (1992): *Translating Literature: Practice and Theory in a Comparative Literature Context*. – New York: Modern Language Association of America.
LEÃO, Rodrigo de Souza (2010): "Entrevista com Paulo Henriques Britto", *Jornal de Poesia*. Available at http://www.revista.agulha.nom.br/r2souza06c.html (last accessed in December 2010).

LEVINE, Suzanne J. (1992): "Translation as (Sub)version: On Translating Infante's Inferno". In: VENUTI, Lawrence (ed.): *Rethinking Translation: Discourse, Subjectivity, Ideology*. – London and New York: Routledge, 75-85.
LYONS, John (1968/1995): *Introduction to Theoretical Linguistics*. – Cambridge: Cambridge University Press.
LYOTARD, Jean-François (1979/2006): "Education and its Legitimation through Performativity" (transl. Geoffrey BENNINGTON and Brian MASSUMI). In: CROME, Keith & WILLIAMS, James (eds.): *The Lyotard Reader and Guid*. – New York: Columbia University Press, 92-103.
LYOTARD, Jean-François (1988/1997): *The Postmodern Explained*. Transl. ian PEFANIS, Morgan THOMAS and Don BARRY. – Cambridge: Cambridge University Press.
MACHADO, Cassiano E. (2002): "Contos Iniciais de Willian Faulkner Ganham 1ª Edição no Brasil", *Folha de São Paulo*, 6th July 2002.
MCGOWAN, John (1991): *Postmodernism and its Critics*. – Ithaca: Cornell University Press.
MENDONÇA, Ana Waleska P. C. (2000): "A Universidade no Brasil", *Revista Brasileira de Educação* 14, 131-150.
MILTON, John (1998): *Tradução: Teoria e Prática*. – São Paulo: Martins Fontes.
MILTON, John (2002): *O Clube do Livro e a Tradução*. – São Paulo: EDUSC.
MOATI, Raoul: *Derrida/Searle: Déconstruction et Langage Ordinaire*. – Paris: Presses Universitaires de France – PUF.
MOOSBURGER, Théo (2008): *Tradução comentada dos versos 1-609 do épico bizantino* Vasileios Digenis Akritis. MA thesis. – Florianópolis: Universidade Federal de Santa Catarina.
MOUNIN, Georges (1963/1998): *Les Problèmes Théoriques de la Traduction*. – Paris: Gallimard.
NELKIN, Dorothy (1996): "The Science Wars: Responses to a Marriage Failed", *Social Text* 46/47, p. 93-100.
NEWMARK, Peter (1981/1986): *Approaches to Translation*. – Oxford: Pergamon Press.
NIETZSCHE, Friedrich W. (1873): "Über die Wahrheit und Lüge im außermoralischen Sinne". Available at http://www.textlog.de/455.html (last accessed in May 2008).
NIRANJANA, Tejaswini (1992): *Siting Translation: History, Post-Structuralism and the Colonial Context*. – Berkeley: University of California Press.
NORD, Christiane (1988/2005): *Text Analysis in Translation: Theory, Methodology, and Didactic Application of a Model of Translation-oriented Text Analysis*. Transl. Christiane NORD and Penelope SPARROW. – Amsterdam: Atlanta, Rodopi.
NORRIS, Christopher (1982/2002): *Deconstruction: Theory and Practice*. – London and New York: Routledge.
NORRIS, Christopher (2000): *Deconstruction and the Unfinished Project of Modernity*. – London: Athlone Press.
O Globo. "Flip Abre Inscrições para Oficina com Paulo Henriques Britto", 6th May 2009.
OSIMO, Bruno (2000): Prefazione. In: TOROP, Peeter: *La Traduzione Totale*, ed. Bruno OSIMO. – Módena: Guaraldi Logos, 2000.

PERRONE-MOISÉS, Leyla (2006): "Derrida e as Ciências Humanas". In: OTTONI, Paulo & FERREIRA, Élida (eds.): *Traduzir Derrida: Políticas e Desconstruções*. – Campinas: Mercado das Letras, 23-30.

PESSÔA, André Vinícius: "Uma Ética da Composição em *Trovar Claro*, de Paulo Henriques Britto". In: *Fórum de Literatura Brasileira Contemporânea (UFRJ)*. Available at http://andreviniciuspessoa.blogspot.com/2009/06/uma-etica-da-composicao-em-trovar-claro.html (last accessed in December 2010).

PETRY, Simone C. (2009): "Teoria do Escopo sob a Luz da Desconstrução: O Papel do Tradutor". In: XVIII Seminário do CELLIP, Ponta Grossa, 2009, 1-10.

PINTO, Manuel da Costa (2003): "Rodapé: 'Macau' é Província de Simetrias Poéticas", *Folha de São Paulo*, 20th September 2003.

PIUCO, Narceli (2008): Corinne ou l'Italie *de Mme de Staël: Da Adaptação à Retradução Estrangeirizante*. MA thesis. – Florianópolis: Universidade Federal de Santa Catarina.

POPPER, Karl R. (1959/2002): *The Logic of Scientific Discovery*. – London and New York: Routledge.

PYM, Anthony; GRIN, François; SFREDDO, Claudio & CHAN, Andy L. J. (2012): *Studies on Translation and Multilingualism: The Status of the Translation Profession in the European Union*. European Commission, July 2012.

Random House Webster's Unabridged Dictionary. Random House Inc., 1999, v3.0 for Windows™ Systems.

RASMUSSEN, Ingvill & LUDVIGSEN, Sten (2009): "The Hedgehog and the Fox: A Discussion of the Approaches to the Analysis of ICT Reforms in Teacher Education of Larry Cuban and Yrjö Engeström", *Mind, Culture, and Activity* 16 (1), 83-104.

REIß, Katharina & VERMEER, Hans J. (1984): *Grundlegung einer allgemeinen Translationstheorie*. – Tübingen: Niemeyer.

RODRIGUES, Cristina C. (1999): *Tradução e Diferença*. – São Paulo: UNESP.

ROUSTANG, François (1976/1986): *Dire Mastery: Discipleship from Freud to Lacan*. Transl. Ned LUKACHER. – Arlington: American Psychiatric Publishing Inc.

RUMELHART, David (1979/1993): "Some Problems with the Notion of Literal Meanings". In: ORTONY, Andrew (ed.): *Metaphor and Thought*. – Cambridge: Cambridge University Press, 71-82.

RÚJULA, Álvaro de (2008): "What is a 'theoretical physicist'?". Available at http://public.web.cern.ch/public/en/people/Theorists-en.html (last accessed in August 2013).

SADOCK, Jerrold (1979/1993): "Figurative Speech and Linguistics". In: ORTONY, Andrew (ed.): *Metaphor and Thought*. – Cambridge: Cambridge University Press, 42-57.

SARUP, Madan (1988/1993): *An Introductory Guide to Post-Structuralism and Postmodernism*. – Harlow: Pearson Education Ltd.

SAUSSURE, Ferdinand de (1986): *Course in General Linguistics* (transl. Roy HARRIS). – Illinois: Open Court.

SCHNITZLER, Arthur (1961/2006): *Arthur Schnitzler: Traumnovelle und andere Erzählungen*, (ed. Michael SCHEFFEL). – Frankfurt am Main: S. Fischer Verlag GmbH.

SEARLE, John R. (1975): “Indirect Speech Acts”. In: COLE, Peter & MORGAN, Jerry L. (eds.): *Syntax and Semantics* vol. 3. – New York: Academic Press, 59-82.
SEARLE, John R. (1979): *Expression and Meaning: Studies in the Theory of Speech Acts.* – Cambridge: Cambridge University Press.
SHIRKY, Clay (1995): *Voices from the Net.* – London: Pearson Education Ltd.
SNELL-HORNBY, Mary (2006): *The Turns of Translation Studies: New Paradigms or Shifting Viewpoints?* – Amsterdam: John Benjamins.
SNELL-HORNBY, Mary (2010): “Is Translation Studies Going Anglo-Saxon? Critical Comments on the Globalization of a Discipline”. In: GILE, Daniel et al. (eds.): *Why Translation Studies Matters.* – Amsterdam: John Benjamins, 97-104.
SPIVAK, Gayatri C. (1976/1997): Translator's Preface. In: DERRIDA, Jacques: *Of Grammatology.* – Baltimore: Johns Hopkins University Press, ix-lxxxviii.
STÄEL, Madame de (1821/2004): “De l'Esprit des Traductions”, *Clássicos da Teoria da Tradução* 2, 140-153.
STEINER, George (1975/1998): *After Babel: Aspects of Language and Translation.* – London: Oxford University Press.
TIRKKONEN-CONDIT, Sonja; MÄKISALO, Jukka; JÄÄSKELÄINEN, Riitta; KALASNIEMI, Mirja & KUJAMÄKI, Pekka (2002): “Do we Need a Shared Ground?”, *Target: International Journal of Translation Studies* 13 (2), 339-343.
TOROP, Peeter (1995/2000): *La Traduzione Totale.* Transl. Bruno OSIMO. – Módena: Guaraldi Logos. Available at www.logos.it (last accessed in May 2007).
VENUTI, Lawrence (1986): “The Translator's Invisibility”, *Criticism 28*, 179-212.
VENUTI, Lawrence (ed.) (1992): *Rethinking Translation: Discourse, Subjectivity, Ideology.* – London and New York: Routledge.
VENUTI, Lawrence (1995): *The Translator's Invisibility: A History of Translation.* – London and New York: Routledge.
VERMEER, Hans J. (1997): Vorwort. In: Wolf, Michaela (ed.): *Übersetzungswissenschaft in Brasilien: Beiträge zum Status von „Original“ und Übersetzung.* – Tübingen: Stauffenburg, p. 9-11.
WITTGENSTEIN, Ludwig (1953): *Philosophische Untersuchung.* Available at http://www.geocities.jp/mickindex/wittgenstein/witt_pu_gm.html#LocalLink-c60 (last accessed in August 2013).
WOLF, Michaela (ed.): *Übersetzungswissenschaft in Brasilien: Beiträge zum Status von „Original“ und Übersetzung.* – Tübingen: Stauffenburg.
ZWISCHENBERGER, Cornelia (2013): *Qualität und Rollenbilder beim simultanen Konferenzdolmetschen.* – Berlin: Frank & Timme.

Appendix

BRITTO (1999, 12 – his English translation):

This difference might be represented in terms of the diagram presented in the first section of this paper. Given a structure of intertextuality of the form

$$\begin{array}{ccccccccc} \ldots \rightarrow & t_c^M & \rightarrow & t_d^N & \leftarrow & t_e^O & \rightarrow \ldots \\ & \uparrow & & \downarrow & & & \\ \ldots \leftarrow & t_h^P & \rightarrow & t_i^Q & \leftrightarrow & t_j^R & \rightarrow \ldots \end{array}$$

containing a text t_c^M and given another structure (intersecting the previous one at t_i^Q, for textuality is a three-dimensional space at the very least)

$$t_{i1}^Q \rightarrow t_{i2}^Q \rightarrow t_{i3}^Q \rightarrow \ldots \rightarrow t_i^Q$$

where each t_{in}^Q is a successive version in the process of elaboration of t_i^Q, the *final version*, then we may say that t_i^Q is a *translation* of t_c^M if and only if:

(i) M (the language in which t_c^M is written) and Q (the language in which t_i^Q is written) are different;

(ii) t_c^M was written before t_i^Q.

(iii) there is between t_c^M (and t_c^M only) and t_i^Q a *controlling relation*, defined as successive interventions of t_c^M in the series t_{i1}^Q, t_{i2}^Q, ... t_i^Q with the effect of approximating t_{in}^Q to t_c^M as to choice of lexical items, metrical patterns, syntactic patterns, etc.

BRITTO (2002, 9):

APPENDIX

The still explosions on the rocks,	*a*	*4*	8	L	[s], [ʒ], [z]	
the lichens, grow	*b*	*2*	4	S	[z], [g], [r]	
by spreading gray, concentric shocks.	*a*	*4*	8	L	[s], [g], [k], [r]	
They have arranged	*c*	*2*	4	S	[r]	
to meet the rings around the moon, although	*b*	*5*	10	L	[m], [r], [ð]	
within our memories they have not changed.	*c*	*5*	10	L	[ð], [m], [r]	
And since the heavens will attend	*d*	*4*	8	L		
as long on us,	*e*	*2*	4	S		
you've been, dear friend,	*d*	*2*	4	S		
precipitate and pragmatical;	*f*	*4*	9	L	[p], [t]	precipitate, pragmatical
and look what happens. For Time is	*e*	*3*	8	L	[t]	
nothing if not amenable.	*f*	*4*	8	L	[n]	amenable
The shooting stars in your black hair	*g*	*4*	8	L	[b]	
in bright formation	*h*	*2*	5	S	[b], [f]	
are flocking where,	*g*	*2*	4	S	[f]	
so straight, so soon?	*i*	*2*	4	S	[s]	
— Come, let me wash it in this big tin basin,	*h*	*6*	11	L	[l], [ʃ], [b]	
battered and shiny like the moon.	*i*	*4*	8	L	[b], [ʃ], [l]	

9

BRITTO (2002, 10):

Os liquens — silenciosas explosões	*a*	*4* 10	L	[ʃ], [s], [z]	
nas pedras — crescem e engordam,	*b*	*3* 6	S	[ʃ], [k], [s], [g]	
concêntricas, cinzentas concussões,	*a*	*5* 10	L	[k], [s], [ʃ],	
Têm um encontro marcado	*c*	*3* 6	S	[t], [k], [d]	
com os halos ao redor da lua, embora	*b*	*4* 10	L	[k], [l], [d]	
até o momento nada tenha mudado.	*c*	*5* 10	L	[t], [m], [n], [d]	
E como o céu há de nos dar guarida	*d*	*5* 10	L	[k], [d]	dar ↕
enquanto isso não se der,	*e*	*4* 8	L	[k], [d]	der ↑
você há de convir, amiga,	*d*	*3* 8	L	[k]	↕
que se precipitou;	*f*	*3* 6	S	[k], [p], [t]	↓
e eis no que dá. Porque o Tempo é,	*e*	*4* 8	L	[k], [d], [p], [t]	dá tempo ↕
mais que tudo, contemporizador.	*f*	*4* 10	L	[k], [t], [d],	con*tempo*rizador
No teu cabelo negro brilham estrelas	g	*5* 10	L	[e]	
cadentes, arredias.	*h*	*3* 6	S	[e]	
Para onde irão elas	g	*3* 6	S		
tão cedo, resolutas?	*i*	*3* 6	S	[e]	
— Vem, deixa eu lavá-lo, aqui nesta bacia	*h*	*5* 11	L	[e], [v], [l], [b], [s]	
amassada e brilhante como a lua.	*i*	*4* 10	L	[s], [b], [l^{j}], [l]	

10

BRITTO (2006a, 65-68):

	RITMO	P	S	C	R	OBS.
Across' the floor' flits' the mechan'ical toy',	˘/|˘/|[illegible]|˘/|[illegible]	5	11	L	*a*	[aj] [k], [ʃ], [s], [b], [l]
fit' for a king' of sev'eral cen'turies back'.	[illegible]|˘/|˘/|˘/|˘/	5	10	L	*B*	
A lit'tle cir'cus horse' with real' white' hair'.	˘/|˘/|˘/|˘/|[illegible]	5	10	L	*c*	
His eyes' are glos'sy black'.	˘/|˘/|˘/	3	6	C	*b*	
He bears' a lit'tle danc'er on his back'.	˘/|˘/|˘/|[illegible]|˘/	5	10	L	*B*	
She stands' upon her toes' and turns' and turns'.	˘/|[illegible]|˘/|˘/|˘/	5	10	L	*d*	[ow] [s], [t], [r], [b]/[p] rimas femininas em b & c
A slant'ing spray' of ar'tifi'cial ros'es	˘/|˘/|˘\\|˘/|˘/|˘	5	11	L	*E*	
is stitched' across' her skirt' and tin'sel bod'ice.	˘/|˘/|˘/|˘/|˘/|˘	5	11	L	*f*	
Above' her head' she pos'es	˘/|˘/|˘/|˘	3	7	C	*e*	
anoth'er spray' of ar'tifi'cial ros'es.	˘/|˘/|˘\\|˘/|˘/|˘	5	11	L	*E*	
His mane' and tail' are straight' from *Chi'rico*'.	˘/|˘/|˘/|˘/|˘/	5	10	L	*g*	[ej] [ow] [k], [ʃ], [p]/[b], [t], [l]
He has' a for'mal, me'lancho'ly soul'	˘\\|˘/|˘‖/|˘\\|˘/	5	10	L	*H*	
He feels' her pink' toes' dan'gle toward' his back'	˘/|˘/|[illegible]|˘/|˘/	5	10	L	*b*	
along' the lit'tle pole'	˘/|˘/|˘/	3	6	C	*h*	
that pier'ces both' her bo'dy and her soul'	˘/|˘/|˘/|[illegible]|˘/	5	10	L	*H*	
and goes' through his', and re'appears' below',	˘/|˘/‖˘\\|˘/|˘/	5	10	L	*g*	[ow] [aw],[b]/[p], [k] falta rima por repetição rima visual *g-h* ritmo ternário
un'der his bel'ly, as a big' tin' key'.	[illegible]|˘/‖[illegible]|˘/|[illegible]	5	10	L	*i*	
He *can'ters* three' steps', then' he makes' a bow',	˘/|˘/|[illegible]|˘/|˘/	5	10	L	*h*	
can'ters again', bows' on one' knee',	[illegible]	[illegible]	[illegible]	[illegible]	*i*	
can'ters, then *clicks*' and stops', and looks' at me.'	[illegible]‖˘/|˘/‖˘/|˘/	5	10	L	[illegible]	
The danc'er, by this' time', has turned' her back'.	˘/|[illegible]|\\/|˘/|˘/	5	10	L	*b*	[aj] [t], [ʃ], [s]
He' is the more' intel'ligent by far'.	[illegible]|˘/|˘/|[illegible]|˘/	5	10	L	*J*	
Fac'ing each' oth'er rath'er des'perately —	[illegible]|\\/|˘/|˘/|[illegible]	5	10	L	*i*	
his eye' is like' a star' —	˘/|˘/|˘/	3	6	C	*j*	
we stare' and say', "Well', we have come' this' far'."	˘/|˘/‖[illegible]|˘/|[illegible]	5	10	L	*J*	

Tabela 1

	RITMO	S	C	R	OBS.
E' um brinque'do de cor'da dig'no de um rei'	[illegible]	11	L	*a*	enjambement
de uma ou'tra e'ra: cava'lo e bailari'na.	[illegible]	11	L	*B*	
Um cava'lo de cir'co, de o'lhos ne'gros,	[illegible]	10	L	*c*	
bran'co no pê'lo e na *cri'na*.	[illegible]	7	C	*b*	
Sobre e'le vai' monta'da a bailari'na.	[illegible]	10	L	*B*	
Na *pon'ta* dos *pés*', e'la rodopi'a.	[illegible]	10	L	*c*	[p] enjambement
Tem' um ra'mo de flo'res ar'tificiais'	[illegible]	11	S	*D*	
na sai'a e no cor*pe*'te de ouropel'.	[illegible]	10	L	*e*	
So'bre a cabe'ça, traz'	[illegible]	6	C	*d*	
um ou'tro ra'mo de flo'res ar'tificiais'	[illegible]	12	L	*D*	
A cau'da do cava'lo é pu'ro Chiri'co.	[illegible]	11	L	*f*	
E' formal' e melancó'lica a sua al'ma.	[illegible]	11	L	*G*	
E'le sen'te em seu dor'so a per'na le've	[illegible]	10	L	*h*	
da bailari'na cal'ma	[illegible]	6	C	*g*	
em tor'no da has'te que a perfu'ra, cor'po e al'ma,	[illegible]	13	L	*G*	
e lhe atrave'ssa o cor'po, sai'ndo por fim'	[illegible]	12	L	[illegible]	[p] falta rima por repetição; rima entre 1 e 5 ritmo trocaico
sob seu ven'tre como u'ma cha've de la'ta.	[illegible]	11	L	*j*	
E'le dá' três' *pa'ssos*, faz' uma mesu'ra,	[illegible]	11	L	*k*	
an'da mais' um *pou'co*, do'bra uma das *pa'tas*,	[illegible]	11	[illegible]	*j*	
an'da, esta'la, *pá'ra* e o'lha para mim'.	[illegible]	11	L	[illegible]	
A dançari'na, a e'ssa altu'ra, está' de cos'tas	[illegible]	12	L	*l*	[p]
O cava'lo é o mais' argu'to de'les dois'.	[illegible]	11	L	*M*	
Entreolha'mo-nos, com cer'to deses*pe*'ro,	[illegible]	12	L	*n*	
e dize'mos depois':	[illegible]	6	C	*m*	
"E', até' aqui' chega'mos nós' dois'."	[illegible]	10	L	*M*	

Tabela 2

<table>
<tr><th>RITMO</th><th>P</th><th>S</th><th>C</th><th>R</th><th>OBS.</th><th>RITMO</th><th>S</th><th>C</th><th>R</th><th>OBS.</th></tr>
<tr><td>˘ / | ˘ / | [illegible] | ˘ / | [illegible]</td><td>5</td><td>11</td><td>L</td><td>a</td><td rowspan="5">[aj]
[k], [ʃ], [s], [b], [l]</td><td>/ ˘ / ˘ ˘ / ˘ / ˘ ˘ /</td><td>11</td><td>L</td><td>a</td><td rowspan="5">enjambement</td></tr>
<tr><td>[illegible] | ˘ / | ˘ / | ˘ / | ˘ /</td><td>5</td><td>10</td><td>L</td><td>B</td><td>˘ / ˘ / ˘ ‖ ˘ / ˘ ˘ ˘ /</td><td>11</td><td>L</td><td>B</td></tr>
<tr><td>˘ / | ˘ / | ˘ / | ˘ / | [illegible]</td><td>5</td><td>10</td><td>L</td><td>c</td><td>˘ ˘ / ˘ ˘ / ‖ ˘ / ˘ /</td><td>10</td><td>L</td><td>c</td></tr>
<tr><td>˘ / | ˘ / | ˘ /</td><td>3</td><td>6</td><td>C</td><td>b</td><td>/ ˘ ˘ / ˘ ˘ /</td><td>7</td><td>C</td><td>b</td></tr>
<tr><td>˘ / | ˘ / | ˘ / | [illegible] | ˘ /</td><td>5</td><td>10</td><td>L</td><td>B</td><td>˘ / ˘ / ˘ / ˘ ˘ ˘ /</td><td>10</td><td>L</td><td>B</td></tr>
<tr><td></td><td></td><td></td><td></td><td></td><td></td><td></td><td></td><td></td><td></td><td></td></tr>
<tr><td>˘ / | [illegible] | ˘ / | ˘ / | ˘ /</td><td>5</td><td>10</td><td>L</td><td>d</td><td rowspan="5">[ow]
[s], [t], [r], [b]/[p]
rimas femininas
em b & c</td><td>˘ / ˘ ˘ / ‖ / ˘ ˘ ˘ /</td><td>10</td><td>L</td><td>c</td><td rowspan="5">[p]
enjambement</td></tr>
<tr><td>˘ / | ˘ / | ˘ \ | ˘ / | ˘ / | ˘</td><td>5</td><td>11</td><td>L</td><td>E</td><td>/ ˘ / ˘ ˘ / ˘ \ ˘ ˘ /</td><td>11</td><td>S</td><td>D</td></tr>
<tr><td>˘ / | ˘ / | ˘ / | ˘ / | ˘ / | ˘</td><td>5</td><td>11</td><td>L</td><td>f</td><td>˘ / ˘ ˘ ˘ / ˘ ˘ ˘ /</td><td>10</td><td>L</td><td>e</td></tr>
<tr><td>˘ / | ˘ / | ˘ / | ˘</td><td>3</td><td>7</td><td>C</td><td>e</td><td>/ ˘ ˘ / ˘ ‖ /</td><td>6</td><td>C</td><td>d</td></tr>
<tr><td>˘ / | ˘ / | ˘ \ | ˘ / | ˘ / | ˘</td><td>5</td><td>11</td><td>L</td><td>E</td><td>˘ / ˘ / ˘ ˘ / ˘ \ ˘ ˘ /</td><td>12</td><td>L</td><td>D</td></tr>
<tr><td></td><td></td><td></td><td></td><td></td><td></td><td></td><td></td><td></td><td></td><td></td></tr>
<tr><td>˘ / | ˘ / | ˘ / | ˘ / | ˘ /</td><td>5</td><td>10</td><td>L</td><td>g</td><td rowspan="5">[ej]
[ow]
[k], [ʃ], [p]/[b], [t], [l]</td><td>˘ / ˘ ˘ ˘ / ˘ / ˘ ˘ /</td><td>11</td><td>L</td><td>f</td><td rowspan="5"></td></tr>
<tr><td>˘ \ | ˘ / | ˘ ‖ / | ˘ \ | ˘ /</td><td>5</td><td>10</td><td>L</td><td>H</td><td>/ ˘ / ˘ ˘ ˘ / ˘ ˘ ˘ /</td><td>11</td><td>L</td><td>G</td></tr>
<tr><td>˘ / | ˘ / | [illegible] | ˘ / | ˘ /</td><td>5</td><td>10</td><td>L</td><td>b</td><td>/ ˘ / ˘ ˘ / ˘ / ˘ /</td><td>10</td><td>L</td><td>h</td></tr>
<tr><td>˘ / | ˘ / | ˘ /</td><td>3</td><td>6</td><td>S</td><td>h</td><td>˘ ˘ ˘ / ˘ /</td><td>6</td><td>C</td><td>g</td></tr>
<tr><td>˘ / | ˘ / | ˘ / | [illegible] | ˘ /</td><td>5</td><td>10</td><td>L</td><td>H</td><td>˘ / ˘ ˘ / ˘ ˘ ˘ / ˘ ‖ / ˘ /</td><td>13</td><td>L</td><td>G</td></tr>
<tr><td></td><td></td><td></td><td></td><td></td><td></td><td></td><td></td><td></td><td></td><td></td></tr>
<tr><td>˘ / | ˘ / ‖ ˘ \ | ˘ / | ˘ /</td><td>5</td><td>10</td><td>L</td><td>g</td><td rowspan="5">[ow]
[aw],[b]/[p], [k]
falta rima por repetição
rima visual g-h
ritmo ternário</td><td>˘ ˘ ˘ / ˘ / ˘ ‖ ˘ / ˘ ˘ /</td><td>12</td><td>L</td><td>[illegible]</td><td rowspan="5">[p]
falta rima por repetição;
rima entre 1 e 5
ritmo trocaico</td></tr>
<tr><td>[illegible] | ˘ / ‖ [illegible] | ˘ / | [illegible]</td><td>5</td><td>10</td><td>L</td><td>i</td><td>˘ ˘ / ˘ ˘ \ ˘ / ˘ ˘ /</td><td>11</td><td>L</td><td>j</td></tr>
<tr><td>˘ / | ˘ / | [illegible] | ˘ / | ˘ /</td><td>5</td><td>10</td><td>L</td><td>h</td><td>[illegible]</td><td>11</td><td>L</td><td>k</td></tr>
<tr><td>[illegible]</td><td>4</td><td>8</td><td>M</td><td>i</td><td>[illegible]</td><td>11</td><td>[illegible]</td><td>l</td></tr>
<tr><td>[illegible] ‖ ˘ / | ˘ / ‖ ˘ / | ˘ /</td><td>5</td><td>10</td><td>L</td><td>[illegible]</td><td>[illegible]</td><td>11</td><td>L</td><td>[illegible]</td></tr>
<tr><td></td><td></td><td></td><td></td><td></td><td></td><td></td><td></td><td></td><td></td><td></td></tr>
<tr><td>˘ / | [illegible] | \ / | [illegible] ˘ / | ˘ /</td><td>5</td><td>10</td><td>L</td><td>b</td><td rowspan="5">[aj]
[t], [ʃ], [s]</td><td>˘ ˘ ˘ / ‖ ˘ / ˘ / ‖ ˘ / ˘ /</td><td>12</td><td>L</td><td>l</td><td rowspan="5">[p]</td></tr>
<tr><td>[illegible] | ˘ / | ˘ / | [illegible] | ˘ /</td><td>5</td><td>10</td><td>L</td><td>J</td><td>˘ ˘ / ˘ / ˘ / ˘ / ˘ /</td><td>11</td><td>L</td><td>M</td></tr>
<tr><td>[illegible] | \ / | ˘ / | ˘ / | [illegible]</td><td>5</td><td>10</td><td>L</td><td>i</td><td>˘ ˘ ˘ / ˘ ˘ ‖ ˘ / ˘ ˘ ˘ /</td><td>12</td><td>L</td><td>n</td></tr>
<tr><td>˘ / | ˘ / | ˘ /</td><td>3</td><td>6</td><td>C</td><td>j</td><td>˘ ˘ / ˘ ˘ /</td><td>6</td><td>C</td><td>m</td></tr>
<tr><td>˘ / | ˘ / ‖ [illegible] | ˘ / | [illegible]</td><td>5</td><td>10</td><td>L</td><td>J</td><td>/ ‖ ˘ / ˘ / ˘ / ˘ \ /</td><td>10</td><td>L</td><td>M</td></tr>
<tr><td colspan="6">TOTAL DE SILABAS: 237</td><td colspan="5">TOTAL DE SILABAS: 256 (271)</td></tr>
</table>

Tabela 3

Cadernos de Literatura em Tradução, n. 7, p. 53-69

CIRQUE D'HIVER	CIRQUE D'HIVER
Across the floor flits the mechanical toy,	É um brinquedo de corda digno de um rei
fit for a king of several centuries back.	de uma outra era: cavalo e bailarina.
A *little* circus horse with *real* white hair.	Um cavalo de circo, de olhos negros,
His eyes are *glossy* black.	branco no pêlo e na crina.
He bears a *little* dancer on his back.	Sobre ele vai montada a bailarina.
She stands upon her toes and turns and turns.	Na ponta dos pés, ela rodopia.
A *slanting* spray of artificial *roses*	Tem um ramo de flores artificiais
is *stitched* across her skirt and tinsel bodice.	na saia e no corpete de ouropel.
Above her head she poses	Sobre a cabeça, traz
another spray of artificial *roses*.	um outro ramo de flores artificiais.
His *mane* and tail are straight from Chirico.	A cauda do cavalo é puro Chirico.
He has a formal, melancholy soul.	É formal e melancólica a sua alma.
He feels her *pink toes dangle toward* his back	Ele sente em seu dorso a perna leve
along the *little* pole	da bailarina calma
that pierces both her body and her soul	em torno da haste que a perfura, corpo e alma,
and goes through his, and reappears below,	e lhe atravessa o corpo, saindo por fim
under his belly, as a *big* tin key.	sob seu ventre como uma chave de lata.
He *canters* three steps, then he makes a bow,	Ele dá três passos, faz uma mesura,
canters again, bows on one knee,	anda mais um pouco, dobra uma das patas,
canters, then clicks and stops, and looks at me.	anda, estala, pára e olha para mim.
The dancer, by this time, has turned her back.	A dançarina, a essa altura, está de costas.
He is the more intelligent *by far*.	O cavalo é o mais arguto deles dois.
Facing each other rather desperately –	Entreolhamo-nos, com certo desespero,
his eye is like a star –	e dizemos depois:
we stare and say, "Well, we have come this far."	"É, até aqui chegamos nós dois."

Tabela 4

BRITTO (2008a, 31):

Finding is the first Act	[illegible]	4	*a*	Primeiro Ato é achar,	[illegible] 2-4-7	7	*a*
The second, loss,	[illegible]	2	*b*	Perder é o segundo Ato,	[illegible] 2-5-7	7	*b*
Third, Expedition for	[illegible]	4	*c*	Terceiro, a Viagem em busca	[illegible] 2-5-7	7	*c*
The "Golden Fleece"	[illegible]	2	*b'*	Do "Velocino Dourado"	[illegible] 4-7	7	*b'*
Fourth, no Discovery -	[illegible]	4	*d*	Quarto, não há Descoberta -	[illegible] 1-4-7	7	*d*
Fifth, no Crew -	[illegible]	3	*e*	Quinto, nem Tripulação -	[illegible] 1-3-7	7	*e*
Finally, no Golden Fleece -	[illegible]	4	*f*	Por fim, não há Velocino -	[illegible] 2-4-7	7	*f*
Jason - sham - too.	[illegible]	3	*e*	Falso - também - Jasão.	[illegible] 1-4-6	6	*e*

TRANSKULTURALITÄT – TRANSLATION – TRANSFER

Bd. 1 Cornelia Zwischenberger: Qualität und Rollenbilder beim simultanen Konferenzdolmetschen. 434 Seiten. ISBN 978-3-86596-527-1

Bd. 2 Sarah Fünfer: Mensch oder Maschine? Dolmetscher und maschinelles Dolmetschsystem im Vergleich. 150 Seiten. ISBN 978-386596-548-6

Bd. 3 Dörte Andres/Martina Behr (Hg.): Die Wahrheit, die reine Wahrheit und nichts als die Wahrheit. Erinnerungen der russischen Dolmetscherin Tatjana Stupnikova an den Nürnberger Prozess. 242 Seiten. ISBN 978-3-7329-0005-3

Bd. 4 Larisa Schippel (Hg.): Magda Jeanrenaud: Universalien des Übersetzens. 332 Seiten. ISBN 978-3-86596-444-1

Bd. 5 Sylvia Reinart: Lost in Translation (Criticism)? Auf dem Weg zu einer konstruktiven Übersetzungskritik. 438 Seiten. ISBN 978-3-7329-0014-5

Bd. 6 Sophia Scherl: Die deutsche Übersetzungskultur in der zweiten Hälfte des 18. Jahrhunderts. Meta Forkel-Liebeskind und ihre Übersetzung der *Rights of Man*. 152 Seiten. ISBN 978-3-7329-0020-6

Bd. 7 Thomas Kammer: Basiswissen für Dolmetscher – Deutschland und Spanien. 204 Seiten. ISBN 978-3-7329-0035-0

Bd. 8 Dorothee Jacobs: Basiswissen für Dolmetscher – Deutschland und das Vereinigte Königreich Großbritannien und Nordirland. 192 Seiten. ISBN 978-3-7329-0036-7

Bd. 9 Sophia Roessler: Basiswissen für Dolmetscher – Deutschland und Italien. 212 Seiten. ISBN 978-3-7329-0039-8

Bd. 10 Annika Selnow: Basiswissen für Dolmetscher – Deutschland und Frankreich. 192 Seiten. ISBN 978-3-7329-0040-4

Bd. 11 Julia Richter: Translationshistoriographie Perspektiven und Methoden. 186 Seiten. ISBN 978-3-7329-0937-7

Bd. 12 Alice Leal: Is the Glass Half Empty or Half Full? Reflections on Translation Theory and Practice in Brazil. 334 Seiten. ISBN 978-3-7329-0068-8

Bd. 13 Kristina Werner: Zwischen Neutralität und Propaganda – Französisch-Dolmetscher im Nationalsozialismus. 130 Seiten. ISBN 978-3-7329-0085-5

Bd. 14 Larisa Schippel/Magda Jeanrenaud/Julia Richter (Hg.): „Traducerile au de cuget să îmblînzească obiceiurile ...". Rumänische Übersetzungsgeschichte – Prozesse, Produkte, Akteure. 368 Seiten. ISBN 978-3-7329-0087-9

Frank & Timme

TRANSKULTURALITÄT – TRANSLATION – TRANSFER

Bd. 15 Elena Kalašnikova (Hg.): „Übersetzer sind die Wechselpferde der Aufklärung". Im Gespräch: Russische Übersetzerinnen und Übersetzer deutscher Literatur. 254 Seiten. ISBN 978-3-7329-0097-8

Bd. 16 Dörte Andres/Martina Behr (eds.): To Know How to Suggest ... Approaches to Teaching Conference Interpreting. 260 Seiten. ISBN 978-3-7329-0114-2

Bd. 17 Tatiana Bedson/Maxim Schulz: Sowjetische Übersetzungskultur in den 1920er und 1930er Jahren. Die Verlage *Vsemirnaja literatura* und *Academia*. 182 Seiten. ISBN 978-3-7329-0142-5

Bd. 18 Cécile Balbous: Das Sprachknaben-Institut der Habsburgermonarchie in Konstantinopel. 90 Seiten. ISBN 978-3-7329-0149-4

Bd. 19 Cornelia Zwischenberger/Martina Behr (eds.): Interpreting Quality: A Look Around and Ahead. 334 Seiten. ISBN 978-3-7329-0191-3

Bd. 20 Mehmet Tahir Öncü: Basiswissen für Dolmetscher – Deutschland und die Türkei. 232 Seiten. ISBN 978-3-7329-0154-8

Bd. 21 Marc Orlando: Training 21st century translators and interpreters: At the crossroads of practice, research and pedagogy. 158 Seiten. ISBN 978-3-7329-0245-3

Bd. 22 Christian Trollmann: Nationalsozialismus auf Japanisch? Deutsch-japanische Beziehungen 1933–1945 aus translationssoziologischer Sicht. 154 Seiten. ISBN 978-3-7329-0281-1

Bd. 23 Ursula Gross-Dinter (Hg.): Dolmetschen 3.0 – Einblicke in einen Beruf im Wandel. 226 Seiten. ISBN 978-3-7329-0188-3

Bd. 24 Lieven D'hulst/Carol O'Sullivan/Michael Schreiber (eds.): Politics, Policy and Power in Translation History. 256 Seiten. ISBN 978-3-7329-0173-9

Bd. 25 Dörte Andres/Julia Richter/Larisa Schippel (Hg.): Translation und „Drittes Reich". Menschen – Entscheidungen – Folgen. 352 Seiten. ISBN 978-3-7329-0302-3

Bd. 26 Julia Richter/Cornelia Zwischenberger/Stefanie Kremmel/Karlheinz Spitzl (Hg.): (Neu-)Kompositionen. Aspekte transkultureller Translationswissenschaft. 404 Seiten. ISBN 978-3-7329-0306-1

Bd. 27 Barbara den Ouden: Translation und Emotion: Untersuchung einer besonderen Komponente des Dolmetschens. 438 Seiten. ISBN 978-3-7329-0304-7

Bd. 28 Larisa Schippel/Cornelia Zwischenberger (eds.): Going East: Discovering New and Alternative Traditions in Translation Studies. 540 Seiten. ISBN 978-3-7329-0335-1

FSC
www.fsc.org
MIX
Papier | Fördert
gute Waldnutzung
FSC® C083411

Kolibri 360 GmbH
Ferdinand-Jühlke-Straße 7
99095 Erfurt, Deutschland
produktsicherheit@kolibri360.de

Druck:
CPI Druckdienstleistungen GmbH
im Auftrag der
Kolibri 360 GmbH
Ferdinand-Jühlke-Str. 7
99095 Erfurt